W9-BCL-172

Historical Portrait of the Progress of Ichthyology, from Its Origins to Our Own Time

FOUNDATIONS OF NATURAL HISTORY

Foundations of Natural History is a series from the Johns Hopkins University Press for the republication of classic scientific writings that are of enduring importance for the study of origins, properties, and relationships in the natural world.

PUBLISHED IN THE SERIES

Materials for the Study of Variation: Treated with Especial Regard to Discontinuity in the Origin of Species,
by William Bateson, with a new introduction by Peter J. Bowler and an essay by Gerry Webster

Nicholas Copernicus: On the Revolutions,
translated by Edward Rosen

Nicholas Copernicus: Minor Works,
translated by Edward Rosen

Problems of Relative Growth, by Julian S. Huxley,
with a new introduction by Frederick B. Churchill and an essay by Richard E. Strauss

Fishes, Crayfishes, and Crabs: Louis Renard's "Natural History of the Rarest Curiosities of the Seas of the Indies,"
edited by Theodore W. Pietsch

Historical Portrait of the Progress of Ichthyology, from Its Origins to Our Own Time, by Georges Cuvier,
edited by Theodore W. Pietsch,
translated by Abby J. Simpson

Mme Lizinka de Mirbel pinx
Richomme

HISTORICAL PORTRAIT OF THE PROGRESS OF ICHTHYOLOGY,
from Its Origins to Our Own Time

GEORGES CUVIER

Edited by Theodore W. Pietsch

Translated by Abby J. Simpson

THE JOHNS HOPKINS UNIVERSITY PRESS
Baltimore and London

This book has been brought to publication with the generous assistance of the University of Washington.

FRONTISPIECE: Baron Georges Cuvier (1769–1832). Engraving by Richomme after a painting by Mme de Mirbel. Courtesy of M.-L. Bauchot, M. Ducreux, and the Bibliothèque Centrale, Muséum National d'Histoire Naturelle, Paris.

Printed in the United States of America on acid-free paper
04 03 02 01 00 99 98 97 96 95 5 4 3 2 1

The Johns Hopkins University Press
2715 North Charles Street
Baltimore, Maryland 21218-4319
The Johns Hopkins Press Ltd., London

Library of Congress Cataloging-in-Publication Data will be found at the end of this book.
A catalog record for this book is available from the British Library.

ISBN 0-8018-4914-4

To Marie-Louise Bauchot,
for her expert guidance in the production of this work
and for her many years of outstanding service to the world
ichthyological community

This science, in common with every other, is the work of many persons, each in his own field, and each contributing a series of facts, a series of tests of the alleged facts of others, or some improvement in the method of arrangement. As in other branches of science, this work has been done by sincere, devoted men, impelled by a love of this kind of labor, and having in view, as the only reward they asked, a grateful remembrance of their work.

— David Starr Jordan, 1905

CONTENTS

LIST OF ILLUSTRATIONS

LIST OF TABLES

PREFACE

Jean Léopold Nicolas Frédéric Cuvier, better known by his adopted name, Georges Cuvier, was a colossal figure in biology during the first quarter of the nineteenth century. Born on 23 August 1769 at Montbéliard, a small town in the present French department of Doubs, close to the Swiss border, he was educated at the Academy of Stuttgart. Too young to compete successfully for a position more to his liking, in July 1788 he accepted a post as tutor to the sons of the comte d'Héricy in Normandy. Living by the sea, with ample free time at his disposal, he walked the beaches and shores, collecting plants and animals, gathering fossils, and studying anatomy. It was here that his interest in natural history blossomed, and it soon became directed toward ichthyology and comparative anatomy (Adler 1989, 18).

In 1795 Etienne Geoffroy Saint-Hilaire (1772–1844), having seen some of Cuvier's unpublished works, invited him to come to Paris to work at the Muséum National d'Histoire Naturelle. In April 1795, at the age of twenty-five, Cuvier was named professor of natural history at the Ecoles Centrales in Paris (Bauchot, Daget, and Bauchot 1990, 7), and in 1796 he became a member of the Institut de France (Académie des Sciences). In 1800 he was named professor at the Collège de France, replacing Louis Jean Marie Daubenton (1716–99), and in 1802, on the death of Jean Claude Mertrud (b. 1728), he took the chair of comparative anatomy at the Muséum National d'Histoire Naturelle in Paris, where he remained until his death on 13 May 1832.

After his arrival in Paris in 1795, Cuvier quickly established worldwide connections, building a vast correspondence with travelers and other naturalists. In many parts of the world naturalists rivaled one another to assist him with notes, manuscripts, drawings, and particularly specimens. For many years all zoological materials were deposited at the Muséum National d'Histoire Naturelle (Jordan 1902, 440, 1905, 400). With these collections as a foundation, Cuvier made substantial contributions to comparative anatomy, zoological classification, and the study of the fossil record. After completing his four-volume *Règne animal distribué d'après son organisation, pour servir de base à l'histoire naturelle des animaux et d'introduction à l'anatomie comparée* in 1816, a publication that David Starr Jordan (1902, 439, 1905, 400) described as no less important than Linnaeus's *Systema naturae,* Cuvier directed his attention almost entirely to ichthyology. In the monumental *Histoire naturelle des poissons,* conceived soon after 1820 (Günther 1880, 18), Cuvier and his most able student, Achille Valenciennes, summarized nearly all that was known about fishes up to that time. Consisting of twenty-two volumes comprising 11,253 pages of text and 650 plates, and published between 1828 and 1849, the *Histoire naturelle* contains descriptions of 4,055 nominal species, of which 2,311 were new to science (Bauchot, Daget, and Bauchot 1990, 5, 19). No other work, either by a single author or by several, has made such a great contribution to the advancement of the systematics of fishes. The work of these two scientists thus constitutes the historical foundation of modern ichthyology, the indispensable basic reference that must be consulted by all fish systematists (Gill 1872, 41–43; Monod 1963, 24–32; Pietsch 1985, 59–62). (For details of Cuvier's life and work, see Jardine 1834; for an assessment of the major features of his zoological theories and practice, see Coleman, 1964; for a thorough overview of the sources for the *Histoire naturelle des poissons,* including biographical sketches of all the collectors and donors mentioned in its pages, see Bauchot, Daget, and Bauchot 1990.)

The first volume of the *Histoire naturelle des poissons,* published in 1828 and written solely by Cuvier, begins with a historical account of the progress of ichthyology from its earliest beginnings through the first quarter of the nineteenth century. This is followed by a detailed description and illustrations of the external and internal anatomy of the European perch *(Perca fluviatilis)* as a basis for a comparative anatomy of all fishes. Finally, the volume closes with a lengthy statement of Cuvier's philosophy of animal classification and an outline of the classification to be followed throughout the succeeding volumes of the series. The second volume begins with detailed descriptions of species, each large taxonomic group preceded by a discussion of its major distinguishing characters. This continues through volume 22. Every species known or previously cited was said to be separately described or at least mentioned in the appropriate chapter.

But it is Cuvier's "Tableau historique des progrès de l'ichtyologie, depuis son origine jusqu'à nos jours" (Historical portrait of the progress of ichthyology, from its origins to our own time), the first part of volume 1 of the *Histoire naturelle des poissons,* that is of interest here: a chronological survey of the science of fishes, carried back beyond Aristotle to the Egyptians, Phoenicians, and Carthaginians, it is an affirmation of Cuvier's vast encyclopedic knowledge, his complete command of the scientific and popular literature, and in analyzing each work, his genius in recognizing the originality of its contribution to ichthyology. Authors, explorers, collectors, and illustrators, all those who called attention to the fishes of Europe and of distant lands and seas, are cited and placed in their historical context (Bauchot, Daget, and Bauchot 1990, 19). The narrative is augmented by 560 footnotes that supply precise biographical and bibliographical information on the persons cited. Although there are two other notable accounts of the history of ichthyology, by Albert Günther in *An Introduction to the Study of Fishes,* published in 1880, and by David Starr Jordan, published by the American Association for the Advancement of Science in 1902 (published again in slightly modified form in *A Guide to the Study of Fishes,* 1905), these are much condensed, general summaries, and their authors owe a large and unacknowledged debt to Cuvier (Wheeler 1958, 192).

Originally a continuous narrative, Cuvier's history falls conveniently into five parts and sixteen chapters. Following a brief introduction, Part 1 reaches back to ancient times, linking our earliest knowledge of fishes with the quest for sustenance: the contributions of the Egyptians, Phoenicians, and Carthaginians, and those of the Greeks and the Romans. Part 2 tells of the great Renaissance ichthyologists Belon, Rondelet, and Salviani and summarizes the significant anatomical contributions of the sixteenth and seventeenth centuries. The beginnings of modern ichthyology are outlined in part 3: the collaboration of Ray and Willughby, and the contributions of their immediate successors and of travelers in the Dutch colonies of the East Indies, in China and Japan, and in the Caribbean. Part 4, "The Foundations of Modern Classification," takes the history from the brilliant and innovative work of Artedi and Linnaeus through the contributions of the great and lesser ichthyologists of the eighteenth century—Gronovius, Forsskål, Pallas, Bloch, and others. Part 5, centering on the era of great collection building in the early nineteenth century, begins with the work of Lacepède and summarizes the contributions of Risso, Rafinesque, Geoffroy Saint-Hilaire, Lesueur, de Blainville, and many more, finishing with Cuvier himself and a summary of the materials assembled for the great *Histoire naturelle des poissons.*

Following the history, we have included an English translation of Cuvier's final chapter of volume 1 of the *Histoire naturelle des poissons,* his philosophy of animal classification: "Distribution méthodique des poissons en familles naturelles et en divisions plus élevées" (Method of classification of fishes into natural families and higher divisions). This discussion provides an in-depth philosophical basis for understanding the criticism or praise Cuvier extended to those who went before him. The 560 notes that augment Cuvier's text, originally placed at the bottoms of the pages, are here gathered together following each chapter of the narrative. Twenty-two classifications of fishes (called *distributions* or *méthodes* by Cuvier), proposals spanning nearly three centuries from Belon, Artedi, and Linnaeus to Oken and Cuvier himself, originally long footnotes, are here set up as tables and placed at appropriate points in the text. Reference material added for this volume includes literature cited, illustration credits, and an index.

In preparing the translation of the text, we made every effort to portray Cuvier as accurately as possible; yet at the same time we felt it necessary throughout to strike a balance between Cuvier's often convoluted, sometimes flowery writing style and one more palatable to modern tastes—but in the body of the text, nothing has been knowingly altered or added. The notes, by contrast, have been rather heavily edited: in many cases the supporting documentation Cuvier provided has been augmented, and additional notes have been added; wherever possible dates of events, of publication of various works, and of births and deaths have been supplied when not included by Cuvier; given names of individuals have been added; and errors have been corrected wherever found. Cuvier's text is a running narrative without chapters; for ease of reading we have inserted part titles and chapter titles. Except for these added titles, and minor insertions that do not add content or alter the meaning of Cuvier's text, all editorial additions are in brackets; minor errors, however—such as misspellings, erroneous dates, and missing or incorrect volume or page numbers—are corrected silently. Names of periodicals, which Cuvier usually wrote out in full and always gave in French regardless of national origin, are here abbreviated, as they appear in the literature cited, and the original languages are restored. Page numbers Cuvier provided for articles appearing in periodicals, all available to the reader in the literature cited, have been eliminated from the notes. Because the precise meaning of Cuvier's taxonomic nomenclature is not always certain, the French vernacular names he gave for various fish taxa have been retained in parentheses following the modern interpretation. For biographical information on eighteenth- and early nineteenth-century French travelers, explorers, and naturalists, we have relied heavily on the information provided by Bauchot, Daget, and Bauchot (1990) in their "L'Ichtyologie en France au début du XIXe siècle: *L'Histoire naturelle des poissons* de Cuvier et Valenciennes." In developing the bibliography, we attempted to cite all editions of works mentioned by Cuvier; with very few exceptions, the references listed in the literature cited are based on personal examination of publications and manuscripts found primarily in the libraries of the British Museum (Natural History), the British Library, and the Bibliothèque Centrale du Muséum National d'Histoire Naturelle, Paris.

Many people have generously given their help with this project, but we owe special thanks to Marie Louise Bauchot of the Muséum National d'Histoire Naturelle, Paris, and Alwyne Wheeler of the British Museum (Natural History), London, both now retired. In addition to immediate and always thorough responses to innumerable queries, they devoted many hours to critical reviews of the manuscript. Their great knowledge of the subject and eager willingness to contribute helped to greatly improved the final draft. To them we are truly grateful.

Many others assisted by providing hospitality during visits to their institutions or by otherwise allowing access to manuscripts, illustrative materials, and published literature: Carol Gokce, Paul Cooper, and Nick Stead, General Library, British Museum (Natural History), London; Nigel Merrett, Gordon Howes, and Oliver Crimmen, Department of Zoology, British Museum (Natural History), London; the kind and always helpful staff of the Reading Room and North Library of the British Library, London; C. M. Craig, the Royal College of Surgeons of England, London; Yves Laissus, Monique Ducreux, and V. Van de Ponseele, Bibliothèque Centrale, Muséum National d'Histoire Naturelle, Paris; Roland Bauchot, Université de Paris, Paris; Pierre Janin and Jacqueline Lanson, Bibliothèque Nationale, Paris; Delphine Allannic, Musée de la Marine, Palais de Chaillot, Paris; Joelle Defaÿe, Muséum de Nice, Nice; M. Boeseman, L. B. Holthuis, and Ingrid Henneke, Rijksmuseum van Natuurlijke Historie, Lieden; Augusta R. André de la Porte, Stedelijk Museum de Lakenhal, Leiden; Jørgen Nielsen and Torben Wolff, Zoological Museum, University of Copenhagen; Christine Karrer, Zoologisches Institut und Zoologisches Museum, Hamburg; Costas Papakonstantinou, National Centre for Marine Research, Athens; Nina J. Root and Debra B. Colchamiro, the Library, American Museum of Natural History, New York; Eveline Nave Overmiller, Library of Congress, Washington, D.C.; Susan L. Jewett and Lisa F. Palmer, Division of Fishes, National Museum of Natural History, Smithsonian Institution, Washington, D.C.; David Burgevin, Photographic Services, National Museum of American History, Smithsonian Institution, Washington, D.C.; Ellen B. Wells and Leslie Overstreet, Smithsonian Institution Libraries, Washington, D.C.; Melinda K. Hayes, Allan Hancock Library of Biology and Oceanography, University of Southern California, Los Angeles; Kraig Adler, Cornell University, Ithaca; Melanie Wisner and Roger E. Stoddard, the Houghton Library, Harvard University, Cambridge; and at the University of Washington, Seattle: Gary L. Menges, Special Collections Division, University of Washington Libraries; Colleen M. Weum, Health Sciences Library and Information Center; Patricia L. McGiffert, Health Sciences Photography; Stanley A. Schockey, Classroom Support Services; and Anna S. McCausland, Interlibrary Borrowing Service.

In addition to the detailed critical reviews provided by M. L. Bauchot and A. Wheeler, the full manuscript was read by J. Richard Dunn and James W. Orr, both of the University of Washington. Parts of the manuscript were critiqued by William D. Anderson Jr., College of Charleston, and Brooks M. Burr, Southern Illinois University, Carbondale. To all these people we extend our grateful thanks.

Thanks are also extended to Alice M. Bennett, whose editing skills greatly improved the manuscript; and to Paula M. Mabee (San Diego State University), coeditor of Foundations of Natural History, for recommending our work for the series. At the Johns Hopkins University Press, we thank Robert J. Brugger, history editor, Barbara B. Lamb, managing editor, Terry Schutz, production editor, Miriam L. Kleiger, manuscript editor, and Margaret B. Gillespie, editorial assistant, for skillfully directing the publication of this volume, and Anita Walker Scott for supervising its design and production. Finally, we are extremely grateful to Richard T. O'Grady, former science editor at the Johns Hopkins University Press, for his expert guidance through the process of review and final acceptance of the manuscript by the Press.

The bulk of the editorial and bibliographic work was done in the fall of 1992 during a three-month sabbatical leave from the University of Washington, for which the editor is extremely grateful. The opportunity to devote uninterrupted time to this project made the final product possible. For this, as well as for support in many other ways, we thank the University of Washington and its administrators, especially Robert C. Francis and Ellen K. Pikitch, respectively former and present directors for research, Fisheries Research Institute, School of Fisheries; Marsha L. Landolt, director, School of Fisheries; G. Ross Heath, dean, College of Ocean and Fishery Science; and Dale E. Johnson, associate dean, Academic Programs and Research, the Graduate School.

HISTORICAL PORTRAIT OF THE PROGRESS OF ICHTHYOLOGY, FROM ITS ORIGINS TO OUR OWN TIME

Georges Cuvier

Introduction

Natural history is a science of facts, and the number of facts it embraces is so great that no one person can gather or verify the totality that make up even one of its branches. Therefore natural history cannot be studied fruitfully except by consulting all the authors who have dealt with it and by comparing the evidence brought by the authors and by nature. But to consult the authors fruitfully, to be able to determine the degree of confidence owed to each of them, even to distinguish between what they report from their own observations and what they have gleaned from the writings of their predecessors, it is necessary to know the circumstances that governed their work, the times they lived in, the condition in which they found the science, and the facilities procured for them either through their personal position or through the help of friends, patrons, or students. These details, placed in chronological order and according to the order in which they are connected, constitute the history of the science, the foundation needed for any work that wishes to show the whole subject—the basis without which it would be impossible to make comprehensible the discussion of what is called synonymy, or the concordance of the names of the species, which itself is indispensable for compiling without confusion and without error what is known of their properties. For this reason we have decided to begin this *Histoire naturelle des poissons* with a history of ichthyology [fig. 1].

We shall try to follow man's knowledge of fishes down through the different ages; to introduce the authors who have written about this part of natural history; and to show the means they had at their disposal for instructing themselves in it, the contribution each made to its progress, the influence their works had on each other, and the continuing usefulness of their works.

We recognize in the progress of ichthyology three principal eras.

HISTOIRE
NATURELLE
DES POISSONS,

PAR

M. LE B.ON CUVIER,

Grand-Officier de la Légion d'honneur, Conseiller d'État et au Conseil royal de l'Instruction publique, l'un des quarante de l'Académie française, Secrétaire perpétuel de celle des Sciences, membre des Sociétés et Académies royales de Londres, de Berlin, de Pétersbourg, de Stockholm, de Turin, de Gœttingue, de Munich, etc.;

ET PAR

M. VALENCIENNES,

Aide-Naturaliste au Muséum d'Histoire naturelle

TOME PREMIER.

A PARIS,

Chez F. G. LEVRAULT, rue de la Harpe, n.° 81;

STRASBOURG, même maison, rue des Juifs, n.° 33;

BRUXELLES, Librairie parisienne, rue de la Magdeleine, n.° 438.

1828.

Like all branches of natural history, for many centuries ichthyology comprised only partial observations. Aristotle, 350 years before our era, began a body of doctrine, but it was weak and, relying only on ideas and rules that were hardly verified, was deprived of any sure means of distinguishing the species. During more than eighteen hundred years, those who wrote about natural history limited themselves to copying Aristotle or commenting on him. But in the middle of the sixteenth century Rondelet, Belon, and Salviani returned to true observation; and in rectifying and extending what Aristotle had written, they gave ichthyology a good basis in descriptions and drawings of a number of well-determined species. At the end of the seventeenth century Willughby and his colleague John Ray were the first to try to classify these species using a method based on distinctive characteristics taken from structure. Finally, in the middle of the eighteenth century, Artedi and Linnaeus completed this undertaking by establishing well-defined genera and assigning to them certain precisely characterized species. Since then, ichthyology has gone forward unobstructed toward perfection, approaching it all the more rapidly as each naturalist engaged in it has applied more ardor and more of that sagacity that discerns what is true and presents the truth in a rational way for rational minds. We are going to demonstrate the history of this progress as far as surviving literature will permit us to trace it.

FIG. 1. *(opposite) Histoire naturelle des poissons,* the title page of volume 1, 1828, containing Cuvier's "Historical Portrait of the Progress of Ichthyology, from Its Origins to Our Own Time." Courtesy of M. Ducreux and the Bibliothèque Centrale, Muséum National d'Histoire Naturelle, Paris.

Part One
THE EARLIEST BEGINNINGS

fishing [fig. 2]: they ate fishes raw, sun dried, or salted;[6] certain districts had no other food.[7] And whether or not the priests ate fishes, it is no less true that they knew the species well enough. Monuments built under their direction reveal in several places very faithful representations of fishes. We can cite in particular the painting of a large fishery, copied by Cailliaud[8] from one of the sepulchral vaults at Thebes, in which more than ten species can be easily distinguished, including damselfishes *(chromis),* Nile perch *(varioles),* elephantfishes *(mormyres),* and catfishes *(silures)* of different kinds.[9] The great *Description de l'Egypte* also shows several species, in particular mullets *(muges)* and elephantfishes *(mormyres),* drawn and colored in a very recognizable fashion.[10]

Also, a certain number of these species were embalmed. We have seen several specimens of the mummified remains of the bynni *(cyprins de l'espèce du binny),*[11] among others, in Passalacqua's collection.[12] This practice probably came from the cult devoted to certain fishes. Strabo[13] assured us that oxyrhynchus, a certain species of elephantfish *(oxyrinque),* and lepidotus *(lépidote)* were revered throughout Egypt.[14] This oxyrhynchus *(oxyrinque)* was particularly revered in the province and town that bore its name, and the town even erected a temple to it.[15]

At Latopolis, the Nile perch *(latos)* was worshiped [fig. 3];[16] at Elephantine, the maeotes *(méote);*[17] at Syene, the phagrus *(phagre)* or phagrorius *(phagrorius).*[18] The last was also probably a cult object in Phagroriopolis, one of the provinces of lower Egypt.[19] Those who believe it is necessary to seek explanations for these bizarre cults find them either in some natural property or in the role that these fishes played in the national mythology. Thus Aelianus believed that phagrus *(phagre)* was worshiped because its arrival announced the proximate flooding of the Nile;[20] but oxyrhynchus *(oxyrinque)* was revered because it was believed to have come from the wounds of Osiris.[21] Plutarch,[22] on the other hand, claimed that phagrus *(phagre),* oxyrhynchus *(oxyrinque),* and lepidotus *(lépidote)* were held in horror by the Egyptians because the fishes devoured the genitals of Osiris when Typho threw his legs into the Nile.[23]

These contradictory interpretations, and perhaps only guesses as well, of puerile customs are scarcely important to our purpose; but we can conclude from them that there were in that country regular observations of fishes, if only for distinguishing between what might be eaten and what must be returned to the river immediately upon capture. We may believe, too, that the practice of holding sacred some species, which probably were raised in the temples much like other animals dedicated to the gods, and the practice of opening them up after death for embalming, caused the men charged with these functions to acquire a more intimate knowledge of their structure and their habits.

Fisheries had much less interest for the Hebrews, who did not live

Part One

THE EARLIEST BEGINNINGS

1 The Egyptians, Phoenicians, and Carthaginians

The study of fishes, originating from their use as food, must have been one of the first sciences acquired, for there is no other nourishment that nature offers in greater abundance or that is easier to obtain. Thus we see that the most primitive peoples and those relegated to the most barren shores are the ones who depend most on fishes. The Greenlanders, Eskimos, and people of Kamchatka are fish eaters, as are the inhabitants of the rocks of the Maldives and of the arid sandy coasts of Makran. It is nature herself that constrains them to this way of living; she does not furnish them with any other resources. In Iceland dried fishes are used for money. For lack of vegetation, fishes are eaten by certain tribes and even by their herds. From earliest times, the Indian Ocean has supported fish eaters on its coasts. Herodotus reported them near the Red Sea,[1] and Nearchus, between India and Persia.[2]

The ease with which they procure their food and their poverty in all other respects have combined throughout the ages to keep fish eaters at the least elevated ranks of civilization. Agatharchides depicted them as the coarsest of humans, strangers to all moral feelings.[3] Their industry is limited to pounding and drying fishes for their winter provisions.[4]

It was probably because they wished to turn men, whom they sought to civilize, away from a mode of living so contrary to agriculture and so little favorable to the development of intelligence that the priests in Egypt inspired in their people a horror of the sea and forbade them to eat fishes. The priestly caste continued to abstain from eating fishes,[5] even when they could no longer stop the people from doing so in a nation where a great river, with its numerous canals and the lakes it flowed into, offered fishes in such prodigious quantity.

Despite the prohibition against priests' eating fishes, it is certain that the common people of Egypt gave themselves up wholeheartedly to fishing [fig. 2]: they ate fishes raw, sun dried, or salted;[6] certain districts had no other food.[7] And whether or not the priests ate fishes, it is no less true that they knew the species well enough. Monuments built under their direction reveal in several places very faithful representations of fishes. We can cite in particular the painting of a large fishery, copied by Cailliaud[8] from one of the sepulchral vaults at Thebes, in which more than ten species can be easily distinguished, including damselfishes *(chromis),* Nile perch *(varioles),* elephantfishes *(mormyres),* and catfishes *(silures)* of different kinds.[9] The great *Description de l'Egypte* also shows several species, in particular mullets *(muges)* and elephantfishes *(mormyres),* drawn and colored in a very recognizable fashion.[10]

Also, a certain number of these species were embalmed. We have seen several specimens of the mummified remains of the bynni *(cyprins de l'espèce du binny),*[11] among others, in Passalacqua's collection.[12] This practice probably came from the cult devoted to certain fishes. Strabo[13] assured us that oxyrhynchus, a certain species of elephantfish *(oxyrinque),* and lepidotus *(lépidote)* were revered throughout Egypt.[14] This oxyrhynchus *(oxyrinque)* was particularly revered in the province and town that bore its name, and the town even erected a temple to it.[15]

At Latopolis, the Nile perch *(latos)* was worshiped [fig. 3];[16] at Elephantine, the maeotes *(méote);*[17] at Syene, the phagrus *(phagre)* or phagrorius *(phagrorius).*[18] The last was also probably a cult object in Phagroriopolis, one of the provinces of lower Egypt.[19] Those who believe it is necessary to seek explanations for these bizarre cults find them either in some natural property or in the role that these fishes played in the national mythology. Thus Aelianus believed that phagrus *(phagre)* was worshiped because its arrival announced the proximate flooding of the Nile;[20] but oxyrhynchus *(oxyrinque)* was revered because it was believed to have come from the wounds of Osiris.[21] Plutarch,[22] on the other hand, claimed that phagrus *(phagre),* oxyrhynchus *(oxyrinque),* and lepidotus *(lépidote)* were held in horror by the Egyptians because the fishes devoured the genitals of Osiris when Typho threw his legs into the Nile.[23]

These contradictory interpretations, and perhaps only guesses as well, of puerile customs are scarcely important to our purpose; but we can conclude from them that there were in that country regular observations of fishes, if only for distinguishing between what might be eaten and what must be returned to the river immediately upon capture. We may believe, too, that the practice of holding sacred some species, which probably were raised in the temples much like other animals dedicated to the gods, and the practice of opening them up after death for embalming, caused the men charged with these functions to acquire a more intimate knowledge of their structure and their habits.

Fisheries had much less interest for the Hebrews, who did not live

FIG. 2. Fishing scene from the Middle Kingdom tomb of Ahanekht, son of Tehutinekht, in the tomb valley at El Bersheh just south of Roda, an island in the Nile River near Cairo (tomb no. 5, outer chamber, inner wall). After Griffith and Newberry 1893, pl. 16.

FIG. 3. Two men carrying a Nile perch, *Lates niloticus,* from the mastaba of Rahotep at Medum, Fourth Dynasty. After Petrie 1892, pl. 12.

FIG. 4. Various fishes depicted on Punic coins from Cadiz. After Heiss 1870, pl. 51.

near the sea and whose country had only one moderate-sized river and two small freshwater lakes, the Dead Sea being too salty to sustain fishes. And yet Moses, at least as a precaution, laid down laws for the use of this food. He forbade the people to eat all fishes that lacked scales or fins,[24] which probably meant in the first case catfishes *(silures)* and in the second the various water snakes, to which was apparently attributed some evil quality.

It was the Phoenicians, dwellers on the coast, who brought fishes to the Hebrews.[25] However, one does not find any positive evidence among the ancients that the Phoenicians fished on a large scale in antiquity. There is no mention of fishes or salteries in the poetic description of their commerce given by Ezekiel;[26] but it is difficult to believe that a navigating people would not also be a fishing people, and if the great establishments of fisheries and salteries that flourished in the times of the Romans on the coasts of Spain did not owe their origin to the Phoenicians, they must at least have been created by the Carthaginians; for tuna *(thon)* and other fishes are commonly seen depicted on Punic medallions from Cadiz and Carteia [fig. 4]. According to Bochart, the name of the village Malaga, a seaport in southern Spain, comes from the Hebrew and Phoenician word *malach,* which means "to salt."[27]

But whatever might have been the knowledge possessed by these early civilizations, it did not contribute to the body of our doctrines today except as bits and pieces that may have been passed down through the works of the Greeks and Romans.

NOTES

1. [Herodotus, *Persian Wars,* bk. 3], "Thalia," chaps. 19 and 20. [Herodotus, a Greek historian born in the Dorian city of Halicarnassus about 484 B.C., died in 425 B.C., is called the "father of history" (Godolphin 1942, 1:xxi).]

2. [Nearchus, in] Strabo *[Geographia],* bk. 15 [pt. 2], chap. 2 [on the country of the "Icthyophagi," or fish eaters (see Jones 1917, 7:131)]. [Nearchus (fl. 320–300 B.C.) was an officer in the army of Alexander the Great, a native of Crete who settled at Amphipolis in Macedonia, who, as leader of various military campaigns, conducted the fleet to the head of the Persian Gulf (325 B.C.) and later (323 B.C.) circumnavigated Arabia from the mouth of the Euphrates to the isthmus of Suez; abstracts of the narrative of his travels were repeated by Strabo and others (*Encyclopaedia Britannica,* 1951, 16:179). For an English translation of Nearchus's description of the country of the "Icthyophagi," see Vincent 1807a.]

3. Agatharchides, in Photius [*Bibliotheca,* cod. 250, v. 31, a collection of criticisms and extracts from 280 works of various authors from antiquity to his time; see Henry 1974, 156–57]. [Agatharchides or Agatharchus, of Cnidus, Greek historian and geographer, lived in the time of Ptolemy Philometor (181–146 B.C.) and his successors. Among other works, he wrote treatises on Asia, Europe, and the Red Sea (see Vincent 1807b); extracts from the last of these were repeated and thus preserved by Photius (*Encyclopaedia Britannica,* 1951, 1:342). Saint Photius (ca. A.D. 820 to ca. 891), patriarch of Constantinople (A.D. 858–67 and 878–86), to whom we are indebted for almost all we possess of the otherwise lost books of many of the ancient writers (see White 1982, 15–47.]

4. Ibid., [v. 34; see Henry 1974, 158]; and Strabo, [*Geographia,* bk. 15, pt. 2, chap. 2].

5. They (the priests) are not allowed to eat fishes (Herodotus, [*Persian Wars,* bk. 2], "Euterpe," chap. 37).

6. Herodotus, [*Persian Wars,* bk. 2], "Euterpe," chap. 77.

7. Ibid., chap. 92. ["Some of these folk live entirely on fish, which are gutted as soon as caught and then hung up in the sun: when dry, they are used as food."]

8. [Frédéric Cailliaud, French explorer, was born at Nantes in 1787. In 1815 he was commissioned to explore the deserts to the east and west of the Nile. Departing from Edfou, he went toward the Red Sea, and in the desert he found the emerald mines of Mount Labarah; he then returned to Egypt by the ancient route of Coptos to Berenice. In June 1818 he visited the Great Oasis, where he made some important archaeological discoveries; then he explored all the known oases to the west of Egypt and ascended the Nile to 10° north latitude. Soon afterward he returned to France, taking residence in his native city of Nantes, where he became conservator of the city's museum and where he died in 1869.]

9. Cailliaud, *Voyage à Méroé* [1823], vol. 2, pl. 75 [paintings from the ancient tombs of Gournah at Thebes, showing scenes of hunting, fishing, harvesting, music, etc.; for further explanation of the plate, see Cailliaud 1826–27, 3:292.].

10. *Description de l'Egypte, Antiquities* [see Description of Egypt 1809–30 (atlas vol. 2, pl. 87, figs. 2–3); Gillispie and Dewachter 1987 (vol. 1, pl. 87, figs. 2–3): paintings of fishes from the fifth Tomb of the Kings (i.e., tomb of Ramses III) at Bybân el-Molouk (Valley of the Kings), Thebes, drawn by the famous botanical artist Pierre Joseph Redouté (b. 1759 at St. Hubert in the Ardennes region of Belgium, d. 1840); see Redouté 1812]. [For more recent studies of fishes depicted on ancient Egyptian tombs and monuments, see Gaillard 1923; Danelius and Steinitz 1967; Gamer-Wallert 1970; Driesch 1983, 1986; Brewer and Friedman 1989.]

11. [The "bynni" or "lepidotus" of Strabo and others is *Barbus bynni* (Forsskål, 1775): see Valenciennes 1842, 174–79; Boulenger 1907, 203–8, text figs. 24–25; Gaillard 1923, 44–49, figs. 28–29; Thompson 1947, 148–49; Gamer-Wallert 1970, 95–98; and Brewer and Friedman 1989, 59.]

12. ["Among the great number of mummies that the scientific explorer M. [Joseph] Passalacqua brought back from the necropolis at Thebes are several fishes belonging, as my father [Etienne Geoffroy Saint-Hilaire; see chap. 13, n. 18] has verified, to the species *Cyprinus lepidotus* [a junior synonym of *Barbus bynni*]: all have been carefully embalmed, wrapped in small bands, and placed in boxes sculpted on the outside and of the same shape as the fishes" (Isidore Geoffroy Saint-Hilaire 1827a, 285). See Passalacqua 1826.]

13. [Strabo; see chap. 3, n. 25.]

14. Strabo, *[Geographia]*, bk. 17, pt. 1, chap. 40 [Jones 1917, 8:109]. [The "oxyrhynchus" of ancient Egypt, a fish of the Nile, sacred to Osiris and said to have sprung from his wounds, refers to two morphologically similar species of *Mormyrus, M. kannume* Forsskål, 1775, and *M. caschive* Linnaeus, in Hasselquist 1762 (see Boulenger 1907, 68; Gaillard 1923, 24–28, figs. 14–17; Thompson 1947, 184; Gamer-Wallert 1970, 91–95; and Brewer and Friedman 1989, 51–52). The "lepidotus" of the ancients is *Barbus bynni* (Forsskål, 1775); see n. 11 above.]

15. Strabo, ibid. ["On the far side of the river lie the city Oxyrynchus and a Nome bearing the same name. They hold in honour the oxyrynchus and have a temple sacred to Oxyrynchus, though the other Aegyptians in common also hold in honour the oxyrynchus"]; Aelianus, *[De natura animalium]*, bk. 10, chap. 46.

16. Strabo, ibid. [chaps. 40, 47]. ["Latos" is the Nile perch, *Lates niloticus* (Linnaeus, 1758), for which the ancient Egyptians had the greatest veneration and of which numerous mummified remains have been found (see Boulenger 1907, 457–58; Gaillard 1923, 81–84, figs. 48–49; Thompson 1947, 144–46; and Brewer and Friedman 1989, 74–75).]

17. Aelianus, *[De natura animalium]*, bk. 10, chap. 19. ["Maeotes" is the name of an unknown Egyptian fish, held sacred at Elephantine because it gave forewarning of the inundation of the Nile (Thompson 1947, 155).]

18. Aelianus, ibid.; Clement of Alexandria says the same thing in *Exhortation to the Greeks*, chap. 2, p. 34. [Clement of Alexandria is Titus Flavius Clemens, a Greek Christian theologian, b. ca. A.D. 150, probably in Athens, d. ca. 212 (Butterworth 1919, xi–xii). "Phagrus" or "phagrorius" of the ancients is the name of an unknown fish: "The Egyptian Phagrus cannot be safely identified" (Thompson 1947, 274–75; see also Gamer-Wallert 1970, 101–7).]

19. Strabo, *[Geographia]*, bk. 17, pt. 1, chap. 26.

20. Aelianus, *[De natura animalium]*, bk. 10, chap. 19. [Claudius Aelianus or Aelian; born at Praeneste in about A.D. 170; he was over sixty years of age when he died (see Scholfield 1958, xi).]

21. Ibid., bk. 10, chap. 46.

22. [Plutarch of Chaeronea; see chap. 3, n. 27.]

23. [Plutarch], *De Iside et Osiride*, bk. 18. ["We are told . . . that notwithstanding all her search, Isis was never able to recover the privy-member of Osiris, which having been thrown into the Nile immediately upon its separation from the rest of the body, had been devoured by the Lepidotus, the Phagrus and the Oxyrynchus, fish which of all others, for this reason, the Egyptians have in more especial avoidance" (see Squire 1744, 23).]

24. Lev. 11:9–12. ["These shall ye eat of all that are in the waters: whatsoever hath fins and scales in the waters, in the seas, and in the rivers, them shall ye eat. And all that have not fins and scales in the seas, and in the rivers, of all that move in the waters, and of any living thing which is in the waters, they shall be an abomination unto you."] The Hebrews rather ill applied this rule to the common eel *(anguille)*, which does not lack scales.

25. Neh. 13:16. ["There dwelt men of Tyre also therein, which brought fishes, and all manner of ware, and sold on the sabbath unto the children of Judah, and in Jerusalem."]

26. Ezekiel 27.

27. [Samuel] Bochart [1599–1667], *Geographia sacra, seu Phaleg* [1712], 167.

2 Aristotle and His Disciples

It is among the Greeks that we find the first foundations of ichthyology, and of all other sciences as well. Actually, it has been claimed that they at first did not consider fishes an important food. Homer's heroes were never served fishes, and even Ulysses wrote that his companions caught fishes only after being pressed by hunger. Some commentators believed that an excuse may be found in the words, "For hunger weighed upon our stomachs"[1]—words used by Menelaus in a like circumstance,[2] but that prove nothing, neither that fishes were undesirable nor that the art of fishing was unknown. The very fact that the companions of Ulysses and Menelaus took fishhooks with them proves the contrary; furthermore, Plato[3] and Athenaeus[4] attributed the abstinence of these warriors from eating fishes to the fear of weakening themselves with too pleasant a diet.

Moreover, Homer spoke in several places about fishing with a hook and with a net. He compared Penelope's pining suitors to fishes in a pile flapping on the beach where the fishermen had just emptied their nets.[5] When Scylla drew six of Ulysses's men into her whirlpool, he depicted them as being like the little fishes to which the fisherman had held out his bait on a long rod.[6] Hesiod placed on the shield of Hercules an alert fisherman, ready to cast his nets over fishes pursued by a dolphin.[7]

How, in fact, could such ignorance and such prohibition, if indeed they ever existed, have continued in force in a country like Greece, invaded everywhere by bays and arms of the sea, where the population was mostly insular and given to navigation from earliest times? We must conclude, on the contrary, that fresh as well as salted fishes were at an early stage perhaps the most important item in the Greek diet. This fact is mentioned unceasingly by the comic poets. Aristophanes,[8] in his works that survive, referred to it twenty times, and Athenaeus cited perhaps two hundred passages and authors (their works now lost) on the subject.

The art of fishing thus became one of the most lucrative and most widespread industries; in favorable locations, great salteries were built, which later became considerable towns. Byzantium and Synope flourished especially for this reason; and it was the great abundance of fishes that gave the port of Byzantium the name the "Golden Horn." Private citizens made rapid fortunes in this commerce, and the ancient comedians made fun several times of a salt-fish merchant named Chaerephilus, who became a citizen of Athens and whose son wasted the fortune that the father had so laboriously amassed.[9]

Different personalities became the objects of satire solely for having loved fishes so excessively. Such were a certain Callimedon, nicknamed the Crawfish *(langouste)*, about whom the comedians would never keep quiet;[10] Philoxenes of Cythera, a poet in dithyrambics who, learning from his doctor that he was about to die of indigestion from eating the greater part of a fish, asked to eat the rest of it first (a funny story so well versified by La Fontaine); the great orators Callias and Hyperides, who loved fishes as much as games of chance; Melanthus the tragedian; and others. Especially cited is Androcides of Cyzicus, a painter whose liking for fishes led him to represent with great care, after nature, the species in the Strait of Scylla, and who was thus the precursor of the great iconographers of our day.[11]

One proof of the great number of species the Greeks succeeded in recognizing is that there remain in their language more than four hundred names for designating fishes, which certainly no other language approaches, and as Buffon said very judiciously: "Doesn't this abundance of words, this richness in fine and precise expressions indicate the same abundance in ideas and knowledge? Doesn't one see that these people, who named many more things than we, as a consequence, knew many more things?"[12]

Naturally, in these circumstances, several writers wrote either about the fishes themselves, on their fisheries, or on the use made of them in the art of cookery and on the precautions that hygiene recommended in their use. One may judge by the numerous citations of Athenaeus that in fact there existed many books on these matters. Unfortunately, Athenaeus did not indicate when each of these cited authors lived, and among the traits that he reported there are rarely any that are useful in identifying the authors. Therefore for some of them it is not easy to determine whether they preceded Aristotle or came after him; nor is it easy to distinguish those whose works were used by Aristotle, this first of naturalists, from those who on the contrary profited from his works. Careful as he was in collecting the books of those who came before him, being one of the first to form a library,[13] it can hardly be doubted that he acknowledged every author who was able to furnish him with interesting facts, and yet he did not cite any of those Athenaeus mentioned as having spoken of fishes, not even those who must have been his contemporaries, such as Archestratus, master of the art of eating, whose gastrology[14] was said to be a guide for Epicurus in his search for pleasure. The only names mentioned in Aristotle's books on animals are those of Aeschylus, Alcmaeon, Ctesias, Diogenes of Apollonia, Herodorus, Herodotus, Homer, Musaeus, Polybius, Simonides, Syennesis,[15] Empedocles, Democritus, and Anaxagoras.[16]

It is true that Aristotle seldom cited any authors but those he wished to refute—a practice too common in our own time—and he has even been accused of ingratitude toward Hippocrates, whose name he did not mention, although he must have borrowed more than one idea from him. On the other hand, I do not believe he did a great wrong to the ichthyologists who preceded him, if in fact there were any before him. Those fragments preserved by Athenaeus that may be attributed to them do not show that they treated their subject methodically or extensively, and we may believe that it was only under the pen of Aristotle that ichthyology, like all other branches of zoology, first took the form of a true science.[17]

This great man, supported by a great prince, collected facts from everywhere, and the facts in his works are so numerous and so new that for several centuries they incurred posterity's disbelief. The personalities described by Athenaeus wondered how Aristotle was able to learn what he wrote in reference to the habits of fishes, their propagation, and other details of their lives that take place in the most hidden recesses of the sea.[18]

Athenaeus himself answered the question; he told us that Alexander gave Aristotle eight hundred talents (more than three million francs)[19] for collecting materials for his *Historia animalium*, to which Pliny added that the king put several thousand men at Aristotle's disposal for hunting, fishing, and observing all that he desired to know.[20]

This is not the place to set forth in detail what Aristotle did with this munificence, to analyze his numerous works of natural history,[21] or to enumerate the immense quantity of facts and laws he succeeded in setting down. We shall not even show the genius with which he laid the foundations of comparative anatomy and established in the animal kingdom, and in several of its classes, a classification based on their structure that was changed very little in succeeding ages. It is only as an ichthyologist that we are to consider him, and in that branch of zoology, even had he treated of none other, one must recognize him as a superior man.

He was completely knowledgeable in the general structure of fishes: "They lack a neck," he said; "their tail is continuous with their body, except in the rays *(raies)*, where it is long and slender; they do not have hands, or feet, or scrotum, or virile member, or teats. One must distinguish them from the marine animals that produce living young, such as the dolphin, which has teats hidden in folds near its vulva. The special character of true fishes is found in the gills and fins: most have four fins [two pairs]; but those with an elongated form, such as freshwater eels *(anguilles),* have only two. Some, like the moray eel *(murène),* lack [paired] fins altogether. Rays *(raies)* swim with their whole wide body. Gills are sometimes supplied with an opercle, but sometimes not, as is the case with the cartilaginous fishes: the gills of some are simple while those of others are double."[22] He even noticed that the swordfish *(xiphias)* has eight gills on each side, which is true in the sense that each of its gills is divided into two combs. "No fish has hair or feathers; most are covered with scales, some have skin either rough or smooth. Their tongue is hard, often armed with teeth, and sometimes so firmly attached that they appear to be devoid of the organ altogether,[23] this because they must swallow rapidly; which is also why their teeth are generally hooked.[24] They lack eyelids. One does not see their ears or their nares,[25] for what serves as nares is in a hidden cavity.[26] Nevertheless, they enjoy the senses of tasting, smelling, and hearing," which the author proved by a number of experiments.[27] "All have blood; all scaly fishes have eggs, but the cartilaginous fishes, except for the anglerfish *(baudroie),* give birth to live young.[28] All have a heart, a liver, and a gallbladder," and in this regard he goes into quite particular and quite true detail on the gallbladder of the dragonet *(callionyme)* and the bonito *(amia);* but he is wrong in denying that fishes have kidneys and a bladder.[29]

"Their intestines vary greatly; there are some, such as the mullet *(muge),* that have a fleshy gizzard like that of birds; others scarcely have an apparent stomach.[30] Blind appendages or ceca adhere near their stomach, quite numerous in some, not so numerous in others. There are even some that have none at all, like most of the cartilaginous fishes.[31] Along the vertebral column extend two organs that serve as testes, the excretory canals of which end at the anus, and that become much larger at the time of spawning."[32]

"Their scales harden with age.[33] Lacking lungs, they have no voice as such, and yet there are several (which he names) that make sounds and a sort of grunting.[34] Like other animals, they need sleep.[35] In most kinds the females are larger than the males. In the sharks *(squales)* and rays *(raies)* the male is distinguished by claspers on both sides of the anus."[36] Not only did Aristotle make numerous observations, from which he deduced such exact rules, but he also showed these different structures in drawings.[37]

As for species, Aristotle knew and named as many as 117, and he wrote about their way of life, their migrations, their friendships and rivalries, the ruses they employ, their loves, the times of their spawning, their egg laying and fecundity, the method of catching them, the times when their flesh is best, with details that one would have difficulty today either contradicting or confirming, so far are modern men from observing fishes the way this great naturalist seems to have done, through his own effort or through his correspondents. It would be necessary to spend several years in the islands of the Archipelago and live there with the fishermen to be able to have an opinion on this subject.[38]

Aristotle, it is true, accepted spontaneous generation, but it must be admitted that he supported this theory with rather specious facts. However, what he said about the difficulty of finding freshwater eels *(anguilles)* in a state of reproduction is well founded, and naturalists of our day scarcely have more certain knowledge of the procreation of this species than the ancients. Even in recent times, one of his most paradoxical assertions has been verified, namely, that a species of *Serranus (channa)* impregnates itself, and that all individuals of the species produce eggs.[39]

What is most regrettable in this mass of such precise information is that the author did not suspect that the nomenclature used in his time would come to be obscure, and that he took no precaution to make recognizable the species he was writing about. This is a general fault among ancient naturalists; one is almost obliged to guess the meanings of the names they used. Even tradition has changed and leads us into error: it is only by very laborious calculations and by bringing together and comparing the characteristics scattered among the authors that one succeeds in getting fairly positive results for some species, but we are still condemned to be ignorant of most of them.

We have purposely given this somewhat extensive discussion of the works of Aristotle on fishes because he is not only the first but also the only one of the ancients to have written about their natural history with a scientific viewpoint and with some genius.

For a while his school followed in his footsteps: Theophrastus,[40] his most worthy student, added some interesting facts to this part of science. In his treatise *On Fishes That Live on Dry Land,* he spoke clearly about the fishes of India, described just recently under the name *Ophicephalus (ophicéphales),*[41] and of the loach *(misgurn, Cobitis fossilis* Linnaeus), which remains alive in the mud after the marsh it inhabits has dried up.

The famous physician and anatomist Erasistratus, grandson of Aristotle, his pupil and also the pupil of Theophrastus, wrote a work on nutrition derived from fishes. Clearchus, another disciple of Aristotle,[42] composed a treatise on the general subject of aquatic animals, of which Athenaeus quoted passages on the flying fish *(adonis* or *exocet)* and on fishes that make sounds.[43] Another treatise on fishes was written by Dorion,[44] who must have lived at about the same time, if he is the same

FIG. 5. Rectangular plate of baked clay decorated with the figures of four marine fishes and two bivalves, the work of a Greek artist of southern Italy, fourth century B.C. After Singer 1959, fig. 3.

man about whom many droll stories are told and who was quoted by Lynceus, a disciple of Theophrastus.[45]

Several works on hygiene can also be attributed to this era and to the influence of the Peripatetic philosophy. I include among them that by Diocles of Carystus, dedicated to Antigonus; the treatise on healthful things for the healthy and the sick by Diphilus of Siphne, a contemporary of Demetrius, son of Antigonus; the treatise on nutrition by Philotimus, disciple of Praxagoras and contemporary of Herophilus; and another on the same subject by Icesius, a disciple of Herophilus. One sees from the quotations of Athenaeus that fishes were often mentioned. If there were any certainty about the time when their authors lived, I could also readily include the poem on men and the sea by Pancratius the Arcadian; the *Halieutica* by Numenius of Heraclea; poems on fishes by Coclus of Argos and by Posidonius of Corinth, a very different man from the great philosopher Posidonius of Apamea; prose essays by Seleucus of Tarsus and Leonides of Byzantium; and the book on salted fishes by Euthydemus of Athens. But all these writers are known to us only from the few words Athenaeus wrote about them.[46]

Still, one may see that fishes were an object of general attention among the Greeks [fig. 5], and it is probable that if the Greeks had continued in the footsteps of Aristotle, ichthyology would have made great progress; but to have more Aristotles, more Alexanders were needed. Positive natural history requires work and expense that a private person without patronage cannot afford. Ptolemy-Lagus, himself a pupil of the philosopher of Stagira [Aristotle], and his son Ptolemy-Philadelphus, throughout their lives[47] gave encouragement to all kinds of science, but their unworthy successors no longer took an interest, and the school they established at Alexandria found it more convenient to cultivate erudition, geometry, and metaphysics than to exert itself in research on the productions of nature. As a natural consequence, the Peripatetic philosophy, especially its emphasis on experiment, fell by degrees into a kind of contempt; the Academy and the Porch [Platonists] were in the ascendant, and scientific observers were ridiculed. The drolleries of Lucian, who shows us a Peripatetic philosopher examining the length of a gnat's life and the nature of the soul in the oyster,[48] were probably told long before him; these kinds of studies became so uncommon that when Apuleius was accused of magic, one of the main arguments used against him was that he engaged in the study of rare and strange fishes.[49]

NOTES

1. [Homer], *Odyssey*, bk. 12, line 332. [Homer, the Greek epic poet, eighth century B.C., according to legend, is the author of the *Iliad* and the *Odyssey*.]

2. Ibid., bk. 4, line 369. ["No man in Homeric times would eat fish when he could get meat" (see Merry 1895, pt. 2, p. 57).]

3. [Plato], *Republic*, bk. 3 [v. 404]. [Plato, born at Athens or Aegina in 427 B.C., became a pupil of Socrates (ca. 470 to ca. 399 B.C.), visited Sicily, formed a school in the garden of Academus in 386 B.C., and died in 347 B.C. (see Lindsay 1954, p. ii).]

4. [Athenaeus], *Deipnosophistae, [or Banquet of the Learned]*, bk. 1, chap. 15. [Athenaeus was an Egyptian, born in Naucratis on the Nile; the age he lived in is uncertain, but most agree that his work, at least the latter portion of it, must have appeared after A.D. 228 (see Yonge 1854, 1:iii).]

5. [Homer], *Odyssey*, bk. 22, v. 384. ["However he found them all weltering in dust and blood, many as the fish dragged forth by sailors from the grey sea in seine nets up on the beach of some bay, where they lie heaped on the sand and languishing for the briny waves, while the sun's shining saps their life away. Just so were the suitors heaped together" (translation of Shaw 1932, 301).]

6. Ibid., bk. 12, v. 251.

7. Hesiod, *Scutum Herculis* ["The Shield of Hercules"], lines 210–15 [see Way 1934, 61]. [Hesiod, born at Ascra, near Mount Helicon, probably flourished during the eighth century B.C.; he is considered the father of Greek didactic poetry (*Encyclopaedia Britannica*, 1951, 11:529).]

8. [Aristophanes, the great comic dramatist and poet of Athens, ca. 448 to ca. 385 B.C. (*Encyclopaedia Britannica*, 1951, 2:347.]

9. Athenaeus, *[Deipnosophistae, or Banquet of the Learned]*, bk. 3, chap. 90; bk. 8, chap. 23 [Yonge 1854].

10. See in Athenaeus, [ibid.], bk. 8, chap. 24, that Callimedon was the butt of endless jokes.

11. Athenaeus, [ibid.], bk. 8 [chap. 25]. [On survivals of ancient Greek zoological illustrations, see Kádár 1978.]

12. [See Buffon 1749–1804, 1:42–43; see also chap. 8, n. 36.]

13. Strabo, *[Geographia]*, bk. 13, pt. 1, chap. 54.

14. Archestratus, from Syracuse or from Gela, was author of a poem on the subject of good living, whose title was variously reported to be *Gastronomia, Hedypathy, Deipnology,* and *Opsopaea.*

15. This is the list of authors connected with the *Historia animalium* that was placed by Friedrich Sylburg [1536–96] at the beginning of his edition of Aristotle's work [published at Frankfurt, 1587].

16. These last [three] authors are cited [by Aristotle] in the books of *De generatione animalium* [e.g., see bk. 4, v. 1].

17. The biography of Aristotle is well known, and only the main points are mentioned here. He was born at Stagirus in Macedonia in 384 B.C. His father, Nichomachus, was physician to Amyntas, the father of Philip of Macedonia, and belonged to a branch of the Asclepiades. He first studied medicine under his father. Orphaned at eighteen years, he went to Athens and supported himself by selling medicinals (Athenaeus called him "Pharmacopole" because of this). He studied under Plato and opened his own school sometime before the death of his master. As early as 356 B.C., Philip appointed him tutor to Alexander. When war erupted between Philip and Athens in 346 B.C., Aristotle left that city and took refuge with Hermias, the prince of Atarna, in Mysia. After the latter was betrayed and killed by the Persians, Aristotle married his sister. In 343 B.C. Alexander, at the age of thirteen, was put in his care. It is believed that he followed Alexander on his expedition to Egypt. In 331 B.C. he went back to Athens and opened a school of his own in a garden known as the Lyceum. After the death of Alexander in 324 B.C., the demagogues supported by the Sophists and the Platonists accused him of impiety and (as he himself explained it) "to prevent the Athenians from committing another crime against philosophy," he withdrew with his disciples to Euboea, where he died in 322 B.C. at the age of sixty-three. [On Aristotle and the beginnings of zoology, see Balme 1970.]

18. [Athenaeus, *Deipnosophistae, or Banquet of the Learned*], bk. 8, chap. 47 [see Yonge 1854].

19. [Ibid.], bk. 9, chap. 58. ["It is said that the Stagirite (Aristotle) received eight hundred talents from Alexander as his contribution towards perfecting his History of Animals" (Yonge 1854, 2:628).]

20. Pliny, *[Naturalis historia]*, bk. 8, chap. 17, v. 44. [Gaius Plinius Secundus, better known as Pliny the Elder, was born at Como in A.D. 23 and died in the famous eruption of Mount Vesuvius in A.D. 79 (Rackham 1938, vii).]

21. Pliny [ibid.], in bk. 8, chap. 17, mentions fifty books [see Rackham 1940, 35], but only twenty-five are extant: nine on the natural history of animals, four on the parts of animals, five on the generation of animals, one on the locomotion of animals, one on the senses and their organs, one on sleep and wakefulness, one on the movement of animals, one on the length of life, brevity and longevity,

one on youth, old age, life and death, and one on respiration. Twelve more are found in another enumeration, albeit quite incomplete, by Diogenes Laërtius [see Yonge 1853, 181–94]: eight on anatomical dissections, one on a selection of dissections, one on composite animals, one on mythological animals, and one on the causes of sterility; this last book is perhaps the one ordinarily included as the ninth volume on the natural history of animals. Also attributed to him is the treatise *De mirabilibus auscultationibus* [On the marvels of hearing], the best edition being that of Johann Beckmann (1739–1811), Göttingen, 1786, in quarto. The *Historia animalium,* which particularly interests us, should be read in the general edition of the works of Aristotle by [Guillaume] du Val [1572–1646], Paris, 1629; or in the special edition by [Julius Caesar] Scaliger [1484–1558], Toulouse, 1619; and especially the one by [Johann Gottlob] Schneider [see chap. 10, n. 10], published at Leipzig, 1811, in four octavo volumes. [Armand Gaston] Camus [1740–1804] published a French edition in 1783 at Paris, two volumes in quarto; the translation is generally accurate, but unfortunately he included notes that are not those of a naturalist. There is a German translation, with notes, by [Christian Friedrich Leberecht] Strack [1784–1852], published at Frankfurt am Main in 1816, one volume in octavo.

22. [Aristotle], *Historia animalium,* bk. 2, chap. 13; and especially *De partibus animalium,* bk. 4, chap. 13.

23. [Aristotle], *Historia animalium,* bk. 2, chap. 13.

24. [Aristotle], *De partibus animalium,* bk. 3, chap. 1.

25. [Aristotle], *Historia animalium,* bk. 2, chap. 13.

26. [Ibid.], bk. 4, chap. 8.

27. Ibid.

28. [Ibid.], bk. 2, chap. 13.

29. [Ibid.], bk. 2, chaps. 15–16.

30. [Aristotle], *De partibus animalium,* bk. 3, chap. 14.

31. [Aristotle], *Historia animalium,* bk. 2, chap. 17; *De partibus animalium,* bk. 3, chap. 14.

32. [Aristotle, *Historia animalium*], bk. 3, chap. 1.

33. [Ibid.], bk. 3, chap. 11.

34. [Ibid.], bk. 4, chap. 9.

35. [Ibid.], bk. 4, chap. 10.

36. [Ibid.], bk. 5, chap. 5.

37. He [Aristotle] designates the parts by letters, as is still done today. See *[Historia animalium],* bk. 3, chap. 1. [Aristotle seems to have been the first to illustrate a biological treatise, referring to diagrams several times in his *Historia animalium.* The figures have long since disappeared, but his descriptions are such that they can often be reconstructed with confidence (see Singer 1959, 25, figs. 12–14; for more on illustrations of Aristotle's zoological works, see Kádár 1978, 30–33).]

38. See [Aristotle, *Historia animalium*], bk. 5, chap. 4; bk. 6, chaps. 10–17; bk. 8, chaps. 2, 13, 20, 30; bk. 9, chap. 37; *De generatione animalium,* bk. 3, chaps. 4–5.

39. [Of the numerous hermaphroditic fishes in the Mediterranean (most of which belong to the families Serranidae, Sparidae, and Centracanthidae), the

most likely one referred to by Aristotle is *Serranus cabrilla* (which still bears the vernacular name *channos*) or, somewhat less likely, *Serranus scriba* (at present called *perdika*); a third species, *Serranus hepatus (channaki),* is even less likely, because sexually mature individuals occupy deep waters (60–120 m), making accurate observations of reproductive behavior next to impossible in Aristotle's time (Lejeune, Boveroux, and Voss 1980; C. Papakonstantinou, pers. comm., 26 June 1991).]

40. Theophrastus, born about 371 B.C. on the island of Lesbos of a fuller named Melanthus, was at first called Tyrtamus. Aristotle named him Theophrastus because of his divine eloquence. He was first a disciple of Leucippus and Plato. Tenderly loved by Aristotle, he succeeded him in his teaching post in 324 B.C.; he had more than two thousand students. He created one of the first botanical gardens. His two main works of natural history are *Enquiry into Plants,* in nine books [see Hort 1916], and *De causis plantarum,* in six books, a kind of plant physiology. He is much better known for his *Characters* [Edmonds 1929], which were translated and so deftly imitated [in 1699] by Jean de La Bruyère [1645–96]. It is said that he lived nearly a hundred years and that all of Athens turned out for his funeral.

41. [The genus *Ophicephalus* (now recognized as a junior synonym of *Channa* Scopoli, 1777) was originally described by Bloch in 1793 (see Bloch 1785–95, 7:137, pl. 358), but Cuvier is here most probably referring to Hamilton's (formerly Buchanan) account of the fishes found in the river Ganges (see Hamilton 1822, 59; see also Gudger 1924).]

42. Athenaeus, *[Deipnosophistae, or Banquet of the Learned],* bk. 7, chap. 1 [see Yonge 1854].

43. Ibid., bk. 8, chaps. 5, 6.

44. Ibid., bk. 7, chap. 17, and several other places in this book.

45. Ibid., bk. 8, chap. 18.

46. Ibid., bk. 1, chap. 22, and elsewhere throughout the book.

47. Alexander died in 324 B.C., Ptolemy-Lagus in 284 B.C., and Ptolemy-Philadelphus in 246 B.C.

48. Lucian, *Vitarum Auctio* [Sale of Creeds, v. 26: "Why, from him (the Peripatetic) you will find out in no time how long a gnat lives, to how many fathoms' depth the sunlight penetrates the sea, and what an oyster's soul is like" (Fowler and Fowler 1905, 204)]. Everyone knows Lucian of Samosata [in Syria], the famous Greek sophist and satirist [ca. A.D. 120 to ca. 180], contemporary [of the Roman emperors] Antoninus Pius, Marcus Aurelius, and Commodus [see Allinson 1905, vii–xxx; Fowler and Fowler 1905, vii–xiv; Levy 1976, xiii–xxiii].

49. Lucius Apuleius, of Madaura in Numidia, also a contemporary of the Antonines [born ca. A.D. 125; the date of his death is unknown (Butler and Owen 1914, vii–xix; Gaselee 1915, v)]. [Platonic philosopher and rhetorician], author of the unusual novel *Metamorphoses* or *Golden Ass* [see Darton 1924], he devotes twenty pages in the first part (chaps. 1–65) of his *Apologia* to justifying his curiosity in the study of fishes and proving it was not for magical operations ["I do not know of fish that are useful for magic . . . therefore, if you do know of such fish, it is clear that you, rather than I, are the magician" *(Apologia,* chap. 30, v. 4;

see Butler and Owen 1914, commentary, 75]. It can be seen from this discourse that he wrote much on the subject of fishes, but nothing has survived. [Numerous scientific works by Apuleius have been lost. He himself mentions his *Quaestiones naturales,* and especially a treatise on fishes titled *De piscibus (Apologia,* chaps. 36, 38, 40; see Butler and Owen 1914, xxvii–xxviii).]

3 Roman Ichthyology and the Silence of the Middle Ages

Although the Romans never favored purely speculative sciences, they were interested in fishes; but it was from an economic standpoint, and also to gratify a love of pleasure that, despite its excesses, could not exhaust the world's riches amassed by the world's oppressors.

Varro[1] and Columella[2] wrote in the first of these two aims. One sees from these authors that at the time of Cicero and Augustus, freshwater fishponds were already common and that the rich built fishponds along the seashore and filled them with seawater, the upkeep of which was very expensive.[3] Licinius Muraena provided the example, which was soon followed by men of the highest nobility, Philippus and Hortensius.[4]

Several of these establishments were of an astonishing magnitude. Gaius Hirrius one day lent Caesar two thousand moray eels *(murènes)* taken from his ponds.[5] Turbots *(turbots)* and soles *(soles),* sea breams *(dorades),* drums *(sciènes),* and every sort of shellfish *(coquillages)* were raised in the saltwater ponds.[6] Each species of fish had its own compartment in these ponds.[7]

These amateurs never complained about expense. Lucullus cut through a mountain near Naples, at immense cost, in order to introduce seawater into one of his fishponds, causing Pompey jokingly to call him Xerxes in a toga.[8] It is said that one master, Vedius Pollio, practiced cruelty to the point of having his slaves thrown into his ponds to feed his fishes.[9] We read elsewhere that goatfishes *(mulles)* were introduced into houses by means of small channels so that the occupants could watch them die before they were eaten,[10] and the extravagant price paid for one of these fish, only because it was longer than the others.[11]

Before long, local species did not suffice for these rich people who were jaded by having all pleasures. A navy admiral, Optatus, commander of the fleet, was employed to stock the Tuscan sea with a species of parrotfish *(scare)*, which until then lived only in the seas of Greece.[12] Meanwhile, the great fisheries and salteries became more extensive: fishes were sought beyond the Pillars of Hercules, and some thousands of men were occupied in providing the capital of the world with fishes.

One can imagine what this abundance might have meant to the science of fishes if the desire for observation and the methods of Aristotle had been preserved; but that being far from the case, no one wrote from nature, and all the works of the Romans, and those of the Greeks who lived under their sway, consisted only of lists taken from Aristotle or one of the other authors of his school. Excerpts from more recent travelers, too often filled with fables, were the only augmentation to natural history at the time. These include the short poem, the *Halieutica*, attributed to Ovid[13] by Pliny, who thought he found in it names of fishes not known to other authors that he thought Ovid must have seen in Pontus, near which he had commenced this work. But despite Pliny's authority on the subject, it is today a matter of dispute whether the poem is Ovid's; what is certain is that three of the names Pliny says are not found elsewhere—orphus *(orphus)*, mormyrus *(mormyrus)*, and chryson *(chryson)* (the last written *Chrysophrys* in the poem)—are all mentioned by Aristotle.

This little poem of 134 lines names fifty-three fishes in all and gives interesting details on the habits of some of them,[14] but they are very probably borrowed from elsewhere. The faculty he attributes to *Serranus (channa)* of conceiving without the male, which Pliny says was a theory advanced only by Ovid, was already to be found in the works of Aristotle,[15] but the philosopher presents it with expressions of doubt: it was more poetic not to doubt, and it must be admitted that, however singular the fact may seem, there are good reasons for giving it credence.[16]

Pliny himself [fig. 6] is well known merely for having assembled in his immense work, without much order or criticism, what he found in the works of Aristotle and other Greeks, or those of more recent Roman historians and travelers.[17] His book 9 is especially devoted to fishes and their natural history. In book 32, chapter 11 he provided a list of the names of aquatic animals, which amounts to 174; but after deleting shellfishes, cetaceans, and other animals that are not true fishes, there remain only 95 or 96, among which one can still suspect some duplication. About 30 do not seem to be cited by Aristotle, not even under their Greek equivalents.

Pliny cited as his authorities for book 9 eight Greek authors and eighteen Latin authors. Of the former, written works have survived for only two, Aristotle and Theophrastus, and for the latter, only three, Cicero, Nepos, and Seneca. Among Pliny's authorities for book 32 he included seven more Greek authors, of which the work of one, Nicander, is extant, and five Latin authors, of which we possess only the poem attributed to Ovid.

FIG. 6.
Gaius Plinius Secundus, better known as Pliny the Elder, A.D. 23–79. After Thevet 1584, 2:612. Courtesy of M. Ducreux and the Bibliothèque Centrale, Muséum National d'Histoire Naturelle, Paris.

This survey of thirty-eight authors does not fail to offer some curious facts, several of which, concerning the fishes of the Indian Ocean, are attributable to the Greeks and are subjects of Alexander or his successors. In the North Sea, for example, the sawfish *(scie)* and sperm whale *(cachalot)* began to receive attention; but one also reads of mermen, mermaids, and sea cows *(hommes, femmes, taureaux de mer)*, all products of the traveler's imagination [fig. 7]. The mullet *(muges)* fishery in the ponds of Languedoc is described, as well as many interesting details on Roman luxury relative to fishes. It would be difficult to recognize several of the species Aristotle mentioned if one did not have the particulars Pliny reports, in addition to what Aristotle said; but often Pliny himself would not be understood if one could not explain him by using passages from Horace, Seneca, Juvenal, and Martial,[18] for everything about gluttony is reported by the moralists and poets, and with much more care than by the naturalists.

Nevertheless, science as such, in the sense of being universal and methodical, is in no way advanced by Pliny's compilation; he took from Aristotle all the generalities and the little that Aristotle provided on the organization of fishes.[19]

FIG. 7. *Anthropomorphes,* mermaid and merman, products of the traveler's imagination. Reproduced from Hendrik Ruysch's 1718 edition of John Jonston's *Historia naturalis,* vol. 1, De Piscibus, pl. 40. T. W. Pietsch collection. Photograph by Patricia L. McGiffert, Health Sciences Photography, University of Washington, Seattle.

Authors of natural history who came after Pliny, and who wrote in Greek—Oppian, Athenaeus, Aelianus—were no better at observation than he, and like his, their works' only scientific merit is that they have preserved passages from the works of authors and travelers now lost.

The *Halieutica* of Oppian[20] is a poem of five cantos on fishing, in which it may be believed that he brought together everything of interest that was to be found in his predecessors. Several facts in it are found in the works of Aristotle, Ovid, Pliny, and others, but there are some that are not to be found anywhere else. All facts are expressed in poetic language, and before making use of them it is best to rid them of the ornamentation with which imagination has embellished them. Judged in this light, the facts contribute to the identification of some species. The fishes Oppian named come to 125; among them are 26 not found in the works of other authors, and about 10 that are found only in Aelianus.

Athenaeus,[21] in a composition as dull as it is improbable, described several erudites coming together for a meal, discoursing on the food served and everything to do with the food, and reciting or reading on this subject a multitude of passages, often very long, taken from more than eight hundred authors of all periods, most of which have escaped oblivion only because they are cited in this composition. Of course natural history is still interested in these passages, and in fact it can use them to determine nomenclature; but of all its branches there is none that can profit more than ichthyology. About 130 fishes are named in it, and among them are 30 that appear for the first time. Poets, historians, physicians, and naturalists are all requisitioned. As one might expect, the quotations often do not clarify anything, but occasionally some shafts of light shine through, or else one could not guess the sense of many of the names used by the ancients. It is here especially that we learn how important an object of interest fishes were in everyday life.

Claudius Aelianus[22] left a treatise of seventeen books on the characteristics of animals. Never has there been a spirit more contrary to method than the one that presided over this compilation, where everything is pell-mell; but the true and valuable facts found in it are extremely numerous. Aelianus certainly had better information than his predecessors on the animals of Africa and India, which is a sign that relations had become easier with these countries. He named approximately 110 fishes, about 40 of which are not in Aristotle but correspond in part with those found in Athenaeus, Pliny, and Oppian.[23]

Ausonius[24] is perhaps the only Latin author writing about fishes who was not a compiler; he described the fishes of the Moselle from his own observations, and he described them not only as a poet but also as a naturalist. He named fourteen of them, all recognizable from what he says about them, most being unknown to the Greeks and Romans, or at least given different names by them. It is here for the first time that we are introduced to the salmon trout *(truite saumonée)*, the common river trout *(truite commune)*, the barbel *(barbeau)*, and other freshwater fishes.

One may add to the works we briefly analyzed above some passages from Strabo,[25] Pausanias,[26] Plutarch,[27] and Apuleius;[28] what Dioscorides and Marcellus Sidétès wrote about fishes used as medical remedies;[29] what Galen[30] took from Philotimus about their qualities as food; and what Oribasius[31] took from Xenocrates[32] on the same subject. In so doing, one can get a fairly complete idea of the knowledge of ichthyology in ancient times, for it is hardly worth the effort to mention here some commentators on the first chapter of the book of Genesis, such as Ambrose,[33] Eustathius,[34] and Pisides,[35] who said nothing about fishes that was original and not commonly known. If I were to mention Philes,[36] I would be talking no longer about the ancients but about a writer from the end of the Middle Ages. Moreover, although a bit more extensive, he like the others is only a copier of the authors from the Golden Age.

Now, from a careful comparison of all these works it seems to me that the ancients distinguished and named about 150 species of fishes, or nearly all the edible species of the Mediterranean Sea; but they did not precisely establish their characteristics and did not even dream of methodically classifying them, so they themselves were constantly confused in their nomenclature. As for the organization of the class of fishes in general, no one after Aristotle had dealt with it. The decadence of the Peripatetic school meant the decline of all direct research on nature;

natural history was no longer a subject except for compilers who understood nothing about the fundamental nature of things, and with regard to this branch of the sciences, the barbarians had nothing to do with it: it no longer existed when they made their invasion. The nine centuries that followed were scarcely more favorable to the science; the monks in their cells, almost the only depository of knowledge during the long sleep of the human spirit, had no means of performing observations. Those among them who showed more curiosity and intelligence were reduced to making excerpts from the imperfect copies that remained to them of Pliny or Aristotle. And the latter for a period of time was unknown to the monks except through translations made not from the Greek but from Arabic.

This character of the ignorant compiler is seen in the chapters that Saint Isidore devoted to natural history in his *Treatise on Origins.* In the chapter where he speaks of fishes,[37] he names some thirty species and searches for the etymology of their names, but in a childish way, sometimes deriving a Greek word from Latin or fixing on the material resemblance of sounds. However, one or two characteristic traits are found here that cannot be found elsewhere.

Albert the Great [fig. 8],[38] worthy of a better century, based his treatise on animals on a broad and regular plan; but in executing it he seems to have had before him only very faulty copies of Pliny and perhaps a Latin version of Aristotle based on some Arabic version. In his chapter on fishes he cited Pliny especially but succeeded in mutilating him, writing, for example, *Tygrius* for *Thynnus, Solaris* for *Silurus,* taking the word *exposita,* which modifies Andromeda, for the name of the fish she was exposed to, and so forth; he changed the meaning of several names and introduced a great number of completely barbarous ones. However, among the sixty-three fishes he talks about there are some he observed himself, such as his second *Alech,* which is the herring *(hareng);* his *Amger,* which is the needlefish *(orphie);* his first *Esox,* which represents an old male salmon *(bécard);* one of his eels *(murènes),* which is the small river lamprey *(lamproie de rivière),* and so on.

His contemporary and a member of his order, Vincent de Beauvais,[39] consulted more sources, and especially an anonymous author of *De la nature des choses,* who is unknown except from Vincent's quotations and who seems to have reported several facts according to his own observations. Also, Vincent had better copies of Pliny than Albert did and made much use of Isidore's *Origins.* His articles on fishes, which are almost as numerous as Albert's but much more extensive and more nearly correct, could very well be earlier than Albert's, the later perhaps even taken from them. What is certain is that their resemblance to each other is such that where they differ from the ancients they must have at least used the

FIG. 8.
Albert the Great, perhaps better known as Albertus Magnus (1193–1280).
After Walsh 1906–17, ser. 2, opposite p. 21.

same sources. Vincent too wrote about the herring *(hareng)* and made mention of the season in which they appear and the practice in his time of salting them and thus sending them great distances.

NOTES

1. Marcus Terentius Varro, who is considered the most erudite of the Romans, was born in 116 B.C. and died in A.D. 28. We mention here only his treatise *De re rustica* [On agriculture; see Store-Best 1912].

2. Lucius Junius Moderatus Columella [flourished about the middle of the 1st century A.D.2], born in Cadiz, contemporary of Claudius, author of a work in twelve books [also titled] *De re rustica* [see Ash 1941; Forster and Heffner 1954–55].

3. Varro, *De re rustica,* bk. 3, chap. 17. ["Those [fishponds] of the one kind, in which water is supplied to our home-fed fishes by the river Nymphs, are kept by men of the people, and are profitable enough; while the other sea-water ponds which belong to the nobles, and get both water and fishes from Neptune, appeal more to the eye than to the pocket, and empty rather than fill the owner's purse" (see Storr-Best 1912, 347).]

4. Pliny the Elder, *[Naturalis historia]*, bk. 9, chap. 80, [v. 170].

5. Ibid., chap. 81 [v. 171]. [According to Rackham's (1940, 279) translation of Pliny, Hirrius provided Caesar with "six thousand lampreys." But use of the term "lamprey" for the Latin *muraena*, as written by Pliny and Seneca, is undoubtedly wrong (M. L. and R. Bauchot, pers. comm., 18 November 1993; see also n. 9 below). Certainly one cannot imagine a human, or anything else for that matter, being devoured so voraciously by lampreys; moray eels are, however, another matter.]

6. Columella, *[De re rustica]*, bk. 8, chap. 16 [see Forster and Heffner 1954–55, 2:405].

7. Varro, *[De re rustica]*, bk. 3, chap. 17 [see Storr-Best 1912, 348].

8. Pliny the Elder, *[Naturalis historia]*, bk. 9, chap. 54. [It is said that Xerxes (519?–465 B.C.), son of Darius the Great, king of Persia (d. 485 B.C.), cut a channel for his fleet through [the isthmus of the peninsula of] Mount Athos (Rackham 1940, 278).]

9. [Lucius Annaeus] Seneca, *De ira* [On anger], bk. 3, chap. 40 ["When one of his slaves had broken a crystal cup, Vedius ordered him to be seized and doomed to die, but in an extraordinary way—he ordered him to be thrown to the huge lampreys [moray eels; see n. 5 above], which he kept in a fish-pond" (translation of Basore 1928, 349)]. Seneca, *De clementia* ["On Mercy"], bk. 1, chap. 18. ["Who did not hate Vedius Pollio even more than his own slaves did, because he would fatten his lampreys [moray eels] on human blood, and order those who had for some reason incurred his displeasure to be thrown into his fish-ponds" (Basore 1928, 409).]

10. Ibid., *Quaestiones naturales*, bk. 3, chap. 18 [see Clarke 1910, 130–32].

11. Ibid., *Epistulae morales ad Lucilium*, 95 [see Campbell 1969].

12. Pliny the Elder, *[Naturalis historia]*, bk. 9, chap. 29 [v. 62].

13. Ovid [Ovidius Naso] needs no introduction; it suffices here to note the year of his birth, 43 B.C.; the year of his exile, A.D. 10; and the year of his death, A.D. 17. It was during these last seven years that his poem *Halieutica* would have been written, if it was indeed written by him [see Richmond 1962].

14. The *Halieutica* was published in the works of Ovid [Richmond 1962] and in the collection *Poetae Latini minores* [Baehrens 1879–83]. Some critics attribute the poem to Gratius Faliscus.

15. [Aristotle], *Historia animalium*, bk. 6, chap. 12.

16. See the treatise by [Filippi] Cavolini [1756–1810] on the reproduction of fishes [1787].

17. Pliny the Elder (Caius Plinius Secundus), one of the most studious and most scholarly men of antiquity, was born at Verona in A.D. 23, studied at Rome, visited the coasts of Africa, served in the Roman armies in Germany, was in Spain during the civil wars that followed upon the death of Nero, and died, commanding the Roman fleets, in A.D. 79 as a result of having observed, with too little caution, the great eruption of Mount Vesuvius. His *Historia naturalis* [first published in Venice in 1469], in thirty-seven books, dedicated to Titus, is his only extant work. He spent a large part of his life gathering materials for it with an unimaginable ardor and perseverance. It is composed of excerpts from more than

two thousand volumes written by authors of whom we possess only about forty volumes.

18. For authors so well known, and whose writings have no direct relation to our subject, we believe we need only note the time when they lived: Horace (Quintus Horatius Flaccus), born at Venusia in 66 B.C., died at age fifty-seven in 9 B.C. Seneca (Lucius Annaeus Seneca), born at Cordoba [Spain] in 4 or 5 B.C. [although some put it as early as 8 B.C. or as late as A.D. 4; see Campbell 1969, 233], put to death in A.D. 65 by order of Nero, his pupil, wrote also on physical science in the seven books of his *Quaestiones naturales;* in it he speaks of fishes that live underground and of the Roman passion for goatfishes *(mulles)* [see Clarke 1910, 129, 130–32]. Martial (Marcus Valerius Martialis) was born at Bilbilis in Celtiberia about the year A.D. 40 and died about the year 100; he wrote mainly under Domitian. Juvenal (Decius Junius Juvenalis), contemporary of Martial, seems to have survived him, but neither the year of his birth nor the year of his death is precisely known [ca. A.D. 60 to ca. 140]; we shall have occasion to cite many passages from these two poets, especially Martial.

19. The best edition of Pliny is the second by [Joannes] Harduin (dated 1723), in two volumes in folio, reproduced by [Johann Georg Friedrich] Franz [1737–89] from 1778 to 1791, in ten volumes in octavo. One might also mention the translation by [Louis] Poinsinet de Sivry [1733–1804], with notes by [Jean Etienne] Guettard [1715–86] and [Anselme Gaëtan] Desmarest [see chap. 13, n. 26], [twelve volumes in quarto, 1771–82]; and for animals, that by [Pierre Claude Bernard] Gueroult [published in 1802], with notes excerpted from [Georges Louis Leclerc de] Buffon, [Jacques Christophe Valmont de] Bomare [1731–1807], and other naturalists. But none of the commentators are knowledgeable enough about natural science to explain their author or to distinguish between what was founded on truth and what was derived from false or exaggerated reports. I have tried another method in my notes on the books of Pliny relative to animals, included in the new edition by [Nicolaus Eligius] Lemaire, Paris, 1827–32.

20. Oppian of Anabarzus in Cilicia was born toward the end of the reign of Marcus Aurelius [r. A.D. 161–80]. His father was a senator of that town who fell into disfavor with Severus and was condemned to exile. Oppian's poetry so much pleased Caracalla that it is said his father was pardoned and he was given a stater of gold for each line of poetry. Thus his date may be fixed at the beginning of the third century. He died of a plague in the town of his birth at about age thirty. The fifth book of his *Cynegetica* [the chase; see Mair 1928, 2–199] and all of his *Ixeutica* are lost, in which he discussed bird hunting, but his *Halieutica* [on fishing] has survived in its entirety [for an English translation, see Jones 1722; Mair 1928, 200–531]. [J. M.] Limes has just published a French translation, Paris, 1817, in octavo. [Jacques Nicolas] Belin de Ballu [1753–1815] translated the *Cynegetica* and published a fine edition in Greek and Latin, Strasbourg, 1786. The best complete edition of Oppian is that of Schneider, Strasbourg, 1776.

21. Athenaeus, author of the *Deipnosophistes,* or *Sophists at Table,* was from Naucratis in Egypt. The supper he recounts, and at which he was present, is supposed to take place at the house of Laurentius, who had been entrusted by Marcus Aurelius with confidential missions; consequently he must have lived in

the second century, and yet he cites Oppian, who did not write until the beginning of the third. Belin de Ballu believes the citation does not come from Athenaeus, but rather comes from the author who abridged the first two books of his works. In fact, we have the first two books in abridged form only; the others exist in their entirety, or mostly so. There are fifteen in all. The best edition has long been that of [Hieronymus] Commelin, 1597, in folio, with translation by [Jacob] Dalechamps and commentary by [Isaac] Casaubon [1559–1614]; but [Johannes] Schweighaeuser has published one that is much to be preferred, Strasbourg, 1801–7, fourteen volumes in octavo. [Jean Baptiste] Lefebvre de Villebrune has published a French translation, Paris, 1789, five volumes in quarto, somewhat imperfect, especially as regards natural history, but far superior to the one by [Michel de Marolles] the abbé de Villeloin, Paris, 1680, one volume, in quarto.

22. The time of Aelianus, author of *De natura animalium,* is not certain. He is generally placed in the second century or the beginning of the third, because some believed him to be the same as one Claudius Aelianus of Praeneste, who was a teacher of rhetoric at Rome after the reign of Antoninus Pius. Others believed him to be a sophist named Aelianus whose life was written by Philostratus and who some say died after Commodus (others say after Elagabalus). Of course it is not impossible that these two are the same man, but neither is said to have written on the subject of natural history.

23. At first there was only an excerpt of Aelianus, translated and arranged in completely different order by [Pierre] Gilles [1490–1555], Lyons, 1535, in quarto. Conrad Gessner [1516–65] published the text and completed the translation in 1556. The finest edition is by Abraham Gronovius [1695–1775], London, 1744, two volumes in quarto. This is the edition (Leipzig, 1784, in octavo) that Schneider has followed [see chap. 10, n. 10], to which he made a number of useful additions, among them an orderly catalog of the animals that appear in the work.

24. Decius Magnus Ausonius, born at Bordeaux [about A.D. 310], tutor to the emperor Gratian, consul in 379, died in 395, has among his poems a short one on the Moselle River, in which he writes about the fishes of that river; it can be found in all editions of his works [see White 1919, 225–63] and in *Poetae Latini minores* [Baehrens 1879–83].

25. Strabo, author of *Geographia,* was born at Amasia in Capadocia about 58 B.C., and died under Tiberius [about A.D. 24; see Jones 1917, 1:xiv–xxi]. In the seventeenth book [pt. 2, chap. 4, of *Geographia*] he names some fishes of the Nile, and in other parts of the same work he talks about fishing for tunas *(thons)* and bonitos *(pelamydes).* [For an English translation of Strabo's *Geographia,* see Jones 1917.]

26. Pausanias [the Traveler], who flourished under [the Roman emperor] Antoninus Pius, compares the fishes of Greece and Egypt in his *Description of Greece* [bk. 4, "Messenia," chap. 34; see Jones and Ormerod 1926, 357–59]. [On the life and works of Pausanias, see Jones 1918.]

27. Plutarch of Chaeronea in Boeotia [Greek biographer and writer], consul and administrative officer in Illyria under Trajan, died in A.D. 140. There are some characteristics in his *Philosophical Essays* that have been taken from [Aristotle's] *Historia animalium.*

28. [On Apuleius, see chap. 2, n. 49.]

29. Pedacius Dioscorides of Anazarba in Cilicia, believed to have lived under Nero [fl. A.D. 50–70], mentions five or six fishes in the second book of his *Materia medica,* but only in connection with their harmful or beneficial medical qualities [e.g., "*Hippocampus* is a little living creature of the sea, which, being burnt & the ashes thereof taken either in Axungia, or liquid pitch, or unguentum Amaracinium, & anointed on, doth fill up the Alopecia with haire"; "The sea Dragon *[Trachinus draco],* being opened & soe applyed, is a cure for ye hurt donne by his prickles" (see Gunther 1934, 93–108)]. Of Marcellus Sidétès [or Sideta, a medical writer, born at Sida in Pamphylia] who was contemporaneous with Antoninus the Pious [A.D. 117–61], we have only a fragment of a poem in which he gives the names of about sixty fishes but no other indication *[Iatrica* or *De remediis ex piscibus* (Remedies derived from fishes), a short verse of 101 lines, published for the first time by Fédéric Morel (1558–1630), 1591, in octavo].

30. Galen [Claudius Galenus] is another of those men too well known to need any introduction save his dates. Born at Pergamum, Asia Minor, in A.D. 129; after studying at Alexandria, he went to Rome in 169, was physician to [the emperor] Marcus Aurelius, and returned to Pergamum after the death of this prince, where he died in about 200. He is the last of the ancient anatomists [for more on Galen, see Brock 1929, xvi–xl; Cole 1949, 42–47]. In his treatise *De alimentorum facultatibus* [On nourishment], bk. 3, chaps. 24–31, he speaks of a fairly large number of fishes relative to the quality of their flesh.

31. Oribasius [of Pergamum] was physician to the emperor Julian in the middle of the third century. In the second book of his *Collecta medicinalia,* after copying Galen's chapters mentioned above, he adds a rather long one (chap. 58), taken from a treatise by Xenocrates on fishes as food, in which are found names and useful characteristics.

32. Who this Xenocrates was is not known for sure. Some suppose, without much foundation, that he is the same man as the Academic philosopher, second successor to Plato.

33. Saint Ambrose, archbishop of Milan, born about 340, died in 397. The first eleven chapters of the fifth book of his *Hexameron* [The six days of creation] are on fishes [see Savage 1961, 159–90].

34. Saint Eustathius [fl. A.D. 325], bishop of Antioch, one of the prelates at the Nicaean Council, in his *Hexameron,* or *Commentary on the Work of the Six Days,* says only a few words about the sawfish *(scie),* the parrotfish *(scare),* the remora *(écheneis),* and the thresher shark *(renard marin).*

35. George of Pisidia, deacon of Constantinople in the seventh century, speaks of the same fishes in a Greek poem that also has the title *Hexameron,* and so on.

36. Manuel Philes, born at Ephesus about 1275, died about 1340, put into political verses some characteristics from the natural history of animals borrowed from Aelianus. His poem, like Aelianus's book, is titled *De animalium proprietate.* The best edition is that of [Joannes] Cornelius de Pauw, Utrecht, 1730, in quarto.

37. [Isidore], *Originum,* bk. 12, chap. 6. Saint Isidore, archbishop of Seville, lived at the end of the sixth century [560–636], at the time of the emperor Maurice

and King Recaredo. He wrote many works on theology, history, and learning. The twelfth book of his *Origins* is the only one of his writings of interest to naturalists. It is found in the edition of his works printed in Paris in 1601, one volume in folio.

38. Albert the Great [Saint Albert, bishop of Ratisbon, also known as Albertus Magnus], from the family of the counts of Bollstedt, was born at Lauingen in Swabia in 1193. After studying at Padua, he taught the philosophy of Aristotle in Paris and acquired fame as a professor. In 1221 he entered the order of Preaching Friars or Dominicans and became provincial of Germany in 1254, then master of the holy palace at Rome, and in 1260, bishop of Ratisbon. He finally returned to the convent, where he died in 1280. His works in the edition of Lyons, 1651, occupy twenty-two large volumes in folio. The sixth volume contains his treatise on animals, and the twenty-fourth book of this treatise [pp. 645–69] is on fishes [see Albertus Magnus 1651]. [For more on Albertus Magnus, see Walsh 1906–17, 2:21–58.]

39. Vincent de Beauvais [or Vincentius Bellovacensis], a Dominican who is presumed to have died in 1256, and whom Albert the Great survived by twenty-four years, compiled a work truly prodigious for the number of objects it comprises, which could be called the encyclopedia of the Middle Ages. His *Bibliotheca mundi sive Speculum majus* is divided into four parts, the first of which, titled "Speculum naturale," contains in one enormous volume in folio the whole of physical science and natural history. It is claimed that the king (either Philip Augustus or Saint Louis [Louis IX]) procured books for him and furnished him with the copyists and assistants necessary for this immense undertaking. It is in the seventeenth book that he treats of fishes. The best edition of his works is that published at Douai in 1624 [see Vincentius 1624].

Part Two

THE SIXTEENTH-CENTURY REVIVAL

4 Belon, Rondelet, and Salviani, Their Contemporaries and Immediate Successors

Better times came. A great movement had been aroused in the human spirit as early as the thirteenth century and in the fourteenth century by such people as Dante, Petrarch, and Boccaccio; the end of the fifteenth century was the moment of its maturity.

Greeks fleeing from Constantinople had spread knowledge of the ancient classics of their nation[1] and in particular had produced better translations of Aristotle;[2] printing had been invented;[3] America had been discovered,[4] and India was being colonized.[5] Learning was reborn, and with it natural history, which at the same time saw an infinitely wider theater become open for research.

Ichthyology was among the first to revive under these happy auspices. The first concern of those who were devoted to the science was to reexamine and explain the extant works of the ancients; in those initial moments, it was there that one hoped to find truth.

At the beginning of the sixteenth century, Massaria attempted to comment on Pliny's book 9.[6] The eloquent Italian historian Giovio, in a work on the subject, took the trouble to research the ancient names of Roman fishes.[7] He described forty-two species, approximately in order of size, and included in the accounts some characteristics that even today are not without interest for naturalists.

Gilles[8] proposed to do the same in his *Traité des noms français et latins des poissons de Marseille.* His articles are shorter but more numerous. He spoke of ninety-three fishes and sometimes had rather good solutions for ancient nomenclature. Moreover, the same author elsewhere rendered a real service in translating Aelianus, putting him in better order and including excerpts from some other authors, which, although still less than perfect, makes the study less tedious.[9] Books 11, 12, and 13 are about fishes. Assembled under each name are the various articles relating to it, which had been scattered about in Aelianus and elsewhere. But because Gilles does not give citations, one cannot readily go back to the sources.

The same virtue and the same fault are to be found in the book by Edward Wotton,[10] *De differentiis animalium.* Using characteristics borrowed from the ancients, the author put them in order and wrote in a uniform style; in a word, he made one book, but he does not cite his sources or does so only here and there. His book 8 is on fishes, but it seems to contain nothing new.

Lonicer,[11] who included some pages on fishes in his *Naturalis historiae opus novum* of 1551, has not even the advantage of having made a good copy of the ancients; his translations of names into modern languages are faulty and his drawings fanciful.

But the three great authors who truly founded modern ichthyology appeared in the middle of the sixteenth century and, what is remarkable, almost at the same time: Belon in 1553, Rondelet in 1554 and 1555, and Salviani from 1554 to 1558. All three, contrary to the compilers who fill our list after Aristotle and Theophrastus, personally saw and examined the fishes they spoke of and saw to it that they were drawn with some exactitude; and yet, too faithful to the spirit of their time, they applied themselves more to finding the names of these fishes in antiquity and composing their natural history from fragments taken from the authors where they believed they had found these names than to describing the fishes in a clear and complete manner, so that were it not for their drawings it would be almost as difficult to determine their species as those of the ancients.

Belon's[12] drawings are the least helpful among those of the three authors, but his conjectures are not necessarily so; because he had traveled in Turkey and Egypt, his writings shed light on nomenclature in use today in the Archipelago, which sometimes aids us in retrieving the nomenclature used by the ancient Greeks. In his *De aquatilibus, libri duo,* he provided drawings of 110 species of fishes, including 22 cartilaginous species and 17 freshwater species, the rest being marine species; and he discussed about 20 other species for which he gives no drawings [see table 1]. Most of these drawings are recognizable, although roughly drawn. Nearly all the marine fishes are Mediterranean, but there are also some species from the Paris market. Some of the drawings are reproduced in his short treatise *Estranges poissons de marins* [1551], and in his *Observations* [1553b] he added a drawing of a fish he believed to be the parrotfish *(scare)* of the ancients, which no one has seen again since his time.[13]

The drawings by Salviani,[14] not so numerous but much finer, are copperplate engravings on a rather large scale [fig. 9]; some have not been surpassed in more recent works. They number ninety-nine; almost

FIG. 9. Frontispiece of Hippolyte Salviani's *Aquatilium animalium historia*, 1554–58. Courtesy of Eveline Nave Overmiller and the Library of Congress, Washington, D.C.

all are of fishes of Italy with some from Illyria and the Archipelago, not counting a few mollusks.

Rondelet[15] is superior to the other two authors in the number of fishes he knew, and even though his drawings, which are wood engravings, do not compare with Salviani's for beauty, they have greater accuracy and are especially remarkable for their characteristic details [fig. 10]. The artist who drew them was assuredly one of the most useful men in ichthyology, and it is regrettable that we do not know his name.[16] There are 197 drawings of marine fishes and 47 freshwater, not counting cetaceans, reptiles, and mollusks. No one before Risso knew Mediterranean

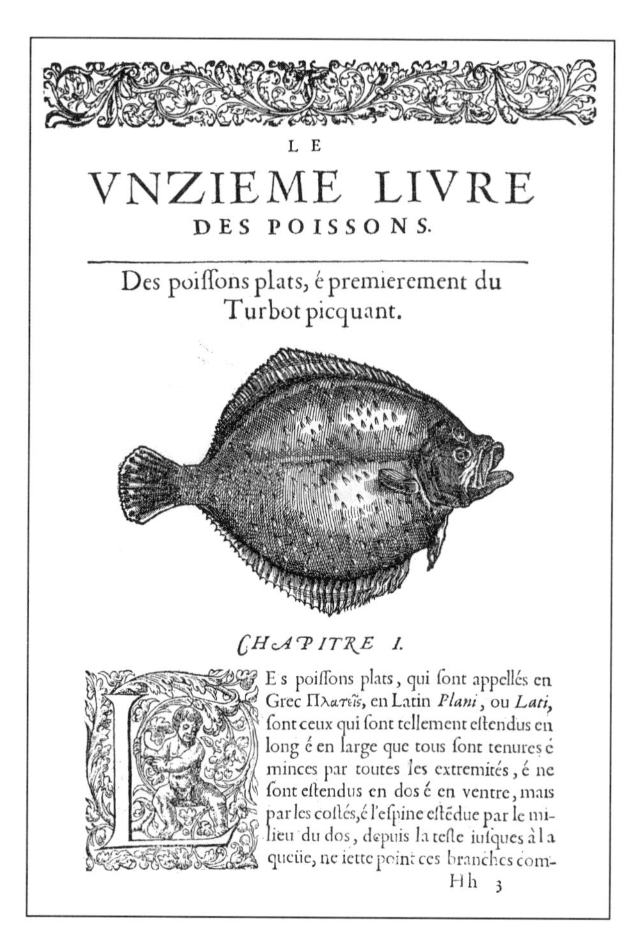

LE

VNZIEME LIVRE

DES POISSONS.

Des poiſſons plats, é premierement du Turbot picquant.

CHAPITRE I.

LEs poiſſons plats, qui ſont appellés en Grec Πλατεῖς, en Latin *Plani*, ou *Lati*, ſont ceux qui ſont tellement eſtendus en long é en large que tous ſont tenures é minces par toutes les extremités, é ne ſont eſtendus en dos é en ventre, mais par les coſtés, é l'eſpine eſtédue par le milieu du dos, depuis la teſte iuſques à la queüe, ne iette point ces branches com-

Hh 3

FIG. 10. The introductory page of book 11 of Guillaume Rondelet's *Histoire entière des poissons,* 1558, pt. 1, p. 245. T. W. Pietsch collection.

fishes as well as Rondelet, and even today it would be impossible to give a natural history of these fishes that would be the least bit complete without consulting him. The reader will see in the course of our work more than one species that Rondelet already knew and that we have been the first to find again. He also often provided observations on their anatomy, which we have been able to verify. Without exactly having a method in the accepted sense of the word today, it can be seen that he had a true sense of genera; he grouped several species with fair accuracy: drums *(sciènes),* wrasses *(labres),* blennies *(blennies),* herrings *(clupes),* mackerels *(scombres),* jacks *(centronotes),* mullets *(muges),* codfishes *(gades),* gurnards *(trigles),* flatfishes *(pleuronectes),* rays *(raies),* sharks *(squales),* eels *(murènes),* minnows *(cyprins),* and trouts *(truites)* are all grouped in his book in such a way that Willughby, and Artedi and Linnaeus after him, had little difficulty in making true genera of them.

FIG. 11. Conrad Gessner (1516–65), an engraving by John Theodore de Bry originally published by Jean Jacques Boissard in his *Icones quinquaginta virorum illustrium*, 1599, pt. 4, pl. 23. After Adler 1989, p. 7.

When these three works were published, Gessner [fig. 11][17] was engaged in the part of his great *Historiae animalium* that deals with aquatic animals. Instead of following the excellent plan he used in the two preceding parts—namely, arranging under certain rubrics the passages of authors from all the ages concerning each species—he inserted articles by Belon and Rondelet, and several by Salviani, as these articles appeared in their books, adding under the heading "Corollary" the passages they had not cited. Such a procedure made this part of his compilation much less useful, because one cannot discern what the ancients said except through the ideas and methods of these moderns. On the other hand, Gessner added to their illustrations, which he copied, many other

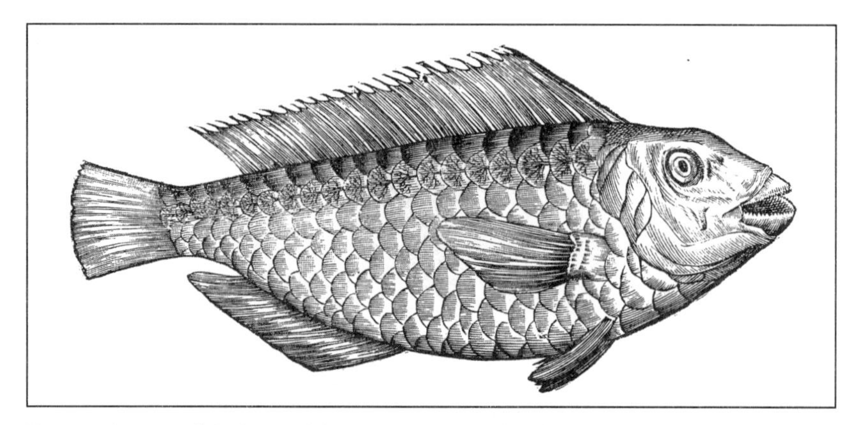

FIG. 12. A parrotfish from Aldrovandi's *De piscibus libri V, et de cetis liber unus,* 1638, p. 8. T. W. Pietsch collection.

illustrations and articles on the fishes of Venice, England, and Germany, which he himself had observed or about which his friends had sent him information. The number of illustrations in the first edition is more than seven hundred, but that includes cetaceans, mollusks, and in general everything that lives in water. There is no attempt at method, and everything is in alphabetical order. Nevertheless, during the rest of the sixteenth and seventeenth centuries, and even part of the eighteenth century, Gessner was the chief authority on all vertebrate animals.

As regards fishes in particular, Aldrovandi[18] and his editor Uterverius hardly did anything but abridge the work of Gessner, reduce it to their own plan, and add to the illustrations they took from it a certain number of new illustrations, among which are in fact several made after nature and that still have some value, although roughly engraved in wood [fig. 12]. Most of the species came from Italian seas, but there are also some from distant countries that were becoming better known.

In fact, discoveries were continuing in the two Indies; colonies were being established there; stories were written that aroused curiosity about the singular natural phenomena to be seen there; scholars formed cabinets and gathered specimens to study at leisure. Little by little, in diverse works there appeared descriptions and drawings, and fishes were not always neglected in these works. Thus André Thevet,[19] in his *Singularitez de la France antarctique,* wrote about the armored catfish *(callichte)* and the hammerhead shark *(marteau).* Léry[20] named several Brazilian fishes, and his nomenclature often agrees with that which Marcgrave used almost a hundred years later. Clusius,[21] in his *Exotica,* made reference to a chimaera *(chimère),* several species of porcupinefishes *(diodons),* a species of boxfish *(ostracion),* and a triggerfish *(baliste).* De Laet[22] in his description of the New World, illustrated a cutlassfish *(trichiure)*, a frogfish (chironecte), a pompano *(gal)*, and other fishes. Nieremberg[23] collected part of the writings of these authors in his *Historia naturae* but also included some accounts taken from manuscript sources.

Something else happened at this time that was most favorable to science: the masters of the new conquests, wishing to know the exact extent of their wealth, sent men of ability to study and describe it. Hernandez in Mexico, at the order of Philip II, made a collection of drawings, with explanations, that would have been of great interest had it been published straightway; but only an excerpt has appeared, years after it was composed, with commentaries that were more perplexing than enlightening;[24] he touched on fishes only briefly and in a very abbreviated manner.

The Dutch achieved in 1637–38, under the leadership of Johann Mauritz, count of Nassau-Siegen, the conquest of northern Brazil. Willem Piso, the count's physician, charged with examining the natural production of the country as it related to public health, had the good fortune of being assisted in this work by a young Saxon medical student, Georg Marcgrave.[25] Of all those who described the natural history of distant lands in the sixteenth and seventeenth centuries, he was assuredly the most intelligent and the most exact, and the one who contributed most to the natural history of fishes. He made known a hundred of them, all new to science at that time, and gave descriptions much superior to those of all the authors who preceded him. The drawings are quite recognizable, despite the fact that they are simple wood engravings; and when some of them were reproduced in Bloch's magnificent work, they were not always reproduced so faithfully. It is here that one sees for the first time the batfish *(malthée; Lophius vespertilio* Linnaeus), the squirrelfish *(holocentrum)*, the cornetfish *(fistulaire)*, the bagrid catfish *(bagres)*, the guitarfish *(rhinobate)*, the driftfish *(pasteur)*, the bonefish *(glossodonte)*, many characins *(characins)*, the trahira *(érythrinus)*, the armored catfish *(loricaire)*, the naked-back knifefish *(carape)*, the sailfish *(istiophore)*, the threadfin *(polynème)*, the toadfish *(batrachus)*, and the tarpon *(mégalope)*, not to mention a multitude of interesting species belonging to genera already known. Piso, in his second edition,[26] added some figures to these, but drawn by another hand and much less accurate.

Conquerors of the Portuguese in the East Indies as in Brazil, the Dutch sent their naturalists there too. Bontius[27] was the first to publish on the fishes of Batavia, but with less precision and with less exact drawings than Marcgrave did for Brazil. Nieuhof [fig. 13][28] added accounts of some species, but only a few.

Becoming solidly established only later in America, it was not until the end of the seventeenth century that the French wrote about the

FIG. 13. Johan Nieuhof (b. 1618). After Churchill and Churchill 1732, frontispiece. Courtesy of Gary L. Menges and the Special Collections Division, University of Washington Libraries, Seattle.

natural history of that part of the world. Du Tertre[29] actually borrowed most of his illustrations from Marcgrave, and Rochefort[30] copied his from du Tertre. Notwithstanding, du Tertre provided valuable observations on some of the species.

Meanwhile, this abundance of foreign natural production did not cause that of Europe to be neglected; on the contrary, it drew new attention to it. Mattioli,[31] in the last editions of his *Commentaire sur Dioscoride*, added accounts of some fishes to those of Gessner and his three predecessors. Imperato[32] mentioned two or three Mediterranean species. Columna[33] and Scilla[34] occasionally described two or three others. Schwenckfelt[35] published a catalog and short descriptions of the fishes of Silesia. Schonevelde[36] composed a fairly exact natural history of the fishes of Holstein and added some species to those Gessner had described. Toward the end of the seventeenth century, Sibbald[37] described some fishes of Scotland. Neucrantz[38] wrote a special treatise on the herring *(hareng)* in which different small species of the same genus are also mentioned.

NOTES

1. After the fall of Constantinople in 1453, and even before, during the wars and calamities that preceded that event.

2. The translation of [Aristotle's] books on animals by Theodorus of Gaza—[b. ca. 1400] a Greek from Thessalonica who went to Italy in 1429 and died in 1478—appeared for the first time at Venice in 1476.

3. Shortly before 1460.

4. In 1492.

5. [The early European settlement of India began with the Portuguese expedition of Vasco da Gama (ca. 1460–1524) who left Lisbon in 1497 and, rounding the Cape of Good Hope, reached the city of Calicut in May 1498.]

6. Franciscus Massaria, *In nonum Plinii de naturali historia librum castigationes et annotationes*, Basle, 1537, in quarto. There is also an edition of Paris, Vascosan, 1542, in quarto, which contains the ninth and the thirty-second books of Pliny.

7. Paolo Giovio [bishop of Nocera], born at Como in 1483, died at Florence in 1552, is well known as one of the most elegant Italian writers. His first work, not so famous as the others, is a Latin treatise on fishes: *De Romanis piscibus libellus ad Ludovicum Borbonium, cardinalem amplissimum*, Rome, 1524, in folio, and 1527, in octavo. There is an Italian translation by [Carlo] Zancaruolo, Venice, 1560, in octavo.

8. Pierre Gilles or Gyllius was born at Alby in 1490, traveled in Italy, and was sent to the Levant by Francis I. Obliged for lack of support to join the army of Soliman II, he bought his way out again, returned by way of Hungary and Germany, and died in Rome at the house of the cardinal d'Armagnac in 1555. His little treatise *De Gallicis et Latinis nominibus piscium Massiliensium* was written before his travels to the Levant and was [first] printed in 1533 [as a summary chapter to his excerpts from Aelianus].

9. This is the book mentioned above [n. 8]: *Ex Aeliani historia per Petrum Gyllium Latini facti, itemque ex Porphyrio, Heliodoro, Oppiano, tum eodem Gyllio luculentis accessionibus aucti, libri XVI, de vi et natura animalium*, Lyons, Gryphe, 1533 and 1535, in quarto.

10. Edward Wotton, physician at Oxford, lived in the first half of the sixteenth century [1492–1555]. His book, titled *De differentiis animalium libri decem* and dedicated to the young King Edward VI, was printed in Paris by Vascosan in 1552, one volume in folio. It is remarkable for its typography.

11. Adam Lonicer [1528–86], *Naturalis historiae opus novum, in quo tractatur de natura et viribus arborum fruticum, herbarum, animantium'q; terrestrium, volatilium et aquatilium*, Frankfurt, 1551, in folio. [His chapter on fishes, titled "De piscium

natura," within which he includes accounts of crustaceans, mollusks, and cetaceans, is found on pp. 300–309.]

12. Pierre Belon, born in the county *(comté)* of Maine in 1517, studied in Germany under Valerius Cordus, traveled in Italy and throughout the Levant, and returned to Paris in 1550. Charles IX lodged him in the château of Madrid in the Bois de Boulogne, and there he was occupied in translating Dioscorides when he was assassinated in the Bois on his way to Paris in 1564. In ichthyology, we have his *Histoire naturelle des estranges poissons marins, avec la vraie peincture et description du daulphin, et de plusieurs autres de son espèce,* Paris, 1551, in quarto; *De aquatilibus, libri duo,* Paris, 1553, in octavo oblong; a French translation of the same work, under the title *La Nature et diversité des poissons, avec leurs pourtraicts, representez au plus Près du natural,* Paris, 1555, in octavo oblong. His *Observations de plusieurs singularitez et choses mémorables, trouvées en Grèce, Asie, Judée, Egypte, Arabie,* etc., Paris, 1553, 1554, and 1555, in quarto, also contains divers articles on fishes. [For more on Belon, see Gudger 1934, 26–28; Cole 1949, 60–62; Allen 1951, 410–12.]

13. Lacepède later described it [this "parrotfish," among the wrasses] under the name *Cheiline scare* [*Cheilinus scarus;* see Lacepède 1798–1803, 3:530].

14. Hippolyte Salviani, from Città di Castello, 1514–72, physician to Cardinal Cervin, who was pope for six weeks under the name Marcellus II, as well as to his successor Pope Julius III, published his *Aquatilium animalium historiae,* in folio, from 1554 to 1558; it was reprinted in Venice in 1600 and 1602. [I have been unable to locate copies of the latter two editions; for more on Salviani, see Gudger 1934, 30–32.]

15. Guillaume Rondelet, born at Montpellier in 1507, son of a pharmacist, was named professor in that town in 1545, traveled with cardinal de Tournon in France, Italy, and the Netherlands, returned to Montpellier in 1551, and died in 1566. He was assisted in the composition of his book on fishes by Guillaume Pellicier, bishop of that town. The first part, *Libri de piscibus marinis,* appeared in Lyons in 1554, in folio. It is divided into eighteen books: the first four treat of generalities; the fifth through the fifteenth describe the different fishes; the sixteenth, cetaceans, turtles, and seals; the seventeenth, mollusks; the eighteenth, crustaceans. The second part, *Universae aquatilium historiae,* is dated 1555 and comprises two books on testaceous species, one book on worms *(vers)* and zoophytes, three books on freshwater fishes, and one on amphibians. There is an abridged French translation of this work, Lyons, 1558, in quarto, titled *L'Histoire entière des poissons,* by Guillaume Rondelet, Esq., etc. [For more on Rondelet, see Gudger 1934, 28–30; Oppenheimer 1936; Cole 1949, 62–72.]

16. [According to Nissen (1951, 83; 1972, 115; see also Baudrier 1964, 172), Rondelet's artist was Georges Reverdi, an "engraver preferred by publishers" at Lyons, born in the region of Dombes in eastern France and active from about 1529, when he worked in Italy, to about 1560 at Lyons. In preparing the woodcuts that illustrate Rondelet's book, Reverdi is said to have worked directly from the author's drawings (M. L. Bauchot, pers. comm., 7 September 1993; see also Kolb 1992, 1:182–87).]

17. Conrad Gessner, the most knowledgeable naturalist of the sixteenth cen-

tury, born in Zurich in 1516, died in 1565, among a multitude of other works has left us a great monument in his *Historiae animalium* [1551–87], printed in Zurich in five books that are usually bound in three volumes in folio. The fourth book, which treats of aquatic animals and is the largest of these volumes—*Historiae animalium liber IV, qui est De piscium et aquatilium animantium natura*—appeared in 1558. An edition not so handsome but more complete was printed at Frankfurt in 1604; another appeared in 1620, as well as an abridgment titled *Nomenclator aquatilium animantium*, Zurich, 1560, with the same figures. [For more on Gessner, see Gudger 1934, 32–36; Allen 1951, 402–3; Wellisch 1975; Adler 1989, 7–8.]

18. Ulisse Aldrovandi, born at Bologna in 1522 of a noble family that still survives, spent his life and fortune in assembling the materials for his great natural history, in thirteen folio volumes [published in Bologna between 1599 and 1688], only four of which he himself published, namely, three on birds [1599–1603] and one on insects [1602]. He died in 1605, at the age of eighty-three. The volume on fishes and cetaceans, drawn up in part from his notes by his successor at Bologna, [Johannes] Cornelius Uterverius, was not published until 1613, but it was reprinted at Bologna in 1638 and 1644 and at Frankfurt in 1623, 1629, and 1640. [For more on Aldrovandi, see Gudger 1934, 36–38; Allen 1951, 403–5.]

19. André Thevet [1504–90], a friar, native of Angoulême, who accompanied [Pierre] Gilles [see n. 8 above] on his journey to Greece in 1550, was with [Nicolas Durand de] Villegaignon [1534–1611] on his expedition to Brazil in 1555 and published his observations at Anvers in 1558, a small book in octavo, with woodcuts. In it he speaks of only two or three fishes [e.g., see his pp. 158–60].

20. Jean de Léry, from La Margalle near Saint-Seine in Burgundy, a protestant minister, born in 1534 [d. 1611], went to Brazil in 1556 at the request of Villegaignon. He published the account of his journey [said to contain the earliest descriptions of Brazilian fishes; see Dean 1923, 275] at La Rochelle in 1578, in octavo. It was reprinted several times [at Rouen, Paris, Geneva, etc.] and has been included in different collections. Chapter 12 of the book [pp. 185–93] treats of fishes and is not without interest. There are no figures.

21. Charles de l'Ecluse (in Latin Carolus Clusius), botanist, born at Arras in 1526, was director of the gardens for emperors Maximilian II and Rudolph II; he died in 1609 while professor at Leiden. The book in question, titled *Caroli Clusii exoticorum libri X*, appeared in Anvers in 1605, one volume in folio.

22. Johannes de Laet, born at Antwerp in 1585, died in 1649, director of the Dutch West India Company, great promoter of geography, editor of Marcgrave [see n. 25 below], and so on, is the author of several works, among them especially his *Novus orbis, seu Descriptionis Indiae occidentalis*, Leiden, 1633, one volume in folio. On pages 570–74 is a chapter titled "Pisces marini Brasiliensium" [in which he provides descriptions of some two dozen species and figures of five species. A hammerhead shark is described and illustrated on p. 576].

23. Juan Eusebio Nieremberg, a Jesuit born in Madrid in 1590 of a family originally from the Tyrol, died in 1658, wrote numerous works. His *Historia naturae, maxime peregrinae, libri XVI*, appeared at Anvers in 1635, in folio. It is a compilation in a pedantic style that does not proclaim its author as having any knowledge of the subject matter; however, it presents articles by authors then in

manuscript form, such as Hernández [see n. 24 below], and others; fishes are presented in the eleventh book.

24. Francisco Hernández de Toledo [1514–78], primary physician to Philip II of Spain [was sent to Mexico in 1570 to collect and explore]. He composed a natural history of that country, embellished with more than twelve hundred paintings of plants and animals. As happens only too often, this work, which cost 60,000 ducats to produce, remained in manuscript, and it is not known what has become of it. Francisco Ximénes [1560?–1620] produced an abridgment without the drawings in Mexico City in 1615, a small volume in quarto. An Italian named Nardo Antonio Recchi, primary physician of the kingdom of Naples, made excerpts of it in ten books, which were acquired by the prince de Cesi [Federico Cesi duca di Acquasparta, 1585–1630] and published in Rome in 1651, in one volume in folio, under the title *Rerum medicarum Novae Hispaniae thesaurus, seu Plantarum, animalium, mineralium Mexicanorum historia, . . .* , with many woodcuts, and with long commentaries by Joannes Terentius [1576–1630], physician at Constance, and Johannes Faber [ca. 1570 to ca. 1640], physician at Bamberg, both of whom were established at Rome, and the famous Fabius Columna [1567–1650]. Even longer commentaries were added to certain illustrations of plants and animals that Recchi had left without description, several of them representing objects foreign to Mexico, even animals of Asia and Africa, which were presented as though they were American. The basis of the text of this part rests on oral assertions by a Capuchin friar named Grégoire de Bolivar, collected by Faber. At the end of the volume are six essays by Francisco Fernández, who can be none other than Francisco Hernández, which seem to me to be his own originals as regards the animal and mineral kingdoms, originals from which Recchi excerpted his ninth and tenth books. When use is made of this bizarre compilation, it is well to ascertain the author of the article consulted; on the other hand, there is very little about fishes in it, and only in the fifth book are there the little essays said to be by Fernández.

25. Willem Piso [1611–78] was sent to Brazil by the Dutch West India Company, directed by De Laet, to serve as physician under Count Johan Maurits of Nassau-Siegen [1604–79], who governed that country from January 1637 to May 1644, and at the same time to gather specimens of the country's natural production. For this purpose De Laet gave him two young German mathematicians and physicians as collaborators: Georg Marcgrave, born at Meissen in 1610, and Heinrich Cralitz. The latter promptly died, but Marcgrave survived the climate and described with care many of the plants and animals; at the same time, he made astronomical and physical observations of all sorts. He died on a voyage to Guinea in 1644. Piso received permission from Count Maurits to have Marcgrave's papers entrusted to De Laet, and his work on natural history was published in Leiden in 1648 in folio, following an essay by Piso on medicine in Brazil, under the title *Historia naturalis Brasiliae.* Marcgrave's work is divided into eight books; the fourth is on fishes. The descriptions are entirely his own, and De Laet only added a few notes. The illustrations are taken from two collections painted at the request of Count Maurits, which he lent to De Laet for the purpose. After returning from Brazil in 1644, Maurits entered the service of Brandenburg and

was made governor of Wesel and grand master of the order of Saint John, and he was elevated to the rank of prince in 1654. He died in 1679, governor of Berlin. These two collections, one in oils and the other in gouache, were arranged by Dr. Christian Mentzel [1622–1701; physician to Friedrich Wilhelm I, elector of Brandenburg]. They were placed in the royal library of Berlin, where they are still kept [these eight hundred or so pictures of Brazilian plants and animals were removed from the Preussische Staatsbibliothek in Berlin during World War II and are at present housed at the Jagiellon Library, Cracow, Poland; see Whitehead 1982; Whitehead and Boeseman 1989, 22, 34–35]. The first collection, by an unknown artist, remained there almost unknown until 1811, when [Johann Carl Wilhelm] Illiger [1775–1813] consulted it to remove doubts raised by Marcgrave's book [see Illiger 1811, vi]. The second collection, which some believe to be by Marcgrave and others by Maurits himself, was brought to the attention of the public by Schneider in 1786 [see chap. 10, n. 10], and Bloch had several drawings copied from it for his great *Ichthyologie* [1785–97], but without seeming to suspect that they were drawn by the prince, and worse yet, adding and deleting and otherwise changing several things quite arbitrarily. We can see, for example, that he has completely distorted the drawing of a squirrelfish *(holocentrum)* to create his *Bodianus pentacanthus* [see Bloch 1785–97, 4:40–42, pl. 225].

These details are taken from the preface to the third part of the great *Naturgeschichte der ausländischen Fische* by Bloch [1785–95, 3:iii–x], and in particular from three articles by [Martin Heinrich Karl] Lichtenstein [see chap. 16, n. 52], included among the memoirs of the Royal Academy of Sciences of Berlin [Lichtenstein 1818, 1819, 1822a, b; see also Lichtenstein 1829]; but we have been fortunate enough to confirm them in part with our own eyes. Valenciennes obtained permission from the conservators of the library to copy these collections, and we are today able to compare them with Bloch's copies and with nature and definitively fix the genera and species to which each fish should be referred. [For more on Piso, see Pies 1981; Whitehead and Boeseman 1989. On Maurits, see Boogaart, Hoetink, and Whitehead 1979. On Marcgrave, see Gudger 1912; Whitehead 1979a, b; Whitehead and Boeseman 1989.]

Marcgrave's works on astronomy, the principal object of his studies, were not so fortunate. Entrusted to Golius [Jacobus Golius, born at The Hague in 1596, died at Leiden in 1667], they have never appeared. There is reason to believe that he preceded the abbé de La Caille [Nicolas Louis La Caille, 1713–62] in determining many of the southern stars.

26. In 1658 Piso published a new, greatly enlarged edition of his essay on the medicine of Brazil, under the title *De Indiae utriusque re naturali et medica.* Marcgrave's work, which in the original edition followed that of Piso, no longer appears in this new edition except in excerpts incorporated into the body of Piso's work, and Piso added his own observations and several new illustrations; but he also deleted much, so that now the one book does not take the place of the other. [Despite this use of Marcgrave's work], it is hard to see how some writers could accuse Piso of plagiarism. On the contrary, he acknowledged Marcgrave at every turn in his two editions.

27. Jacques Bondt or Jacobius Bontius [b. 1599], physician at the city of

Batavia in 1625, died in 1631, was the author of a treatise titled *Historiae naturalis et medicae Indiae Orientalis libri VI.* It was printed in 1658 along with Piso's second edition, a circumstance that determined the title of that edition [see n. 25 above]. The purely medical part had already appeared in 1645 with the treatise by Prosper Alpinus [1553–1617], *De medicina Aegyptiorum.*

28. Johan Nieuhof, born [in 1618] at Bentheim in Westphalia, employed in various capacities by the Companies [East and West] of the Dutch Indies, and for some time governor of Ceylon, perished at Madagascar in 1672. His book [published posthumously under the direction of his brother Hendrik] titled *Gedenkwaerdige zee- en lant-reize door de voornaemste landschappen van West en Oostindien,* Amsterdam, 1682, in folio, contains thirty-eight drawings of fishes, most of them interesting [see the original Dutch edition, Nieuhof 1682, pt. 2, pp. 268–81; or an English translation of Nieuhof's voyages compiled by Awnsham and John Churchill 1732, 305–11]; copies of them are found at the end of Willughby's [1686] plates.

29. Jean Baptiste du Tertre, Dominican monk, missionary to the Antilles, born at Calais in 1610 [died at Paris in 1687], composed a general natural history of the Antilles. The first edition was published in 1654, in one volume in quarto; the second, 1667–71, three volumes in quarto, is much more complete.

30. [Charles de] Rochefort [1605–91], Protestant minister at Rotterdam, borrowed from Du Tertre's first edition the greater part of his *Histoire naturelle et morale des îles Antilles de l'Amérique,* Rotterdam, 1658, in quarto.

31. Pietro Andrea Mattioli, born at Sienna in 1501, physician at Rome, Trent, Gorizia, and Prague successively, died [of the plague] at Trent in 1577, is famous for his commentary on Dioscorides, printed first in Italian at Venice in 1544 and 1548 [the "Privilegio" is dated 1544], and in Latin in 1554 and 1565. This last edition, by [Vincenzo] Valgrisi, is the best: there are very handsome woodcuts of fishes, some of which, however, are taken from Salviani and from Rondelet. This work has been translated and reprinted many times. [For more on Mattioli, see Zanobio 1974.]

32. Ferrante Imperato [1550–1625], Neapolitan physician, author of a natural history in Italian almost entirely about chemistry and mineralogy, printed first in 1599 in folio, then in 1610 in quarto [I have been unable to locate a copy of the 1610 edition], and at Venice in 1672. It contains references to a remora *(écheneis)* and an oarfish *(gymnètre).*

33. Fabius Columna [Fabio Colonna], born at Naples in 1567, his father an illegitimate son of Cardinal Pompeo Colonna, died about 1650, provides a drawing of a eagle ray *(mylobate)* in his "Naturalium rerum observationes," 1606. His "De glossopetris dissertatio," published in 1752, contains interesting details on the teeth of sharks *(squales).*

34. Shark *(squale)* teeth were also an object of study for the Sicilian painter Agostino Scilla [1639–1700], in his work on petrification titled *La vana speculazione disingannata dal senso,* Naples, 1670, a small volume in quarto, of which there is a Latin translation titled *De corporibus marinus lapidescentibus,* Rome, 1752, in quarto. It contains, in addition, an illustration of a hammerhead shark *(marteau)* and another of a rare species of shark *(squale)* with seven gill openings [see Scilla 1670, pl. 28].

35. Caspar Schwenckfelt [1563–1609], physician at Hirschberg, had a natural history of animals of Silesia printed at Lignitz in 1603 in quarto titled *Theriotropheum Silesiae,* filled with good observations but without illustrations. The fifth book [377–456] treats of fishes and indicates several species, all freshwater.

36. Stephan von Schonevelde was a physician at Hamburg. His *Ichthyologia et nomenclaturae animalium marinorum, fluviatilium, lacustrium, quae in Florentissimis ducatibus Slesvici et Holsatiae* [. . . *occurrunt* . . .], Hamburg, 1624, in quarto, is accompanied by seven plates in which several species then new are fairly well represented.

37. Robert Sibbald [1641–1722], physician at Edinburgh, worked for twenty years on his *Scotia illustrata, sive Prodromus historiae naturalis,* Edinburgh, 1684, one volume in folio, in which there are a few mediocre illustrations of fishes. He also published a capital work on cetaceans [Sibbald 1773], but this has nothing to do with our subject.

38. *De harengo, exercitatio medica, in qua principis piscium exquisitissima bonitas summaque gloria asserta et vindicata* by Paul Neucrantz [1605–1671], physician at Rostock, published at Lubeck, 1654, in quarto.

TABLE 1. [Belon, in his *De aquatilibus, libri duo* (1553a), divided aquatic animals into two primary divisions, those provided with blood and those without, corresponding to the modern taxa Vertebrata and Invertebrata. The contents of the first of these were classified according to size and further subdivided by differences in the structure of the skeleton, mode of reproduction, number of limbs, and body form and on the physical character of the habitat occupied. His fishes were classified as follows.

I. Larger fishes or cetaceans
- A. Viviparous cetaceans with bony skeletons (= Cetacea)
- B. Viviparous amphibians
  - 1. With four limbs: seals, hippopotamus, beaver, otter, and other aquatic mammals
  - 2. With two limbs: mermaids, etc.
- C. Oviparous amphibians (= reptiles and frogs)
- D. Viviparous cartilaginous fishes
  - 1. Of an oblong form (= sharks)
  - 2. Of a flat form (= rays and *Lophius*)
- E. Oviparous cartilaginous fishes (= sturgeons and catfishes)
- F. Oviparous cetaceans, with spines instead of bones (= large marine fishes such as tunas, swordfishes, sciaenoids, basses, gadoids, *Trachypterus*)

II. Spinous oviparous fishes of a flat form (= flatfishes)

III. Fishes of a high form (e.g., *Zeus*)

IV. Fishes of a snakelike form (= eels, needlefishes, barracudas)

V. Small oviparous, spinous, scaly, marine fishes
- A. Pelagic forms
- B. Littoral forms
- C. Forms that inhabit rocky localities

VI. Fluviatile and lacustrine fishes

Note: Belon's classification, not provided originally by Cuvier, has been included here by the editor, following the interpretation of Günther (1880, 4).]

5 Anatomical Contributions of the Sixteenth and Seventeenth Centuries

Especially important in laying new foundations in fish science were the observations by the anatomical school, founded in Italy by such men as Vesalius, Eustachius, and Fallopius, which flourished in the sixteenth century and, during the seventeenth, was brought by a happy necessity to the study of animal anatomy.[1] One of its cleverest masters, Fabricius d'Aquapendente, studied fish reproduction and scales in some detail and published an anatomical description of the houndshark *(émissole)*.[2] Casserius, his pupil and successor, made known many interesting facts about the brain and sense organs of these animals, particularly their nares, eyes, membranous labyrinth, and otoliths.[3] Severinus, in his *Zootomia democritea,* while laying the foundations of a general anatomy of animals, also laid those of fish anatomy.[4] In his *Antiperipatias* he sought to prove that fishes breathe the air contained in water, and relying on his ideas on the uniformity of structure in all animals, he even maintained that they have a lung independent of their gills; but he mistook the kidney for this organ. Borelli, in his *De motu animalium* explained the mechanism of swimming and made known the use of the swim bladder [fig. 14].[5] Malpighi described the singular fold in the medullary substance of the optic nerve, which he observed particularly in the swordfish *(xiphias)*.[6] Steno[7] wrote about the brain, eye, and teeth of the shark *(squale)* [fig. 15][8] and gave a good description of the viscera of the ray *(raie)*[9] and the electric ray *(torpille)*,[10] which were also described by Lorenzini.[11]

The spirit of this school was brought to England by Harvey,[12] to Germany by Coiter,[13] to the north of Europe by the Bartholin family,[14] and before long, it spread throughout Europe. The anatomy of fishes profited from it.

A group of young physicians from Amsterdam, among them Blasius[15] and the immortal Swammerdam,[16] addressed themselves to the viscera

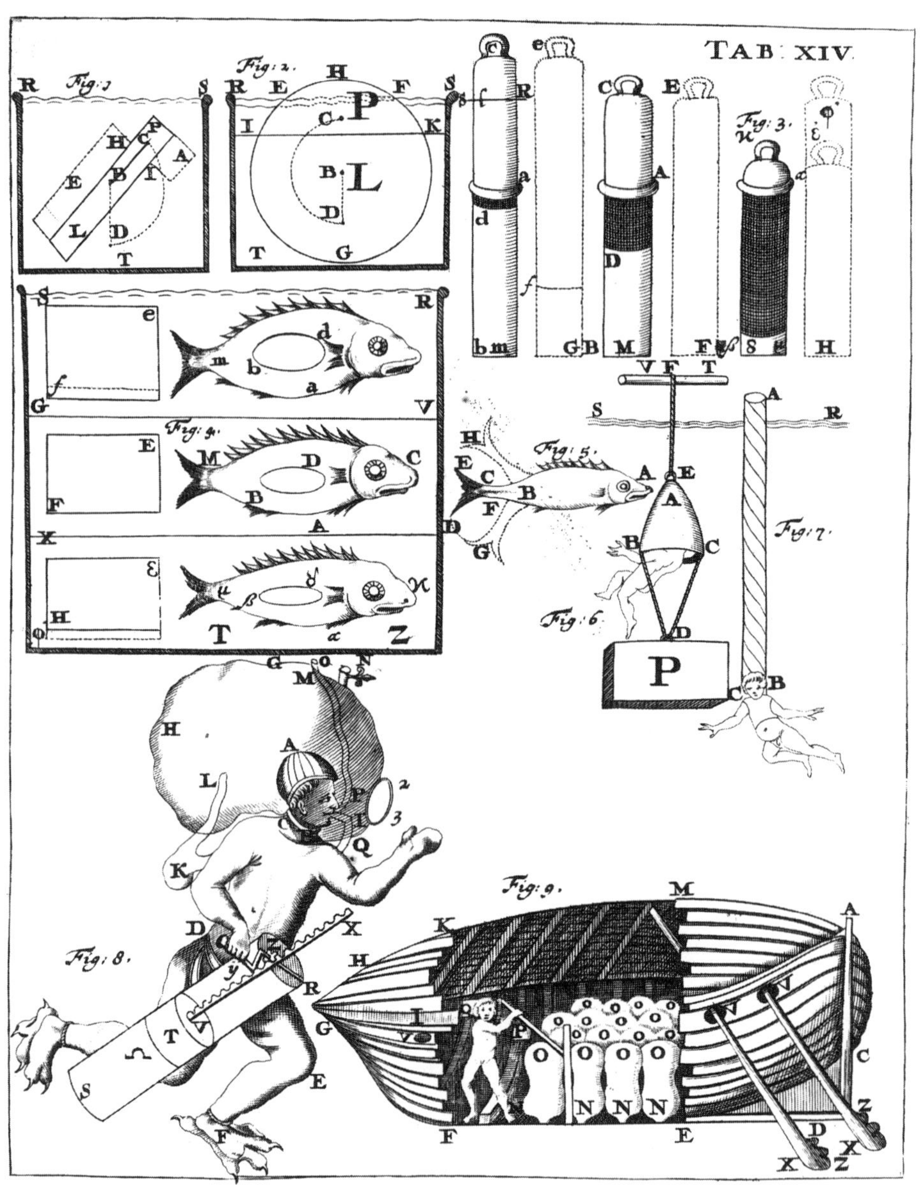

FIG. 14. Giovanni Alfonso Borelli's illustrations of the mechanism of swimming and swim-bladder function in fishes from his *De motu animalium*, 1743, pl. 14. Courtesy of Colleen M. Weum and the Health Sciences Library and Information Center. Photograph by Patricia L. McGiffert, Health Sciences Photography, University of Washington, Seattle.

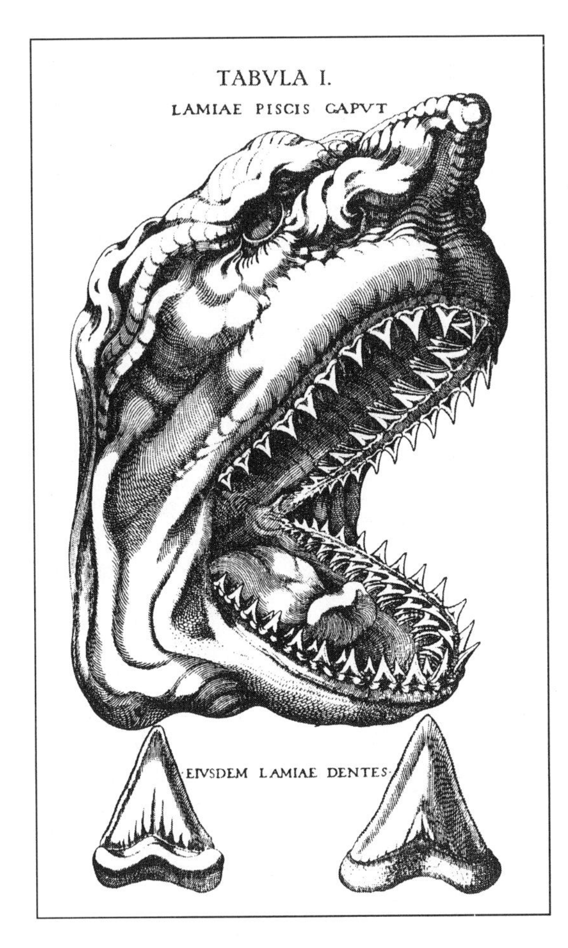

FIG. 15.
Head and teeth of a shark from "Canis carchariae dissectum caput," by Nicolaus Steno, first published in 1667.
After Maar 1910, vol. 2, pl. 1. Courtesy of Colleen M. Weum and the Health Sciences Library and Information Center. Photograph by Patricia L. McGiffert, Health Sciences Photography, University of Washington, Seattle.

of fishes and made known those numerous appendages that take the place of a pancreas in most species, and the canal that in several species joins the swim bladder to the stomach.[17] Even the pores of their skin became the object of a dissertation by Rivinus.[18] Finally Duverney, at the beginning of the eighteenth century, revealed in detail their respiratory organs and the whole mechanism of respiration as well as their branchial circulation and thus completed to a certain point our knowledge of their organization.[19]

Specialized descriptions, in the form of monographs, became very numerous. Boccone wrote his observations on sharks *(squales)*;[20] Vallisneri, on the freshwater eel *(anguille)*.[21] Needham dissected the pike *(brochet)*, the common carp *(carpe)*, and the shad *(alose)* and discoursed on the respiration of fishes and their organs, as well as their swim bladder.[22] Schelhammer and Thomas Bartholin published the anatomy of the sword-

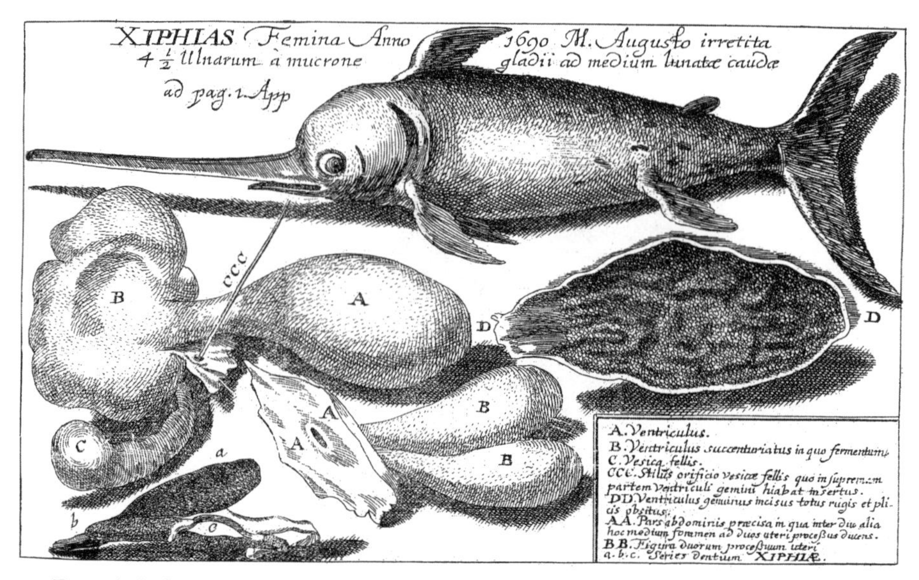

FIG. 16. A dissection of the swordfish, *Xiphias gladius,* from Philipp Jacob Hartmann's "Descriptio anatomico-physica xiphiae, sive Gladii piscis," 1695, pl. 1. T. W. Pietsch collection.

fish *(xiphias)*;[23] the former also described the anatomy of the lumpfish *(lump)* and the cuskeel *(donzelle)*.[24]

In particular, a rather large number of anatomical descriptions of species are to be found in the memoirs of connoisseur-naturalists,[25] including that of the salmon *(saumon)* by Peyer,[26] of the trout *(truit)*[27] and the freshwater burbot *(lote)*[28] by Muralt, the catfish *(silure)* and the swordfish *(xiphias)* by Hartmann [fig. 16],[29] the anglerfish *(baudroie)* by König,[30] and the lamprey *(lamproie)* by Waldschmidt.[31] König also described the unusual stomach of the mullet *(muges)*.[32]

In the memoirs of the physicians at Copenhagen[33] appeared the anatomy of the needlefish *(orphie)* by Borrichius[34] and those of the freshwater eel *(anguille),* the angular rough shark *(squale centrina),* the electric ray *(torpille),* and the lamprey *(lamproie)* by Jacobaeus.[35]

The Académie Royale des Sciences, Paris, which in the beginning intended to describe and dissect the animals in the menagerie at Versailles, and which in fact went rather far in carrying out this enterprise in its "Mémoires pour servir à l'histoire des animaux,"[36] included in this publication anatomical observations on the thresher shark *(squale faux)*.[37] Some of these monographs were included in the collections of Blasius[38] and Valentini.[39] But the most valuable observations at the time on this subject are to be found in the anatomical work of Collins,[40] wherein are seen, in twenty-eight well-engraved plates, the viscera and brains of some twenty fishes.[41] The illustrations of the brains were especially important and had no equal for a long time.

It was desired that all this material be united in one body of work. Toward the middle of the seventeenth century a physician from Silesia, Jonstonus, undertook this task for the whole animal kingdom;[42] but he did it as a simple compiler, without personal knowledge of things and, moreover, completely neglecting anatomy. His volume on fishes in particular is actually nothing more than a rather elegant abridgment of Aldrovandi and the authors who came before Aldrovandi [fig. 17]; the sixth book, on the subject of foreign fishes, is based only on Marcgrave and Nieremberg. Also, he copied the plates of these various authors and of Schonevelde, but to the edition of 1718, titled *Theatrum animalium*,[43] have been added some drawings of fishes of the East Indies, which will be discussed later.

Jonstonus tried to arrange fishes according to a method, but it was a very ill-conceived one. He mingled distinctions based on their habitats with those based on their structure, the latter being ill chosen and inconsistently followed in detail. The fishes ranked under each chapter do not by any means have all the characteristics indicated in the title.

NOTES

1. See [Albrecht von] Haller, *Bibliotheca anatomica*, vol. 1 [1774].

2. Hieronynmus Fabricius [Fabricius d'Aquapendente], born at Aquapendente [1533–1619], student of [Gabriello] Fallopio [1523–62], master of [William] Harvey [see n. 12 below], professor at Padua from 1565 to 1609. [His anatomical description of the houndshark appeared in *De formato foetu*, published in Venice in 1600 (see especially his pls. 31–32). For a facsimile edition, a biography, and commentary on the embryological treatises of Hieronymus Fabricius, see Adelmann 1942. On Fabricius as an anatomist, see Cole 1949, 99–112.]

3. Jules Casserius (Giulio Casseri) of Placentia [b. 1545], died in 1616, successor to Fabricius at Padua, published two works quite remarkable for their time: *De vocis auditusque organis historia anatomica*, Ferrara, 1600, one volume in folio; and *Pentaestheseion, hoc est de quinque sensibus liber*, 1610, in folio. In the first he describes and illustrates on pp. 94–95 the brain, ears, and nares of the pike *(brochet)*; in the second [on p. 292], he shows the eye, but with less exactitude. [For more on Casserius, see Cole 1949, 112–25, fig. 53.]

4. Marcus Aurelius Severinus (Marco Aurelio Severino), born at Tarsia in Calabria in 1580, professor at Naples in 1610, died in that city in 1656. His *Zootomia democritea*, published at Nuremberg under the auspices of Volkamer, 1645, one small volume in quarto, is the first general philosophical treatise on comparative anatomy. It includes notes on the anatomy of a dozen fishes and rough woodcut illustrations of viscera. His *Antiperipatias, seu De respiratione piscium* did not appear until 1661, at Amsterdam, in one small volume in folio without illustrations. [For more on Severinus, see Cole 1949, 132–49, fig. 54.]

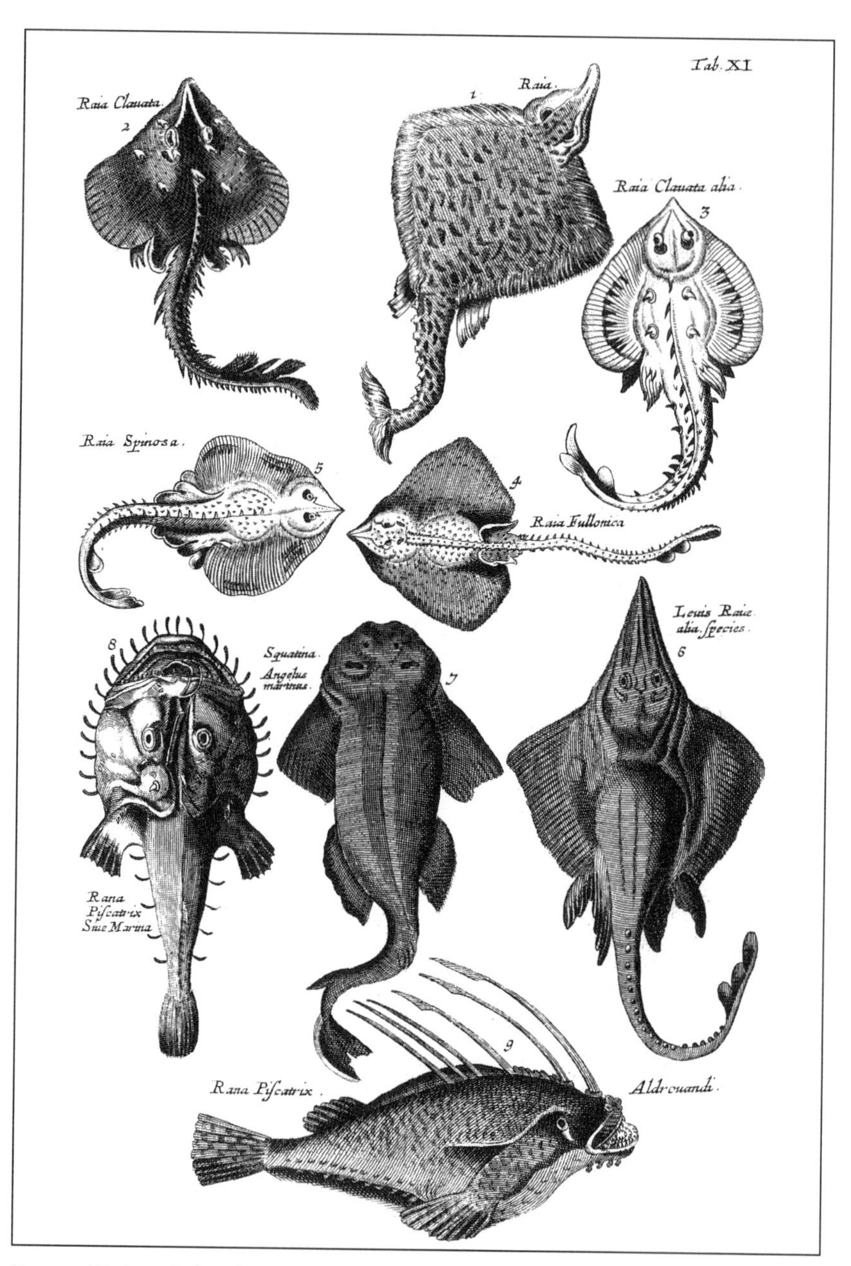

FIG. 17. Various fishes from Hendrik Ruysch's 1718 edition of Joannes Jonstonus's *Historiae naturalis,* vol. 1, *De piscibus,* pl. 11. T. W. Pietsch collection. Photograph by Patricia L. McGiffert, Health Sciences Photography, University of Washington, Seattle.

5. Giovanni Alfonso Borelli, born at Naples in 1608, professor at Pisa and Florence, died in Rome in 1679, was one of the founders of the school of Iatromathematicians [the application of mathematics and mechanics to medicine], who sought to apply exact calculations to physiological phenomena. His *De motu animalium* was published posthumously, Rome, 1680–81, two volumes in quarto. [For more on Borelli and his research on fish locomotion, see Videler 1993, 94–95, fig. 5.1.]

6. Marcello Malpighi, born near Bologna in 1628, professor at Bologna, Pisa, and Messina and first physician to Pope Innocent XII, died at Rome in 1694, includes among his numerous anatomical discoveries the folds in the optic nerve of fishes, which he described in 1664 in the swordfish *(xiphias)* in a letter addressed to Fracastor; it was reprinted in his *Opera omnia botanico-medico-anatomica*, Leiden, 1687, one volume in folio [see Malpighi 1687, 120]. [For more on Malpighi, see Cole 1949, 177–97.]

7. Nicolaus Steno [originally Niels Steensen], born at Copenhagen in 1631, became a Roman Catholic at Florence in 1667, tutor of the grand duke's son, professor of anatomy, bishop in *partibus infidelium* [i.e., appointed over a territory not yet erected into a see], and died at Schwerin in 1686. He was the author of several writings on physiological anatomy published in Denmark, France, and Italy [e.g., see nn. 8–10 below]. [For an account of Steno's life and works, see Maar 1910, 1:i–xxviii; 2:351–59.]

8. [Steno], in a dissertation following his *Elementorum myologiae specimen*, Florence, 1667, in quarto; reprinted by Blasius [1681], 263–72. [For a facsimile of Steno's "Canis carchariae dissectum caput," see Maar 1910, 2:113–39.]

9. [Steno], in "De anatome rajae epistola," attached to his *De musculis et glandulis observationum specimen* [1664]; reprinted by Blasius [1681], 298–302.

10. [Steno], in *Acta Medica et Philosophica Hafniensia*, vol. 2, obser. 89 [1675].

11. Stefano Lorenzini, *Osservazioni intorno alle torpedini*, Florence, 1678, in quarto, and in Latin [in 1680] in *Misc. Curio. Medico-Physica* [see n. 25 below]; reprinted by Valentini [1720], pt. 2, pp. 110–15.

12. William Harvey, born at Folkestone, Kent, in 1578, died in London in 1657, physician to James I and Charles I, illustrious discoverer of the general circulation of the blood, was a student of Fabricius d'Aquapendente [see n. 2 above]. One can see that his discoveries are a continuation of his master's ideas. [For more on Harvey, see Cole 1949, 126–31.]

13. Volcher Coiter, physician [and anatomist] at Nuremberg, was born at Gröningen in 1534 and died in 1576 in France, where he had come as physician with the Germany army. He published a treatise on [bone and] cartilage, Bologna, 1566, and later wrote *Diversorum animalium sceletorum explicationes*, Nuremberg, 1575, in folio. [For more on Coiter, see Cole 1949, 73–83; Allen 1951, 405–6.]

14. The Bartholin family, to which the study of anatomy owes many works, was Danish. Caspar Bartholin, born at Malmø in Scania in 1585, died in 1630, was author of *Institutiones anatomicae* [1632] and *De unicornu* [Padua, 1645, followed by a second edition at Amsterdam, 1678]. Five of his sons produced works. Erasmus (1605–98) wrote on the mineral "Iceland spar" *(cristal d'Islande)* [1669] and on other physicomathematical subjects. Thomas (1616–80), discoverer of the lym-

phatic system, is the one who interests us most. In his essay *De luce animalium,* Leiden, 1647, and Copenhagen, 1669; in his *Historiarum anatomicarum et medicarum rariorum centuriae V. et VI.* (Copenhagen, 1661); and in *Acta medica et philosophica Hafniensia* (1673–80, five volumes in quarto) he has given several facts about the anatomy of fishes—for example, in his "Centuriae V. et VI.," a note by [Joannes] Rhodius [1661, 15–17] describes the color, sometimes red, sometimes green, of the liver of the lamprey *(lamproie).* Caspar II, son of Thomas, published a dissertation [on shark's teeth], *De glossopetris,* Copenhagen, 1704, in quarto, and in 1706 in duodecimo [I have been unable to find a 1706 edition of *De glossopetris*].

15. Gerard Blasius, born near Brugge in 1623, died [in 1692] at Amsterdam, where he practiced medicine, was the author of numerous works on anatomy [e.g., see Blasius 1681]. [For more on Blasius, see Cole 1949, 150–55.]

16. Jan Swammerdam, the celebrated author of *Biblia naturae,* was born at Amsterdam in 1637, and after traveling to France and Germany, died in his home country in 1680. His papers passed through several hands and were finally purchased by [the Dutch physician and chemist Hermann] Boerhaave [born at Voorhout in 1668, died at Leiden in 1738], who published this great work in 1737–38 [there is also a London edition of 1758]. [For more on Swammerdam, see Cole 1949, 270–305; Schierbeek 1967.]

17. These facts and several others are to be found in two small [anonymous] books, one titled *Observationes anatomicae selectiores collegii privati Amstelodamensis,* Amsterdam, 1667, in duodecimo; the other, *Observationum anatomicarum collegii privati Amstelodamensis, pars altera, in quibus praecipue de piscium pancreate ejusque succo agitur,* Amsterdam, 1673, in duodecimo [see Amsterdammers, Collegium Privatum, 1667, 1673; for a facsimile edition with introduction, see Cole 1938; see also Cole 1949, 330–41].

18. August Quirinus Rivinus [1652–1723], "Observatio anatomica circa poros in piscium cute notandos," 1687, 160–62.

19. Joseph Guichard Duverney, born at Feurs in 1648, professor of anatomy in the Jardin du Roi in 1679, died in 1730, was the author of several memoirs on human and comparative anatomy, collected in two volumes in quarto, Paris, 1761. He also published, with Claude Perrault [1613–88], "Mémoires pour servir à l'histoire naturelle des animaux," 1733–34 [see n. 36 below]. His memoir [on respiration and the branchial circulation of fishes] is published in *Hist. Mém. Acad. Roy. Sci.* (Paris) for 1701 [see Duverney 1743] and reprinted in the second volume (2:496–510) of his *Oeuvres anatomiques* [1761]. [Johann Julius] Walbaum [1724–99] had it reprinted in his *Artedius renovatus* [1789], pt. 2, pp. 167–83 [see Duverney 1789].

20. Paolo Sylvio Boccone [Italian naturalist] of Palermo, where he was born in 1633, traveled to France and England, and published at Amsterdam in 1674 his *Recherches et observations naturelles,* one volume, in duodecimo, in which he gives illustrations of the swordfish *(xiphias),* the sand lance *(ammodite),* and observations on sharks *(squales)* and their teeth; reprinted by Valentini [1720], pt. 2, pp. 118–19. In his *Museo di fisica,* Venice, 1697, in quarto, he illustrates the lake whitefish *(lavaret).* He entered the Cistercian order in 1682 and died in 1704.

21. Antonio Vallisneri, born in 1661 in the state of Modena, student of Mal-

pighi, professor at Padua in 1700, died in 1730, was a great observer who wrote on the reproduction of insects, of animals in general, on the chameleon, the ostrich, and so on [see Vallisneri 1733, 1:385–455]. [His observations on the eel are found in the second volume of his collected works (1733), 2:89–95, and reprinted by Valentini (1720), pt. 2, pp. 126–30.]

22. Walter Needham, physician in London [b. 1631?], died 1691, was the author of *De formato foetu*, Amsterdam, 1668, in duodecimo, in which he writes about these subjects in a digression on respiration. The work is titled "De biolychnio & ingressu aeris in sanguinem, item de sanguificatione" [see Needham 1668, chap. 6, pp. 147–201], because he lent credence to the analogy of respiration with combustion, then generally held by the English and also in our day by [A.] Crawford [1748–95] and [Antoine Laurent] Lavoisier [1743–94]. His *Disquisitio anatomica de formato foetu* is reprinted in the *Bibliotheca anatomica* of [Jacob] Manget [1699], 1:687–723; and an excerpt from it on fishes is reprinted by Valentini [1720], pt. 2, pp. 123–24.

23. Günther Christoph Schelhammer, born at Jena in 1649, professor at Helmstedt, Jena, and Kiel, died in Kiel in 1716. It was at Kiel that he dissected the swordfish and the other fishes mentioned in the text [see Schelhammer 1707, 3–17]. The dissertation of Bartholin is reprinted by Blasius [1681], 307–8; that of Schelhammer, by Valentini [1720], pt. 2, pp. 102–8.

24. Another dissertation by Schelhammer [1707; "Lumpus anglorum piscis" is found on pp. 20–23; "Ophidion," on pp. 23–24)], reprinted by Valentini [1720], pt. 2, pp. 108–9.

25. The "Imperial Academy of Inquiring Naturalists," whose members were scattered about Germany, was founded in 1652 by [J. L.] Bausch, a physician at Schweinfurt. The first volume of its *Miscellanea Curiosa Medico-Physica Academiae Naturae Curiosorum, sive Ephemeridum Medico-Physicarum* appeared in 1670 [see Cole 1949, 341–69].

26. *Misc. Curio. Medico-Physica*, decur. 2, vol. 1, obser. 85 [Johann Conrad Peyer (1653–1712), Swiss physiologist, on the internal anatomy of the salmon, 1683]; reprinted by Valentini [1720], pt. 2, pp. 120–21.

27. [Ibid.], decur. 2, vol. 1, obser. 47 [Johann von Muralt (1645–1733), professor of medicine at Zurich, on the anatomy of the trout, 1683b]; Valentini [1720], pt. 2, pp. 121–22.

28. [Ibid.], decur. 2, vol. 1, obser. 46 [Muralt, on the anatomy of the freshwater burbot, 1683a]; Valentini [1720], pt. 2, pp. 132–33.

29. [Ibid.], decur. 2, vol. 7, obser. 40; decur. 3, vol. 2, appendix [Philipp Jacob Hartmann (1648–1707), on the anatomy of the catfish, 1689, and the swordfish, 1695]; Valentini [1720], pt. 2, pp. 101–2.

30. [Ibid.], decur. 3, vol. 2, obser. 139 [Emanuel König (1658–1731), on the anatomy of the anglerfish, 1695]; Valentini [1720], pt. 2, pp. 134–35.

31. [Ibid.], decur. 3, vols. 5–6, obser. 231 [Wilhelm Ulrich Waldschmidt (1669–1731), on the anatomy of the lamprey, 1700]; Valentini [1720], pt. 2, p. 131.

32. [Ibid.], decur. 2, vol. 5, obser. 100 [König, on the stomach of the mullet, 1687].

33. *Acta medica et philosophica Hafniensia*, a collection of observations pro-

duced by Thomas Bartholin and some of his friends; there are five volumes, published from 1673 to 1680 [see Cole 1949, 369–93].

34. Olaus Borrichius, born at Borchen in Denmark in 1626, professor at Copenhagen, chemist and naturalist, died in 1690. [His anatomy of the needlefish *(orphie)* appeared in 1675, in *Acta medica et philosophica Hafniensia*], 2:149–51; reprinted by Valentini [1720], pt. 2, pp. 119–20.

35. Oliger Jacobaeus, Danish naturalist, born at Århus in 1650, married into the Bartholin family, professor at Copenhagen, died in 1701. [In addition to his various anatomical works on fishes (see Jacobaeus 1680a, b, c)], he described the collections of the Royal Museum of Copenhagen.

36. "Mémoires pour servir à l'histoire naturelle des animaux" (vol. 3 of *Mémoirs de l'Académie Royale des Sciences,* Paris, 1699, in three parts) is the work of Perrault and Duverney [see Perrault 1733–34]. The editor, Claude Perrault, physician and famous architect of the colonnade at the Louvre, was born in Paris in 1613, and died in 1688.

37. [Perrault, "Description anatomique d'un renard marin," 1733], reprinted by Valentini [1720], pt. 2, pp. 82–84.

38. Blasius, professor at Amsterdam, mentioned above [n. 15], *Anatome animalium, terrestrium variorum, volatilium, aquatilium, serpentum, insectorum, ovorumque, structuram naturalem ex veterum, recentiorum, propriisque observationibus proponens, figuris variis illustrata,* Amsterdam, 1681, in quarto.

39. Michael Bernhard Valentini, professor at Giessen, was born in that city in 1657 and died in 1729. His *Amphitheatrum zootomicum, tabulis aeneis quamplurimis exhibens historiam animalium anatomicam,* was published at Frankfurt am Main, 1720, two parts in one folio volume. [For more on Valentini, see Cole 1949, 175–76.]

40. Samuel Collins [1618–1710], English physician, attached for a time to the czar of Russia, then to the queen, wife of Charles II of England [Catherine of Braganza, daughter of the king of Portugal]. His work, titled *A System of Anatomy, Treating of the Body of Man, Beasts, Birds, Fish, . . . ,* two volumes in folio, seventy-four plates, London, 1685, is remarkable for its many handsome illustrations of the anatomy of animals. [For more on Collins, see Cole 1949, 156–74.]

41. [The twenty species represented in the work of Collins (see n. 40 above) include] the dory *(dorée),* angelshark or monkfish *(ange),* stingray *(pastenague),* ray *(raie),* houndshark *(émissole),* freshwater burbot *(lote),* common bream *(brème),* perch *(perche),* smelt *(éperlan),* gudgeon *(goujon),* mullet *(muge),* red mullet *(surmulet),* common eel *(anguille),* gurnard *(grondin),* whiting *(merlan),* plaice *(plie),* flounder *(flet),* lamprey *(lamproie), Gadus virescens,* tench *(tanche),* common carp *(carpe),* and anglerfish *(baudroie).*

42. Joannes Jonstonus, born at Lessno or Lissa, in the palatinate of Posen in 1603, of a family originally from Scotland, died in 1675, was an assiduous compiler and published numerous works. His *Historiae naturalis* appeared in installments at Frankfurt am Main [1649–62]: the fishes and cetaceans in 1649, [other aquatic animals in 1650], the birds in 1651, quadrupeds in 1652, insects and snakes in 1653, [and plants in 1662], [seven parts] in folio. It was reprinted in its entirety at Amsterdam in 1657.

43. Hendrik Ruysch [ca. 1663–1727], editor of the third edition of Jonstonus,

published at Amsterdam by the Wetsteins [see Jonstonus 1718], was the son of the famous physician and anatomist Frederik Ruysch [1638–1731], who outlived his son by some four years. [Anatomy, both human and comparative, was Frederik Ruysch's primary interest, but he spared nothing to add additional objects of nature to his collection. He was especially well known for his collection of injected human bodies, which he was able to maintain in a fresh state for long periods; his museum in Amsterdam displayed entire bodies of infants and adults in a state of mummification, all arranged to express dramatic gestures: a "perfect necropolis, all the inhabitants of which were asleep and ready to speak as soon as they were awakened," as described by one visitor (Gonzalez-Crussi 1988, 48). For more on Ruysch and his embalming techniques, see Scheltema 1886; Hazen 1939; Cole 1949, 305–11, 448, 458; Engel 1986, 234–36.]

Part Three

THE BEGINNINGS OF MODERN ICHTHYOLOGY

6 Ray and Willughby and Their Immediate Successors

Ray[1] and Willughby[2] had the honor of being the first to write an ichthyology in which the fishes were clearly described according to nature and classified based on characteristics drawn only from their structure, and in which their natural history was finally rid of all passages from ancient writings—accounts repeated so arbitrarily about the various species by authors in the sixteenth century, so many of which are patently unlikely or unintelligible.

De historia piscium, although it bears only Willughby's name on the title page, is in large part the fruit of their common labors.[3] They had gathered the principal materials for it on a journey to France, Germany, and especially Italy, from 1663 to 1666, during which Willughby described and dissected all the fishes they were able to procure. Ray arranged them under classes and families, based first on the cartilaginous or bony nature of the skeleton, and then on the general shape, the teeth, the presence or absence of pelvic fins, the type of fin rays, soft or spiny, and finally the number of dorsal fins [see table 2]. But failing to keep in mind what was meant by spiny fins and cartilaginous skeleton, he did not always place the species where they should be according to his classification. Thus the sturgeon *(esturgeon)* remains with the bony fishes, while the tuna *(thon)* is left among the fishes without spines in their fins, and so on. Neither are there any well-defined and well-circumscribed genera; nevertheless, relationships between fishes in many places could be readily shown, so that of these connections it took only a few words to form several genera that have since come into use.

As for species, the authors assembled not only those that they saw and described from nature, amounting to 178, but also those of preceding authors, whose descriptions they intercalated with their own, arranging them as much as possible according to the method they had adopted.

FIG. 18. A porcupinefish, *Diodon histrix,* from Francis Willughby's *De historia piscium libri quatuor,* 1686, pl. I.5. Courtesy of Eveline Nave Overmiller and the Library of Congress, Washington, D.C.

One sees in these additions proof of the prodigious care Rondelet had put into his research and of the success he achieved. Willughby was often surprised by the great number he could not find and that had not escaped the naturalist of Montpellier. After Rondelet, Marcgrave furnished the most. Willughby was also very much helped with freshwater fishes by a manuscript by Baldner, a fisherman at Strasbourg;[4] and at the end of the volume, Ray added a supplement taken from Johan Nieuhof[5] and some foreign fishes furnished by Martin Lister.[6] The total number is 420; but the authors occasionally failed to recognize that some species were identical to certain others taken from the works of preceding authors, so there is some duplication, that scourge of natural history that is always ready to intervene as soon as the most severe criticism is not brought to bear on a compilation.

The second volume, titled *Icthyographia* and composed entirely of plates, presents copies of all the illustrations by Salviani, Rondelet, Marcgrave, Clusius, Nieuhof, and other ichthyologists, with a certain number of new illustrations marked with a plus (+) [fig. 18]. They have been well engraved in copperplate, but of course they have only the degree of fidelity of the originals.

The *Synopsis methodica piscium,*[7] a posthumous work by Ray, is scarcely more than an abridgment of the *Icthyographia*, with some supplements taken from du Tertre or accounts provided by Sloane and a Cor-

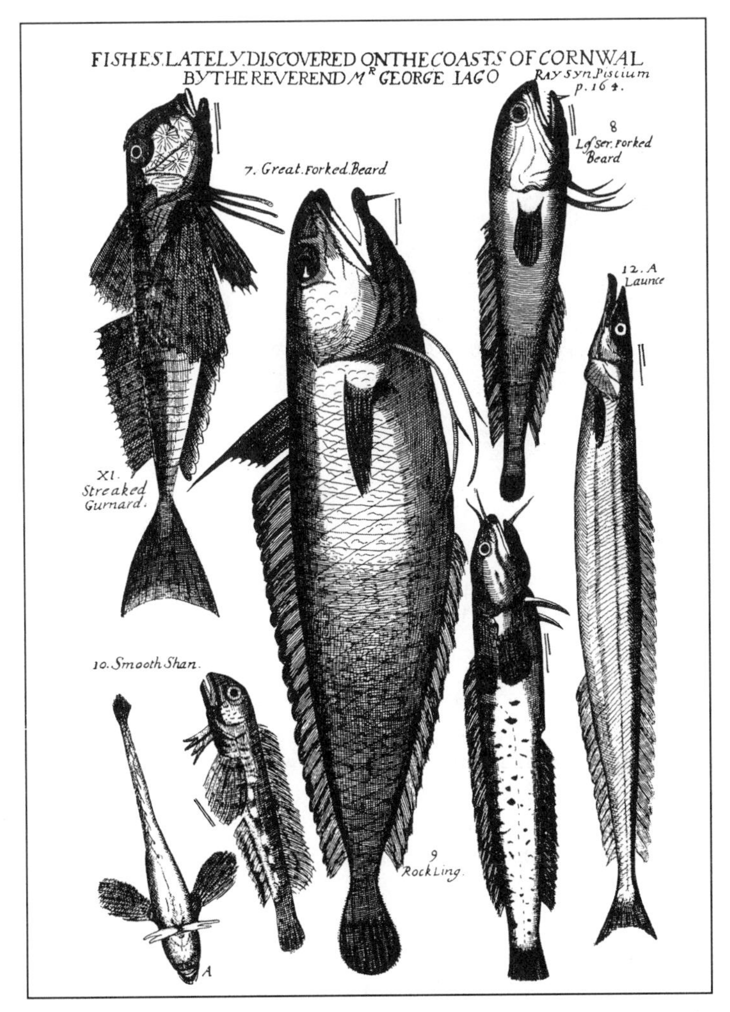

FIG. 19. Fishes from the coasts of Cornwall from John Ray's *Synopsis methodica piscium*, 1713. After Ray 1978.

nish priest named George Jago [fig. 19]. The latter are interesting because they contain information on fishes of Europe that had not yet been described.

The book by Willughby forms an epoch, and a happy one, in the history of ichthyology. Subjected thenceforth to methodical forms, this science could make regular progress, distinguishing new species from old, adding them to the mass while placing them with certitude, and above all it had a fairly complete model for descriptions. Because Willughby had no nomenclature of his own, nor fixed names for his genera,

FIG. 20. *Turdus oculo radiato* from Mark Catesby's *The Natural History of Carolina, Florida and the Bahama Islands,* 1731–43, vol. 2, pl. 22. Courtesy of Eveline Nave Overmiller and the Library of Congress, Washington, D.C.

his influence on the authors immediately succeeding him was scarcely noticeable.

Seldom does one see his influence in the writings of that time, or even long after, on the natural history of the various provinces of England,[8] although some fishes are described; nor even in the writings devoted to the natural history of particular fishes, such as that on the herring *(hareng)* by Dodd.[9]

His influence is more noticeable in writings that describe natural resources in the English colonies, especially in Sloane's publication on the fishes of Jamaica[10] and Catesby's on those of the Carolinas.[11] The former described only thirty-nine fishes, and the illustrations are poorly done from samples poorly dried. The descriptions themselves are based solely on these samples, and most of the species were already to be found in the works of Marcgrave and du Tertre. Catesby provided better drawings, colored according to life, which gives them a validity altogether too uncommon [fig. 20]. There are forty-three fishes, and because they were caught farther north, they are less likely to be found among fishes already described.

Hughes, whose natural history of Barbados was not published until 1750, gave no indication in his writings of being aware of Willughby, even though he writes about some twenty species.[12]

Edwards is hardly to be included among the ichthyologists;[13] only

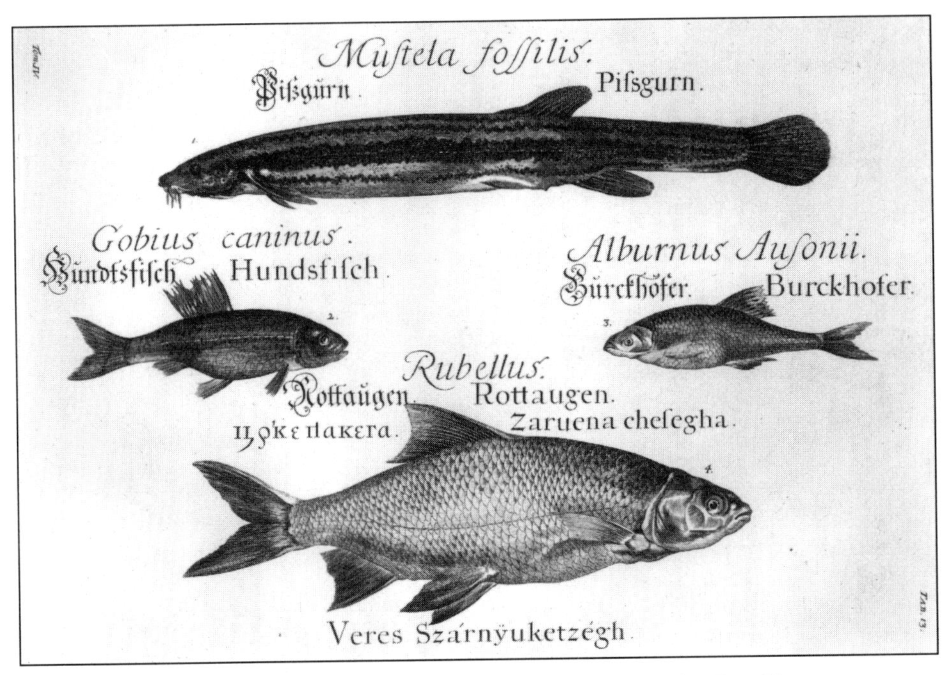

FIG. 21. Fishes of the Danube from Luigi Ferdinando Marsigli's *Danubius Pannonico-Mysicus,* 1726, vol. 4, pl. 13. Courtesy of Eveline Nave Overmiller and the Library of Congress, Washington, D.C.

fourteen fishes are found in his writings, and half of those were already known; therefore this is not the place to note his adherence to Willughby, whom he followed faithfully in his disposition of birds.

Willughby exerted even less authority on the Continent than in England; he was neglected even in works on fisheries or natural history of northern Europe; Zorgdrager,[14] Egede,[15] Anderson,[16] Horrebow,[17] and even Cranz[18] took no notice of him. However, they addressed themselves more to cetaceans and seals than to fishes and did not mention the latter except as targets of large fisheries.

Pontoppidan,[19] who must have known both Artedi and Linnaeus, cited only Willughby, and then only once or twice, so much are these so-called topographic naturalists usually behind in current knowledge. There were exceptions, however: some authors were careful to consult the great English ichthyologist and conform to his classification in their descriptions. Marsigli[20] should be placed in the first rank for his work on the fishes of the Danube, in his natural history of that river, which he assembled with care, described with exactitude, and illustrated with magnificent plates [fig. 21], and to which he added an exact anatomy of the beluga sturgeon *(esturgeon hausen)*. The number of his species is fifty-three, among them a *poécilie* [actually a loach, genus *Barbatula*][21] found nowhere else.

NOTES

1. John Ray or Wray (in Latin Raius), English theologian and one of the great naturalists of the seventeenth century, was born at Black Notley, Essex, in 1627 and died in 1705. He carried the spirit of true method into all the branches of natural history and contributed more than anyone else to the regular progress this science made during the ensuing century. [For more on Ray, see Allen 1951, 417–26; Baldwin 1986; Raven 1986.]

2. Francis Willughby, born in 1635 of an ancient lineage in England, the several branches of which have had or still have peerages, died in 1672. He joined with Ray, his teacher and friend, to work on the natural history of animals. [For more on Willughby, see Allen 1951, 417–26; Welch 1972; Raven 1986.]

3. Willughby's *De historia piscium libri quatuor* was printed at Oxford in 1686, one volume in folio, with one set of 188 plates dating from 1685. Engraving expenses were met by the members of the Royal Society of London; the president, Samuel Pepys [1633–1703], alone had 60 of them printed. [For a facsimile edition, see Willughby 1978.]

4. We have been informed of this work [of Baldner], which still exists in the public library of Strasbourg [Cuvier apparently based his knowledge on a copy of Baldner's manuscript that was later destroyed by fire during the siege of Strasbourg on 24 August 1870; see Allen 1951, 419–20; Allgayer 1991, 2]; it contains mediocre drawings of forty-five freshwater fishes and several other animals. [Another copy of Baldner's work resides in the Bibliothèque Centrale du Muséum National d'Histoire Naturelle, Paris, MS 201: "Poissons de Baldner," a collection of colored drawings of forty-five freshwater fishes and several other animals, with various notes in French, paper, 325 x 240 mm (see Boinet 1914, 32)—this is apparently a copy made for Cuvier of the Baldner original then extant at Strasbourg.]

[Léonard Baldner (Baltner of Cuvier, a spelling that appears to have originated with Ray 1686, in his preface to Willughby's *De historia piscium*), fisherman and conservator of waters *(Wasserzoller)* near Strasbourg, was born in that city in 1612 and died there in 1694. In addition to the Paris copy and the original lost in 1870, Baldner's work, which contains drawings and descriptions of fishes but also birds, quadrupeds, insects, and annelids, is represented by copies at the Municipal Library in Strasbourg, the University Library of Strasbourg, the British Museum (Natural History), and the Staendische Landesbibliothek Kassel. For more on Baldner and his zoological work, see Allgayer 1991.]

5. [On Johan Nieuhof, see chap. 4, n. 28.]

6. [Martin Lister, English naturalist and physician, born at Radclive, near Buckingham, in 1639, died at Epsom in 1712. He was educated at St. John's College, Cambridge, graduating in 1658–59; was elected a fellow of the Royal

Society of London in 1671; and received the M.D. degree at Oxford in 1684. He contributed numerous articles on natural history, medicine, and antiquities to the *Philosophical Transactions of the Royal Society of London* but was best known as a conchologist, a specialty in which he was held in high esteem (*Encyclopaedia Britannica*, 1951, 14:203). For more on Lister, see Cole 1949, 231–45.]

7. Ray, *Synopsis methodica piscium*, London, 1713, one volume in octavo. [For a facsimile edition, see Ray 1978.]

8. [James] Wallace [d. 1688], *An Account of the Islands of Orkney*, 1700; [Charles] Leigh [1662–1701?], *The Natural History of Lancashire, Cheshire, and the Peak, in Derbyshire*, Oxford, 1700, in folio; [John] Morton [rector of Oxendon, 1671?–1726], *The Natural History of Northampton-shire*, London, 1712, in folio; [John] Coker, *A Survey of Dorsetshire*, London, 1732, in folio; Silas Taylor [1624–78], *The History and Antiquities of Harwich and Dovercourt, in the County of Essex*, to which is added an *Appendix Containing the Natural History of the Sea-coast and Country about Harwich* by Samuel Dale [1659–1739], London, 1732, in quarto. One could even extend this judgment to [William] Borlase [1695–1772], *The Natural History of Cornwall*, Oxford, 1758, in folio, in which, however, one sees several interesting fishes, notably the medusafish *(pompile* or *centrolophe)* [see Borlase 1758, 261–74, pls. 26, 27]; and to [John] Wallis [1714–93], *The Natural History and Antiquities of Northumberland*, London, 1769, two volumes. Most of these authors seem to have taken for their model *The Natural History of Oxford-shire* by Robert Plot [1640–96], printed in 1677, Oxford, in folio, rather than the works of Ray and Willughby. Plot [1677, pl. 10, p. 212] provides illustrations of a small lamprey *(lamproie)* and a minnow *(cyprin)*.

9. *An Essay towards a Natural History of the Herring*, by [James Solas] Dodd [1721–1805], London, 1752, in octavo.

10. Hans Sloane, born at Killyleagh in Ireland in 1660, physician to the duke of Albemarle, who was governor of this realm in 1687, died president of the Royal Society of London in 1753. Sloane published his work in English under the title *A Voyage to the Islands of Madeira, Barbados, Nièves, St. Christopher's, and Jamaica*, London, 1707–25, two volumes in folio, with 274 plates.

11. Mark Catesby, born in about 1679 [d. 1749], lived in Virginia from 1712 to 1719 and returned there from 1722 to 1726 at the expense of [Samuel] Dale [1659–1739], [William] Sherard [1659–1728], and Sloane [see Frick and Stearns 1961]. His *Natural History of Carolina, Florida and the Bahama Islands*, in two volumes in folio, London, 1731–43, with 220 plates, at the time surpassed all other publications in the beauty of its illustrations. A German translation appeared [at Nuremberg in six parts] from 1749 to 1770 [a second London edition was published in 1771, and a facsimile edition, with introduction and notes, in 1974]. [For more on Catesby, see Allen 1951, 463–78.]

12. Griffith Hughes [b. ca. 1707], Anglican vicar at St. Lucia on the island of Barbados, published in English a natural history of that island, London, 1750, in folio. Of his twenty fishes, only two are illustrated ["the triangular fish" and "the dolphin" (see Hughes 1750, bk. 10, pp. 299–314, pls. 28, 29)].

13. George Edwards [1694–1773], English painter and librarian of the Royal Society of London, published a series of two collections containing in all 362 plates: *A Natural History of Uncommon Birds, and of Some Other Rare and Undescribed*

Animals, Quadrupeds, Reptiles, Fishes, Insects, etc. in four volumes in quarto, London, 1743–51; and *Gleanings of Natural History, Exhibiting Figures of Quadrupeds, Birds, Insects, Plants, etc.*, in three volumes, 1758–64. His illustrations are quite exact and are among the best of the eighteenth century. In each volume he follows the ornithological method of Willughby. [For more on Edwards, see Allen 1951, 480–86.]

14. Cornelius Gisbert Zorgdrager [b. ca. 1650] author of a very confused work on the whale fishery of Greenland and the cod fishery of Newfoundland, printed in Dutch at Amsterdam in 1720 and [at The Hague in] 1727, and in German at [Leipzig in 1723 and at] Nuremberg in 1750.

15. Hans Egede [1686–1758], Norwegian clergyman [and bishop of Greenland], left for Greenland in 1721 out of religious zeal and lived there until 1736. It is under his direction that the Moravian brothers established a mission there in 1733. His description and natural history of Greenland was printed at Copenhagen, in Danish in 1741, in quarto, and in French in 1763, in octavo. There is an English translation, London, 1745.

16. Johann Anderson, merchant and burgomaster from Hamburg (1674–1743), was author of a natural history of Iceland, Greenland, Davis Strait, and other northern countries, printed in German at Hamburg in 1746, and in French at Paris in two volumes in duodecimo, 1750.

17. Niels Horrebow [1712–60], Danish clergyman, sent to Iceland by the king of Denmark, produced a physical, historical, and so forth, description of that island, printed in Danish at Copenhagen in 1752, [in English at London in 1758], and in French at Paris in 1766, two volumes in duodecimo.

18. David Cranz [1723–77], Moravian missionary, author of a history of Greenland, printed in German at Barby in 1765–70, two volumes in octavo; and in English at London in 1767, in octavo. An excerpt of it is published in [Antoine François Prévost's (1697–1763)] *Histoire général des voyages*, vol. 19, 1770.

19. Erik [Ludvigsen] Pontoppidan, born at Århus in 1698, bishop of Bergen in Norway, died in 1764, published in Danish a natural history of Norway, Copenhagen, 1752–53 in quarto, of which there is an English translation, London, 1755, in folio, and two German translations: Copenhagen, 1753–54, two volumes in quarto; and Flensburg and Leipzig, 1769, one volume in quarto. He writes about fishes, but as a naturalist who is not well educated in the science and who is too credulous. [For a facsimile edition of his *Norges naturlige historie*, see Pontoppidan 1977.]

20. Count Luigi Ferdinando Marsigli [or Marsilli], born 1658, a nobleman of Bologna, officer in the service of Austria in 1682, prisoner in Turkey in 1683, degraded in 1703 for surrendering Fribourg, founder of the Institute of Bologna in 1715, died in 1730. He published in 1726 a description of the course of the river Danube and of the natural production of its waters and its banks, in six volumes, in folio, under the title *Danubius Pannonico-Mysicus* [For more on Marsigli, see Stoye 1993]. The fourth volume, devoted to fishes [and titled *De piscibus in aquis Danubii viventibus*], contains very handsome illustrations of fifty-three species. The anatomy of the beluga sturgeon *(hausen)* is described and illustrated in vol. 6 [pp. 15–17, pls. 9–21].

21. [Obviously this fish cannot be a member of the New World cyprinodontiform family Poeciliidae (as defined by current taxonomy), as indicated by Cuvier. Marsigli (1726, pl. 25, fig. 1) referred to a fish he called *Gobius caninus* and provided an illustration of the same labeled *"Fundulus"* or "nostris Germanis *Grundel*," which is quite clearly the European stone loach *Barbatula barbatula* (Linnaeus, 1758) (A. Wheeler, pers. comm., 6 November 1993; see Valenciennes, 1847, p. 539).]

TABLE 2. This will give the reader an idea of the classification of Ray and Willughby [as used in their *De historia piscium libri quatuor* of 1686]. Please note, however, that for purposes of abridgment we have entered the names of the genera as we know them today, that they named only species, and did not always group them correctly. [See Willughby 1686, 22–25.]

Classis I. CETACEI [cetaceans]				
	Balaena	*Delphinus*	*Phocaena*	
Classis II. CARTILAGINEI [cartilaginous forms]				
LONGI [elongate forms]				
	Squali	*Zygaena*, etc.		
LATI [broad forms]				
	Raiae	*Partinacae*	*Aquilae*, etc.	*Rana piscatrix*
Classis III. OSSEI [bony forms]				
PLANI [depressed forms]				
	Pleuronectes			
NON PLANI [nondepressed forms]				
Anguilliformes [body eel-like]				
	Muraenae	*Taeniae*	*Gunnellus*	*Silurus*
	Anguillae	*Remora*	*Mustela*	*Gobio*, etc.
	Lampetrae			
Corpore contractiore [body contracted]				
Sine ventralibus [pelvic fins absent]				
	Orbes	*Stromateus*	*Xiphias*	
	Balistes	*Acus*		
Cum ventralibus [pelvic fins present]				
Malacopterygii [soft rayed]				
Pinnis dorsalibus 3 [dorsal fins 3]				
	Gadi			
Pinnis dorsalibus 2 [dorsal fins 2]				
	Merlucii	*Truttae*	*Lumpus*	
	Thynni	*Gobii*	*Atherina*, etc.	
Pinna dorsali 1 [dorsal fin 1]				
	Coryphenae	*Argentina*	*Exocetus*	*Cyprini*
	Chaetodontes	*Belone*	*Lucius*	*Cobitis*
	Clupeae	*Saurus*	*Sturio*	
Acanthopterygii [spiny rayed]				
Pinnis dorsalibus 2 [dorsal fins 2]				
	Labrax	*Triglae*	*Trachini*	*Faber* et *Vomer*
	Sphyraena	*Mulli*	*Batrachus*	*Glaucus*
	Mugiles	*Callyonimi*	*Perca*	*Sciaenae*
Pinna dorsali 1 [dorsal fin 1]				
	Spari	*Serrani*	*Cernuae*	
	Labri	*Scorpaenae*	*Gasterostei*	

7 Contributions of Travelers in the East and West

It is scarcely worthwhile to name here some travelers of the era, such as Bosman,[1] Leguat,[2] Labat,[3] De Bruin,[4] Lucas,[5] and Kolb,[6] who were not naturalists at all and produced only a few inexact drawings of fishes and some articles often mixed with anecdotes. But we cannot treat so lightly two collections of fish drawings made in the Dutch colonies in the East Indies, which were used in publications by Ruysch, Valentijn, and Renard.

The first collection was made for Corneille de Vlaming by an unnamed artist [fig. 22];[7] it served as the original for the first part of Renard's fishes of the Indies.[8] The second collection was the work of a man named Samuel Fallours[9] and conformed less to nature. It is presented in the second part of Renard, whose work was not published until 1754, although it had been prepared more than thirty years earlier; but as early as 1718, they were mingled together in a publication by Ruysch, which appeared at the beginning of the third edition of Jonstonus, under the shared title *Theatrum animalium* [fig. 23];[10] and in 1726 Valentijn borrowed illustrations from both publications and included with them a certain number of others in the third volume of his great history of the East Indies.[11]

The descriptions provided by Ruysch and Valentijn are based on the illustrations, and the notes that Valentijn added seem very dubious; but the illustrations, especially those in the first collection, are not at all imaginary, as was believed for a long time. Pallas very early maintained, and with reason, that the illustrations are for the most part done from nature, and every day there comes to us, as proof of the good faith of the artists, one or another of the species that have been represented.[12]

Depending on how clever or meticulous the artist was, it is true that nature has been more or less faithfully reproduced, but almost never are the delicate characters precisely drawn; attention is never paid to the

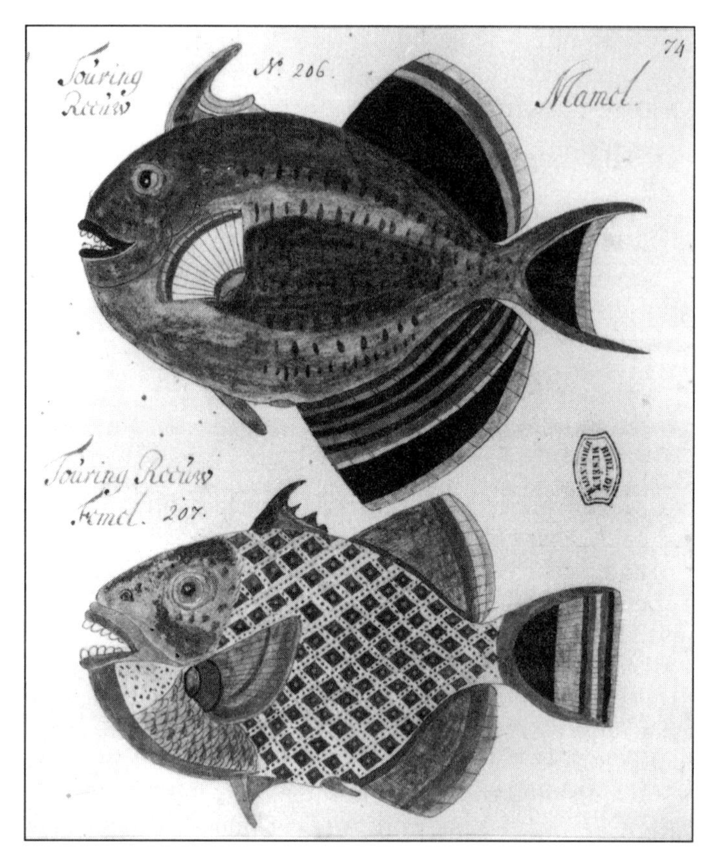

FIG. 22. Triggerfishes from "Zee-Tooneel," a collection of colored drawings made by Isaäc Johannes Lamotius (b. ca. 1645), attributed erroneously by Cuvier to Cornelis de Vlamingh (b. 1678). Manuscript 339, p. 74, figs. 206–7, Bibliothèque Centrale du Muséum National d'Histoire Naturelle, Paris. Courtesy of M. Ducreux and the Bibliothèque Centrale, Muséum National d'Histoire Naturelle, Paris.

number of fin rays or spines. But despite their faults, these collections are still indispensable, either for giving an idea of the natural colors of known species or for helping us recognize new species that travelers are bringing to us daily from those teeming seas.

The number of these illustrations is 459 in Renard, 527 in Valentijn, and 396 in Ruysch; but there is much duplication, and a rather large number of crustaceans should be excluded. One can compare these drawings made in the Indies with the different collections of paintings or engravings coming from China or Japan, and one might even prefer the latter, for they are often as accurate in depicting characteristics as our best European works. Among the published engravings, the *Encyclopédie*

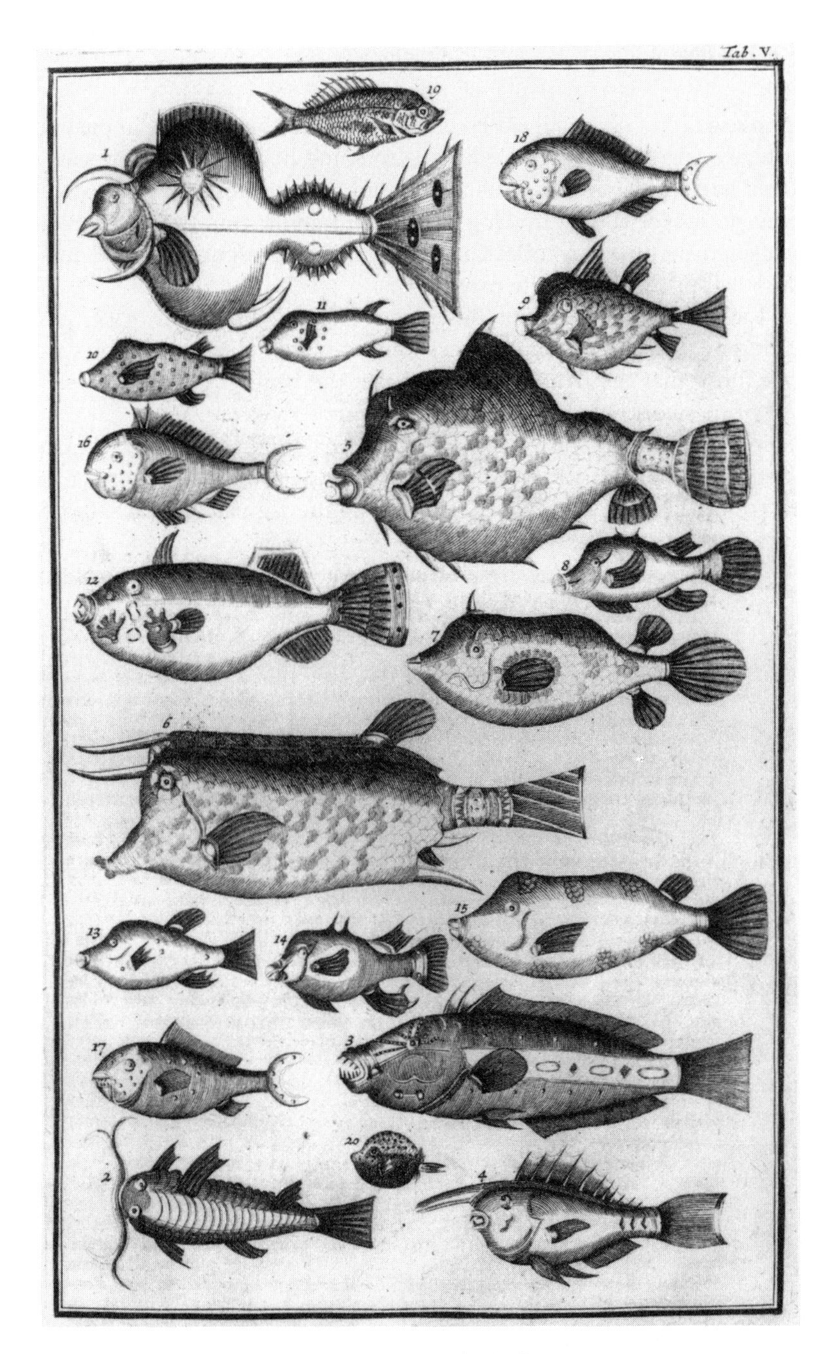

FIG. 23. One of twenty plates that illustrate "Collectio nova piscium Amboinensium," in Hendrik Ruysch's 1718 edition of Joannes Jonstonus's *Historiae naturalis,* vol. 1, pt. 1, pl. 5. T. W. Pietsch collection. Photograph by Patricia L. McGiffert, Health Sciences Photography, University of Washington, Seattle.

japonaise ought especially to be cited,[13] and another volume in particular on fishes, also from Japan, which is in some libraries;[14] the species depicted are quite recognizable, and doubtless these books can give us some notion of the ichthyology of these remote countries. Several unpublished manuscript collections of paintings present even better illustrations, but they are quite rare.[15]

Kaempfer[16] had some of these fishes engraved in his history of Japan, adding details taken from Japanese books and comparisons of the species with those that the Dutch were taking at the time in the Moluccas. This part of his work was copied by Charlevoix.[17]

We should note, however, that all these Chinese and Japanese documents, dull and mixed with fables, are almost useless as far as the texts are concerned; one can profit only from the figures, the artists being much superior to the writers.

But there was one traveler at that time who worked as a true naturalist, to whom ichthyology would have been eminently grateful had his work been published while he was still alive, and that was Father Plumier.[18] His reputation as a superior botanist had been made long before; but he was a good zoologist as well, and both in his native Provence and in the Antilles he made a series of numerous drawings, especially of fishes, remarkable for their detail and accuracy, lacking hardly anything more than the rendering of the correct number of spines, and the expressing of the denticulation of the opercular bones in species where this is masked by the skin while the animal is fresh [fig. 24].[19]

Unfortunately the author, little admired by the ignorant monks[20] among whom he lived, died before publishing this part of his research; his manuscripts lay neglected in his monastery,[21] and only a few excerpts appeared in Feuillée's[22] work and in the journals of a certain Gautier d'Agoty,[23] who was not able to appreciate them. It was not until the end of the eighteenth century that a copy, prepared by the author himself, fell into the hands of Bloch,[24] who used the illustrations in his great work, but sometimes with alterations similar to those to which he subjected the drawings of Prince Maurits.[25]

Another copy of Plumier drawings, rather inexact and too highly colored, made for the great collection of vellums by Aubriet,[26] the famous painter of Tournefort,[27] was in part engraved for the *Histoire naturelle des poissons* by Lacepède;[28] but because the latter did not always realize that these drawings were originally the same as the ones that had already appeared in Bloch's work, there resulted several duplications. The same drawing sometimes occasioned the establishment not only of a species but also of a genus that was purely imaginary.[29]

The carelessness of Plumier's editors should in no way detract from the esteem owed to this respectable and laborious observer. To this day

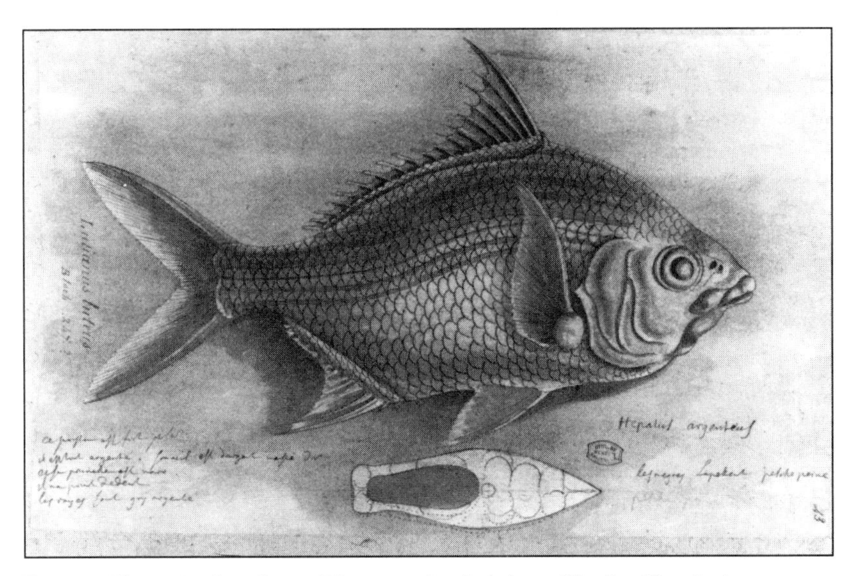

FIG. 24. The striped mojarra, *Diapterus plumieri*, from Charles Plumier's "Poissons et coquilles." Manuscript 31, fig. 43, Bibliothèque Centrale, Muséum National d'Histoire Naturelle, Paris. Courtesy of M. Ducreux and the Bibliothèque Centrale, Muséum National d'Histoire Naturelle, Paris.

we know certain species only through him, and his manuscripts prove that most of the errors that have slipped into publications produced by others of his works were not his.

NOTES

1. Willem Bosman [b. 1672], employed by the Dutch West India Company in Africa, published in 1705 at Utrecht, in octavo, a *Voyage en Guinée*, with more numerous and more exact descriptions and illustrations of animals than are given in most works of this sort. [For more on Bosman, see Tye and Jones 1993, 219–24.]

2. François Leguat was born at Bresse in 1637, expatriated following the revocation of the Edict of Nantes [1685], confined on Rodriguez Island from 1691 to 1693, returned to Europe in 1698, and died in London in 1735. In 1708 he published in London, in octavo, *A New Voyage to the East-Indies by Francis Leguat and His Companions*, in which he provides illustrations of some animals, but they seem to be drawn from memory [rather than from nature]. [The "Succet or Remora" is described and pictured on p. 88.]

3. Jean Baptiste Labat, a Dominican missionary, was born in Paris in 1663, served on the island of Martinique in 1694, returned to Europe in 1706, and died in Paris in 1728. He published four works that are in some respects of interest to natural history: (1) *Nouveau Voyage aux isles de l'Amérique*, Paris, 1722, six volumes in duodecimo (a number of later editions were produced: The Hague in 1724 and

1738, Paris in 1738 and 1742, as well as German and Dutch translations [I have been unable to locate copies of the two 1738 editions]); (2) *Nouvelle Relation de l'Afrique Occidentale, d'après les mémoires de Brue,* Paris, 1728, five volumes in duodecimo (reprinted in 1732 and 1758 [I have been unable to locate copies of these two reprints]); (3) *Voyage du chevalier des Marchais en Guinée et à Cayenne fait en 1725–1727,* Paris, 1730, four volumes in duodecimo (reprinted in 1731 at Amsterdam); (4) *Relation historique de l'Ethiopie occidentale,* part of which was translated into Italian by the Capuchin friar [Giovanni Antonio] Cavazzi [d. ca. 1692]; Paris, 1732, five volumes in duodecimo. [For more on Labat, see Allen 1951, 439.]

4. Cornelius de Bruin [1652–1719], a Dutch painter and author of two travel books in which he presents some objects of natural history, notably some fishes: *Voyage au Levant,* Delft, 1700, one volume in folio; *Cornelis de Bruins reizen over Moskovie, door Persie en Indie,* Amsterdam, 1714, one volume in folio, and another Amsterdam edition *[Voyages de Corneille Le Brun par la Moscovie, en Perse, et aux Indes Orientales],* 1718, two volumes in folio.

5. Paul Lucas, born at Rouen in 1664, died in 1737, made a number of journeys to the Levant and gave an account of three, under the titles *Voyage du Sieur Paul Lucas au Levant,* Paris, 1705, two volumes in duodecimo; *Voyage du Sieur Paul Lucas, fait par ordre du roy dans la Grèce, l'Asie Mineure, la Macédoine et l'Afrique,* Paris, 1712, two volumes; *Voyage du Sieur Paul Lucas, fait en M. DCCXIV, etc. par ordre de Louis XIV dans la Turquie, l'Asie, Sourie, Palestine, Haute et Basse Egypte, etc.,* Rouen, 1719, three volumes. These works, often reprinted, contain some accounts of the fishes of the Nile.

6. Peter Kolb, born in 1675 at Wunsiedel in Bayreuth, was sent to the Cape Town in 1704, returned to Europe in 1712, and died in 1726. In 1719 he published in German an account of his journey, Nuremberg, three parts in folio, which was printed again in German in two volumes, Amsterdam, 1727. There is a much abridged French translation, Amsterdam, 1741, three volumes in duodecimo. He speaks of some fishes, but they were not well observed. The illustrations seem to have been done in Europe after his return [for more on Kolb, see Rookmaaker 1989, 29–30].

7. Cornelis de Vlamingh [b. 1678] had been boatswain for the Dutch East India Company in Bengal and brought the fleet back as admiral in 1715. The original collection of his paintings is now at the [Bibliothèque Centrale du] Muséum National d'Histoire Naturelle, Paris [MS 339]; it is titled *Zee-Tooneel, verbeeldende een wonderbare verscheidenhiet van zwemmende en kruipende Zee-Dieren* [Sea scene, showing a marvelous variety of swimming and crawling sea animals] and contains a statement that it was drawn from nature at Vlamingh's orders and under his supervision. [Despite this attribution to Vlamingh, all available evidence indicates that the "Vlamingh Drawings" were made by Isaäc Johannes Lamotius (ca. 1645 to after 1717), a Dutch governor of the Indian Ocean island of Mauritius (see Pietsch and Holthuis 1992). For more on Vlamingh, see Schilder 1976, 1985; Pietsch 1995.]

8. [Renard's] *Poissons, écrevisses et crabes, de diverses couleurs et figures extraordinaires, que l'on trouve autour des Isles Moluques, et sur les côtes des Terres Australes, . . . ,* Amsterdam, 1754, in folio. Louis Renard [born at Charlemont, 1678 or 1679,

died at Amsterdam, 1746], agent of the king of England at Amsterdam, began preparing his publication in 1718 or 1720, but his death delayed for a considerable time the definitive publication, which did not take place until 1754, under the auspices of Aernout Vosmaer [1720–99; see Pieters 1980]. Renard provided [in his book] a certificate from Frederik Julius Coyett [born at sea ca. 1680, died at Batavia in 1736], to the effect that the drawings in the first part had been done in the house of his father, Balthasar Coyett [born on Formosa probably ca. 1650, died at Batavia in 1725], governor of Ambon. This assertion can be reconciled with the title of Vlamingh's collection [see n. 7 above] only by supposing that two copies were made, one for Vlamingh and another for Coyett. [For more on Louis Renard and his *Poissons, écrevisses et crabes*, see Pietsch 1984, 1986, 1991, 1993, 1995; Pietsch and Rubiano 1988.]

9. Samuel Fallours, minister to the sick at Ambon, who also returned [to Holland] in 1715, acknowledged himself to be the author of the second collection in a letter, also published by Renard at the beginning of his *Poissons, écrevisses et crabes* [1719, 1754]. [For more on Fallours, see Pietsch 1986, 1991, 1995.]

10. Hendrik Ruysch [see chap. 5, n. 43] attributes these drawings, which are essentially the same as the ones in the two parts of Renard, to one person alone: "Quae, ibi," he says, "conciones ad populum habebat, et caeteras res quae pertinent ad religionem per aliquot annos curabat" [Who held services there for the people, and took care of other matters related to religion for some years; see Ruysch 1718, vol. 1, in the dedicatory epistle to Hermann Boerhaave, par. 1, lines 27–29]; a designation that seems to refer to Fallours [see n. 9 above], which makes me believe that the copy acquired by the Wetsteins, and published by Ruysch, was a copy made by Fallours, not only of his own drawings that make up the second part of Renard, but also of those that had been done previously, either for Coyett or for Vlamingh. The latter are much better than his own drawings. [For more on Hendrik Ruysch, see Engel 1986, 236; Pietsch 1995, fig. 49.]

11. François Valentijn, Protestant minister at Ambon, born at Dordrecht in 1666, died [at The Hague] in 1727, made his first stay in the East Indian archipelago in 1685–94 and another in 1706–14. He is the author of a great Dutch work, in five volumes in folio, printed at Dordrecht and Rotterdam, 1724–26, titled *Oud en nieuw Oost-Indiën*. In the third volume, which describes [the Moluccan island of] Ambon, he wrote about natural history, but much of it is supposition and written as if by a man who is a complete stranger to the science. [For more on Valentijn, see Haan 1902; Serton 1971, 3–9; Beyers 1977, 796; Engel 1986, 281; Rookmaaker 1989, 30–31; Pietsch 1991, 7–8; 1995. For a full account of the relation between the fish drawings that appear in the publications of Ruysch 1718, Renard 1719, and Valentijn 1724–26, see Pietsch 1995.]

12. [A breakdown of the number of identifiable Renard figures indicates that those of vol. 1 are significantly more accurate than those of vol. 2: of the 219 figures in vol. 1, 152 can be identified to species, 51 to genus, and 11 to family; of the 241 figures in vol. 2, 122 can be identified to species, 48 to genus, and 31 to family. Thus, only about 2 percent of the figures of vol. 1, and about 17 percent of those of vol. 2 are beyond recognition at the taxonomic level of family or below (see Pietsch 1995).]

13. The *Encyclopédie japonaise,* [formerly] in the king's library [but now in the Department of Manuscripts, Oriental Section, Bibliothèque Nationale, Paris, MSS japonaise 340], consists of a number of small quarto volumes [eighty-one original fascicles bound in thirteen volumes]; the part on fishes contains illustrations of seventy-nine species.

14. The other work is [known to us from two copies: one] in the [Bibliothèque Centrale du] Muséum National d'Histoire Naturelle, Paris [MS 2321; see Boinet 1914, 81]; and [the other, formerly] in the library of the late Joseph Banks [but now in the collections of the East India Office, Blackfriars Road, London (pressmark 16116.d.17)]. Lacepède cited it under the inexact title *Manuscrit chinois.* It is a slender volume in folio [205 x 290 mm], containing [descriptions in Japanese and] many illustrations of aquatic animals, including eighty-three of fishes, engraved on wood and colored, in large part like those in the *Encyclopédie japonaise.* Jean Pierre Abel Rémusat [1788–1832] kindly deciphered for us some of the articles in these books; we feel confident in citing these drawings, because the ones of known species have taught us to appreciate the fidelity of the others.

[The fishes depicted in the Paris copy of the *Manuscrit chinois* are identified with French vernacular names added in pencil by Cuvier. The Banks copy at the East India Office is described by Yu-ying Brown (1988, 23, 55, pl. 51) under the title *Umi no sachi* (Boon of the seas): A haiku anthology on fishes and shellfishes (but also including a squid, a jellyfish, an octopus, sea cucumbers, and several turtles) illustrated by Katsuma Ryūsui (1711–96); two volumes in one, 1762; block printed in color, with powdered mica to give sheen to the fishes.]

15. We have consulted several of these [manuscript] collections of paintings, among them a very handsome one in the library of the Muséum National d'Histoire Naturelle, Paris, which Lacepède also used; it contains fifty-four sheets in transverse folio, superbly drawn. Dussumier [see chap. 16, n. 49] recently lent us another collection that contains twenty-four even more careful drawings. The duke of Rivoli possesses a superb manuscript, brought back from Japan by the late [Isaac] Titsingh [1744–1812], in which the Japanese names are added in European transliteration; he was kind enough to show us this volume, which contains drawings of thirty-one fishes.

[The Lacepède manuscript cannot now be identified (although there is some evidence that it is MS 2321: "Collection of fishes, mollusks, and crustaceans, engraved and illuminated in Japan, with their Chinese and Japanese names, and the French names added in pencil by Cuvier," two volumes of fifty-seven double folio sheets bound in one, measuring 290 x 200 mm; see Boinet 1914, 81) but the Dussumier and Titsingh collections are still present in the collections of the Bibliothèque Centrale du Muséum National d'Histoire Naturelle, Paris. The Dussumier manuscript (MS 5037), titled "Twenty-four fishes painted at Canton. Monsieur J. J. Dussumier," is actually a collection of twenty-four folio pages (measuring 435 x 540 mm) containing illustrations of twenty-five fishes, thirteen of which are identified with reference to the text of the *Histoire naturelle des poissons.* The Titsingh manuscript (MS 396), titled "Poissons du Japon," is an oblong volume of twenty-three folio sheets (measuring 415 x 280 mm) bearing, in addition to fishes, illustrations of crustaceans, sea cucumbers, and turtles (see Boinet 1914, 83).]

16. Engelbert Kaempfer was born at Lemgo, in the principality of Lippe, in 1651. He traveled to Persia in 1684, embarked in 1688 with a Dutch fleet, which was sailing in the Persian Gulf, arrived in 1689 at Batavia, and from there went to Japan. He left Japan about the end of 1691, returned to Europe two years later, and died in 1716, physician to the count of Lippe. He published in 1712 his *Amoenitatum exoticarum,* in five books, and left a manuscript in German titled "Ecclesiastical and Secular Natural History of Japan," which was later acquired by Hans Sloane, translated into English by [Johann Gaspar] Scheuchzer [1702–29], and printed in London in 1727. There is a French translation produced at The Hague, 1729, two volumes in folio [and one in Dutch published at Amsterdam, 1733, one volume in folio]. Plates 11–14 show aquatic animals, among which are twelve species of fishes.

17. Pierre François Xavier de Charlevoix, a Jesuit, born at Saint-Quentin in 1682, died at La Flèche in 1761. In his *Histoire du Japon,* Paris, 1736, two volumes in quarto, he copies Kaempfer but does not name him except when he refutes him. He places the article on fishes in a supplement at the end of the second volume. [On Charlevoix, see Allen 1951, 503–4.]

18. Charles Plumier, born at Marseilles in 1646, entered the order of the Minims [a Catholic monastic order founded in Italy in 1435 by Saint Francis of Paula; see Whitmore 1967] in 1662, taught by Boccone [see chap. 5, n. 20] in Italy, friend of Tournefort [see n. 27 below] and [Pierre Joseph] Garidel [1658–1737], made his first voyage to Martinique and neighboring islands, and even the continent [but see below], in 1688–89, and returned there twice again with government missions. He died in 1704 at the port of Santa Maria near Cadiz when he was preparing to depart for Peru.

[The localities in the New World visited by Plumier are provided in detail by Urban (1920, 5). Some accounts (e.g., Duvau 1823, 94) indicate that Plumier collected plants on the North American continent, but Urban (1920, 5; see also Fournier 1932, 54) adamantly denies this possibility: "Das amerikanische Festland ist von Plumier nicht berührt worden. Alle Identifizierungen seiner Tafeln mit Arten, die nur auf dem Kontinente vorkommen, sind deshalb irrig."]

In addition to his *Description des plantes de l'Amérique* (Paris, 1693, in folio), his *Nova plantarum Americanarum genera* (Paris, 1703, in quarto), his *Traité des fougères de l'Amérique* (Paris, 1705, in folio), and the fascicles published by [Johannes] Burmann [professor of botany at the Athenaeum Illustre and at the Amsterdam Medical Garden, 1707–79] at Amsterdam (1755–60), Plumier left a large quantity of manuscripts, which had remained in the library of the Minims in the Place Royale, Paris, and are now deposited in the king's library and the [Bibliothèque Centrale du] Muséum National d'Histoire Naturelle, Paris. An account of them was published by [Auguste] Duvau [1771–1831], *Biographie Universelle,* vol. 35 [1823]. [For more on Plumier, see Urban 1920; Fournier 1932; and Whitmore 1967.]

19. The drawings of fishes are now in the [Bibliothèque Centrale du] Muséum National d'Histoire Naturelle, Paris, bound in three volumes of different sizes, one titled "Poissons, oiseaux, lézards, serpens et insectes" [MS 24], which contains 157 illustrations of fishes; the second, "Poissons d'Amérique" [MS 25], contains 100; and the third, "Poissons et coquilles" [MS 31], has 80. But several

figures are repeated, and many fishes are [not from America at all but] from our own waters in France. One may still see in most of them the pinholes used for pouncing the drawings, probably for the copy that was used by Bloch [in the preparation of his great works on fishes].

20. It is worth noting the contemptuous tone Labat [see n. 3 above] uses in speaking of a man [Plumier] who was in every regard much superior to himself (e.g., in his *Nouveau voyage aux isles de l'Amérique* and elsewhere): ["I hardly know any man easier to deceive than that good priest. He had a marvelous talent for drawing plants and he was able to produce excellent works of that sort, if he was cloistered, but because he desired to come out of the cloister, he fell into an infinity of blunders, of which the one I mention here is not even one of the most considerable" (Labat 1722, 1:287–88)].

21. [Antoine Laurent] de Jussieu [1748–1836] assured me that the monks [at the convent of the Minims at the Place Royale in Paris] used them as stools for seating themselves near the fire.

22. Louis Feuillée [a monk of the order] of the Minims [see Whitmore 1967], was born at Mane near Forcalquier in 1660. He traveled as an astronomer to the Levant in 1699, to the Antilles and New Spain in 1703, to Peru and Chile, 1708–11, and died in 1732. In his *Journal des observations physiques mathématiques et botaniques*, Paris, 1714, two volumes in quarto, and the remainder in a third volume published in 1728, he included many things pillaged from the papers of Plumier, his confrere, but he took only a few articles on fishes.

23. Jacques Gautier d'Agoty [born at Marseilles in 1710, died at Paris in 1781], painter, [engraver, physician, anatomist], and author of numerous works embellished with color plates that were executed by a process of his own. He included several drawings by Plumier in his *Observations sur l'histoire naturelle, sur la physique et sur la peinture*, 1752–55, six volumes. Many more are reproduced in a continuation of this series produced by his son [Arnaud Eloy Gautier D'Agoty, d. 1771], with the help of [F. V.] Toussaint: two volumes, 1756–57 [each with a slightly different title]. *Observations sur la Physique* [called the *Journal de Physique* after 1794] by [l'Abbé] Rozier, [J. A. Mongez], and [Jean Claude de] La Métherie is itself a continuation of the former.

24. Bloch gives an account of this manuscript in the preface to the third part of his great *Naturgeschichte der ausländischen Fische* [1785–95, 3:iii–viii]. It seems that it was prepared by Plumier [see n. 18 above] himself in the hope of having it printed in Holland. A Frenchman in the service of Prussia took it to Berlin, where it was sold at auction. Its title was *Zoographia Americana, pisces et volatilia continens, auctore R. P. C. Plumier.* It was composed of 169 pages in folio but contained more than just fishes, so that Bloch took from it only thirty-four illustrations for his work; three more were used by Schneider in Bloch's posthumous work, the *Systema ichthyologiae* [see Bloch and Schneider 1801]. It is not known what happened to it at the sale of Bloch's books [see Karrer 1980, 185–88; Wells 1981, 8].

25. For example, Bloch [1785–95, 2:146, pl. 175] deliberately changed the shape of the head of the fish called *"vive"* on Martinique, which is a tilefish of the genus *Malacanthus (malacanthe)*, in order to make it look like a dolphinfish *(coryphène);* he called it *Coryphaena plumieri* (see Bloch and Schneider 1801, 298–99).

Lacepède (1798–1803, 4:427, pl. 8, fig. 1) reproduced the drawing more accurately, almost as it appears in Plumier's manuscript but, on Bloch's authority, still left it among the dolphinfishes *(coryphènes).*

26. [Claude Aubriet, born at Châlons-sur-Marne about 1655, became painter to the king, the king's cabinet, and the king's garden in 1700; later (1706–35) he held the appointment of painter of miniatures to the king (see n. 28 below). He died at Paris in December 1742.]

27. [Joseph Pitton de Tournefort, the celebrated French botanist, professor of botany at the Jardin du Roi, 1656–1708.]

28. It appears that Aubriet [see n. 26 above], who was paid by the page to continue the great collection—begun in 1640 for Gaston d'Orléans, brother of Louis XIII, and now deposited at the [Bibliothèque Centrale du] Muséum National d'Histoire Naturelle, Paris [see Laissus 1967]—took originals wherever he could. It appears also that he knew of Plumier's drawings, but he painted his copies according to the descriptions only, or even according to his imagination. There is nothing to prove that he worked under the supervision of the original author. Lacepède had thirty-seven of these drawings by Aubriet engraved [for use in his *Histoire naturelle des poissons,* 1798–1803].

29. For example, *Harpe bleu-doré [Harpe caeruleo-aureus]* of Lacepède [1798–1803, 4:427], pl. 8, fig. 2, is the same as Bloch's [1785–95, 5:20], pl. 258, *Sparus falcatus;* [Lacepède's, 1798–1803, 3:542], pl. 33, fig. 1, *Cheilodiptère chrysoptère [Cheilodipterus chrysopterus]* is the same as Bloch's [1785–95, 6:66, pl. 306], *Sciaena plumieri,* and so on.

Part Four

THE FOUNDATIONS OF MODERN CLASSIFICATION

8 Artedi and Linnaeus, Their Contemporaries and Immediate Successors

It was not until the first third of the eighteenth century that there appeared the work destined finally to give the natural history of fishes a truly scientific form, completing what Willughby and Ray had started—namely, the ichthyology of the Swede Peter Artedi.[1]

Passionately interested from early childhood in studying fishes, and born with a true genius for classification, this naturalist promptly perceived that only Willughby had described these animals well; but he also took note that English ichthyology had not entirely achieved its goal, for lack of having established its genera, of having designated them by fixed and convenient names, or of having assigned to its species short and comparable characters based on their structure.

He worked thenceforth without rest to fill this lacuna in the science. [The result was his *Ichthyologia, sive Opera omnia de piscibus,* published in 1738, a book consisting of five parts: "Bibliotheca ichthyologica," "Philosophia ichthyologica," "Genera piscium," "Synonymia nominum piscium," and "Descriptiones specierum piscium."] After presenting in the "Bibliotheca" a list of the authors who had written about fishes before him, he analyzed in the "Philosophia" all the interior and exterior parts of these animals, created a precise terminology for the different forms these parts might take, drew up for himself some rules for the nomenclature of the genera and species, and finally, subdivided the class more accurately than Willughby had done. His orders are based solely on the consistency of the skeleton, the opercula of the gills, and the nature of the rays of the fins, disregarding habitat and anything else foreign to structure; he named them Acanthopterygii, Malacopterygii, Branchiostegi, and Chondropterygii. Mention is not made here of his Plagiuri, which contained the cetaceans. The order Branchiostegi, ill defined and poorly composed, cannot survive; but the other three orders are natural, and nothing that has been tried since has been able to replace them.

In the "Genera piscium" he defined for each genus an invariable substantive name and positive, clear-cut characters, based in general on the number of rays in the membrane of the gills [i.e., branchiostegal rays] (whose importance he was the first to notice), on the relative position of the fins, on their number, on the parts of the mouth where the teeth are situated, on the structure of the scales, and even on internal parts such as the stomach and the appendages of the cecum. These genera, numbering forty-five [see table 3], are so well formed that almost all are valid today, and the subdivisions that have had to be introduced by the ever increasing number of species have very rarely been such that it was necessary to separate them from each other. Thirteen additional genera were not formally established but simply mentioned in the appendixes of this and the following part of Artedi's work;[2] of these thirteen, Linnaeus accepted three, and several others have been revived by his successors. Under each genus is found a list of species known well enough so that the author believed he could classify them with their definitions and by short descriptions.

In the "Synonymia nominum piscium," arranged under each species with great erudition, are listed all the articles of earlier authors who mentioned the species, the drawings of the species, and the names that have been applied to them. Artedi listed even the Greek and Latin names, but according to Rondelet's ideas rather than his own research. In the list he included 274 species of fishes, rejecting species when their existence or characteristics did not seem to him to be well enough established. In the appendix he added 17 others as belonging to the genera indicated; finally, in his "Descriptiones specierum piscium," he described the 72 species he was able to see for himself, according to his terminology and with as much detail as clarity.

Nothing like this had existed before in ichthyology, and even though Artedi constantly kept Willughby's work before him in composing his book, it is nonetheless true that he brought science a great step forward and vastly surpassed his predecessor.

The author was not fortunate enough to publish his work himself; but he found an editor worthy of him in his friend from youth, the famous Linnaeus, who retrieved the manuscripts from the hands of his landlord and devoted nearly a year of his time to revising, completing, and preparing them for printing. He published them at Leiden in 1738, but as early as 1735 he used them for his account of the fishes in the first edition of his *Systema naturae,* which appeared in Leiden that year as three large single-page tables [see table 4].

Linnaeus,[3] who himself later became such a great authority in ichthyology, at first did not dare to depart from the footsteps of a friend who, in this science, had been his master. But in the second edition of his *Systema naturae* (1740a), he had the great good sense to provide the number of fin rays for each species. This observation, imitated by his successors, produced unheard-of advantages for ichthyology, not exactly for determining species, but for recognizing the natural genera and subgenera to which each species should be attributed. Often it is the only guide that can lead us through the many confused and incomplete descriptions that fill the literature.

In the sixth edition of the *Systema naturae* (1748a),[4] Linnaeus added only two genera to those of Artedi, *Aspredo (asprèdes)* and *Callichthys (callichtes),* both of which (in the tenth edition) he later suppressed. The ninth edition, reprinted at Leiden under the supervision of Gronovius,[5] received only the new genera that that editor had just established in his *Museum ichthyologicum* [see table 5]: *Silurus (silures), Solenostomus (solénostomes), Gymnogastrus (gymnogastres), Charax (charax), Uranoscopus (uranoscopes), Atherina (athérines), Plecostomus (plécostomes), Polynemus (polynèmes), Mystus (mystes), Holocentrus (holocentres),* and *Callorhyncus (callorhynques).* Yet most of these new genera had been indicated earlier in Artedi's supplements or in the manuscript of the third volume of Seba, prepared by Artedi and of which Gronovius had knowledge.

It was only in the tenth edition of the *Systema naturae,* published in 1758–59 [fig. 25], that Linnaeus, trusting to his own resources, created a new ichthyological classification, divided some genera and combined others, gave the species some common names and characteristic phrases, and added several to those that Artedi had accepted as sufficiently verified.

The most appropriate of the changes in the general distribution was to separate the cetaceans from the fishes, with which they had been grouped since the time of the ancients. Aristotle had already remarked that the cetaceans are warm blooded, that they breathe with lungs, that they give birth to live young and nurse them; in short, that their whole interior structure is that of a viviparous quadruped. Ray and Artedi repeated these characters, yet they continued to classify the cetaceans with the fishes. Brisson[6] was the first to separate them and make them a class apart, which he placed immediately after that of the viviparous quadrupeds; Linnaeus in turn united the cetaceans and the viviparous quadrupeds to form his class Mammalia.

Not so felicitously, he placed Artedi's chondropterygian fishes next to the reptiles under the name Amphibia nantes. It is not easy to understand how he could suppose that they have lungs, especially because he left the sturgeon *(esturgeon)* there and added the anglerfish *(baudroie),* which Artedi had placed in his Branchiostegi.

Linnaeus carried this transgression against natural order much further in his twelfth edition, when he joined the rest of Artedi's Branchio-

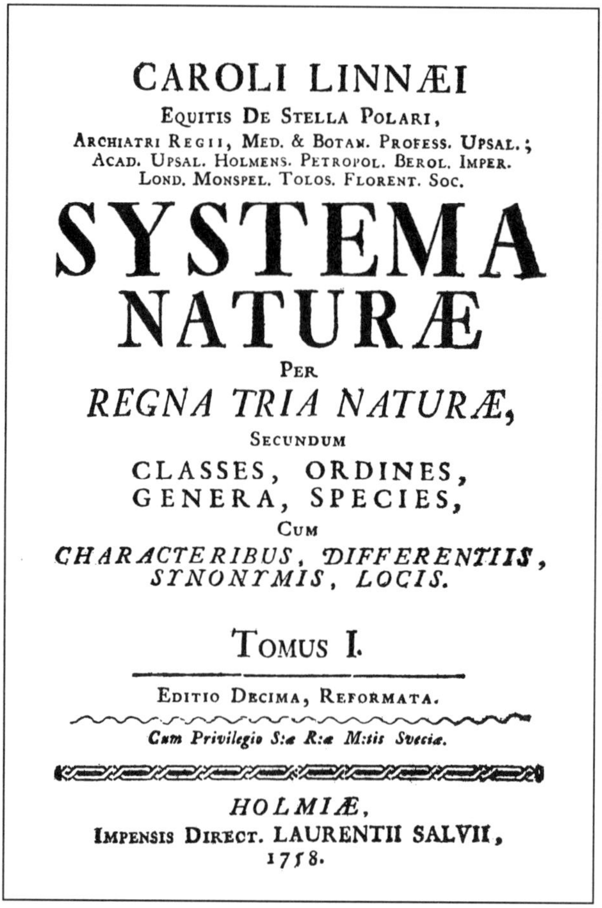

CAROLI LINNÆI
EQUITIS DE STELLA POLARI,
ARCHIATRI REGII, MED. & BOTAN. PROFESS. UPSAL.;
ACAD. UPSAL. HOLMENS. PETROPOL. BEROL. IMPER.
LOND. MONSPEL. TOLOS. FLORENT. SOC.

SYSTEMA
NATURÆ
PER
REGNA TRIA NATURÆ,
SECUNDUM
CLASSES, ORDINES,
GENERA, SPECIES,
CUM
CHARACTERIBUS, DIFFERENTIIS,
SYNONYMIS, LOCIS.

TOMUS I.

EDITIO DECIMA, REFORMATA.

Cum Privilegio S:æ R:æ M:tis Sveciæ.

HOLMIÆ,
IMPENSIS DIRECT. LAURENTII SALVII,
1758.

FIG. 25.
The title page of the tenth edition of Carolus Linnaeus's *Systema naturae*, 1758–59, vol. 1.
T. W. Pietsch collection.

stegi to the Amphibia nantes, namely, the boxfishes *(coffres)*, the puffers *(tétrodons)*, and even the pipefishes and their allies *(syngnathes)*, which Artedi had placed in his Malacopterygii.

In my opinion, it was not a better change (although the change was to remain in effect for a long time) to abandon the division of bony fishes into Acanthopterygii and Malacopterygii (which had been in effect since the time of Willughby) and replace it with a distribution based on the presence or absence of pelvic fins and their position relative to the pectorals. Nothing interferes more with the true correlations of genera than to recognize as orders the Apodes, Jugulares, Thoracici, and Abdominales: the swordfish *(xiphias)*, for example, is separated from the mackerels *(scombres)*, and the barracuda *(sphyrène)*, which is almost a perch, is placed among the pikes *(brochets)*, and so on.

In the twelfth edition of the *Systema naturae,* Linnaeus suppressed some of the genera of Artedi and Gronovius: *Holocentrus (holocentres)* was included with the perches *(perches); Anableps (anableps),* included with *Cobitis (cobites); Coregonus (corrégones), Osmerus (osmères),* and *Charax (charax),* all included within *Salmo (saumons);* and *Aspredo (asprèdes), Callichthys (callichtes),* and *Mystus (mystes),* included with *Silurus (silures).* At the same time he divided other genera, separating *Tetraodon (tétrodons)* and *Diodon (diodons)* from *Ostracion (ostracions), Callionymus (callionymes)* from *Trachinus (vives),* and *Mullus (mulles)* from *Trigla (trigles);* and he added entirely new genera—*Mormyrus (mormyres), Centriscus (centrisques),* and *Pegasus (pégases)*—so that the total comes to fifty-seven. Moreover, he changed some of the names Gronovius had assigned: *Plecostomus (plécostomes)* of the Dutch ichthyologist became *Loricaria (loricaires); Solenostomus (solénostomes)* became *Fistularia (fistulaires); Gymnogaster (gymnogastres), Trichiurus (trichiures);* and *Callorhyncus (callorhynques), Chimaera (chimères).*

The number of species recognized by Linnaeus came to 414—some taken from publications printed after Artedi, such as the work of Edwards,[7] the second volume of Catesby,[8] the natural history of Jamaica by Browne,[9] and especially the dissertations of Johann Friedrich Gronovius and the *Museum ichthyologicum* of Laurens Theodorus Gronovius [fig. 26]; other species were observed by Linnaeus himself on his travels[10] and personally examined in various collections.[11] Still others were procured for him by students[12] whom he sent to many foreign regions; the most zealous of his students in ichthyology at the time were Hasselquist,[13] Osbeck,[14] and Löfling.[15]

The twelfth edition of the *Systema naturae,* published from 1766 to 1768, was enriched by several additional species and good citations taken from the *Zoophylacium* of L. T. Gronovius [fig. 27][16] and from the third volume of the description of the collections of Seba,[17] a volume valuable for its drawings of foreign fishes, superior to all others up to that time, the text of which had been prepared in 1734 and 1735 by Artedi, although it was not published until 1759, at the expense and under the supervision of Gaubius.[18]

Linnaeus also profited in this edition from the natural history of Aleppo by Russell;[19] from the natural history of perches of the Danube by Schaeffer;[20] from the first descriptions of the fishes in the museum of St. Petersburg published by Koelreuter;[21] from descriptions published by a society formed at Trondheim by Bishop Gunner;[22] and handwritten observations, since published, on the swordfish *(espadon)* by Kölpin.[23] He even collected citations from older books, the description of the museum of Gottorp by Olearius[24] and the *Gazophylacium* by Petiver.[25] Finally, he included some fishes from the Carolinas, for which he was obliged to

FIG. 26. Laurens Theodorus Gronovius (1730–77) and his two sons, Johannes and Samuel. From a portrait by Isaac L. la Fargue van Nieuwland, 1775. Courtesy of Augusta R. André de la Porte and the Stedelijk Museum de Lakenhal, Leiden, Holland.

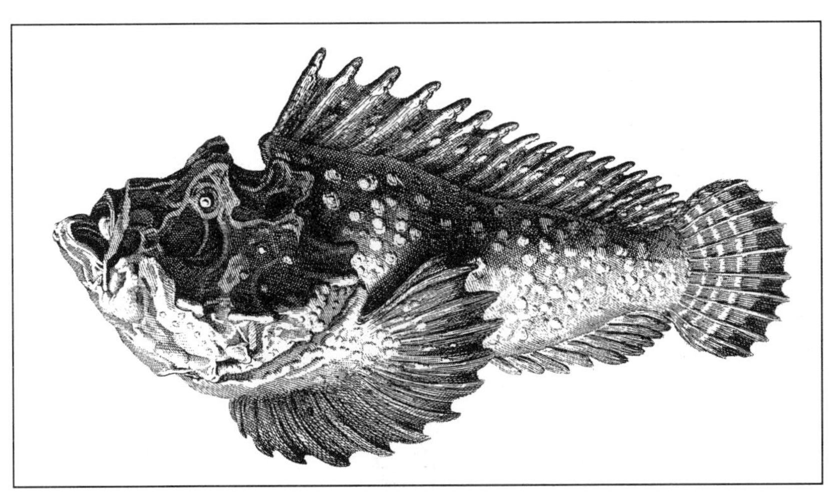

FIG. 27. A stonefish, *Synanceia horrida,* from Laurens Theodorus Gronovius's *Zoophylacium Gronovianum,* 1763–81, pt. 1, pl. 11. T. W. Pietsch collection.

Alexander Garden,[26] and he used two more from the same source in the appendix of his *Mantissa plantarum.*[27] Had he known about them, he could have taken material from Meyer,[28] Hill,[29] Knorr,[30] De Nobleville and Salerne,[31] and especially from Kramer,[32] who had followed his classification methods and where he would have found a new genus, *Poecilia* [actually *Umbra*].[33] A multitude of those special descriptions called monographs would also have served him well, but their obscurity or small significance caused him to neglect them.

What is remarkable, and what is not due to neglect, is that Linnaeus never cited the ichthyological fascicles of Klein,[34] even though they offered new species, fine drawings, and some groups that could have formed the basis for good genera; but the names were poorly composed, and the distribution of the genera was ill defined and not very natural [see table 6]. Moreover, Linnaeus, who had been violently attacked by Klein,[35] apparently wished to avenge himself against him as he did against Buffon [fig. 28][36]—the only vengeance appropriate to a true scientist, if indeed even that was appropriate, the vengeance of not mentioning him.

The distribution of orders was not at all improved in the twelfth edition of the *Systema naturae;* on the contrary, Linnaeus, as mentioned above, still placed all the members of Artedi's Branchiostegi in the Amphibia nantes. To the genera already accepted he added *Cepola (cépoles),* indicated by Artedi under the name *Taenia (taenia); Teuthis (teuthis),* which corresponds to *Hepatus (hépatus)* of Gronovius; and *Amia (amia)* and *Elops (élops),* which he described from fishes sent by Garden, bringing the

FIG. 28. Georges Louis Leclerc, comte de Buffon (1707–88). Courtesy of M.-L. Bauchot, M. Ducreux, and the Bibliothèque Centrale, Muséum National d'Histoire Naturelle, Paris.

number of genera to sixty-one. He entirely ignored the new genera erected by Gronovius in the *Zoophylacium* [see table 7], although there were quite good ones, which have since been recognized again by Bloch, by Lacepède, or by us, such as *Gonorhynchus (gonorhynchus), Leptocephalus (leptocéphalus), Eleotris (éléotris), Amia (amia)* of Gronovius, which is the same as *Apogon (apogon), Mastacembelus (mastacembelus),* and *Umbra (umbra),* which is a killifish *(fondule),* and so on [see table 8].

Linnaeus also ignored several remarkable species from the *Zoophylacii Gronoviani,* in particular *Cynodus cauda bifurca,* and such, which is the same as *Cheirodactylus (chéirodactyle)* of Lacepède, so that the total number of species comes to only 477. But numerical augmentation is of least consideration in the works of this illustrious naturalist: the precision of the characters, the convenience of a well-established terminology, easily remembered common names given to each species, and the binary nomenclature introduced into ichthyology as in all the rest of the system of nature were very important advantages. These advantages are what gave Linnaeus the preeminence that was granted to him in one way or another by every naturalist in his time and proved by the nearly universal adoption of his nomenclature, even to the almost exclusive use of his classification, however imperfect and artificial it may have been.

If some writers, such as Duhamel du Monceau,[37] continued to follow old classifications, it was out of ignorance rather than by a premeditated design of resisting the revolution taking place. For ichthyology particularly, the true naturalists writing immediately after Linnaeus either were in entire submission to him or had nothing original enough or even good enough in the changes they proposed to win over his followers.

Pennant,[38] in his *British Zoology,* even though he had the good sense to rejoin the Amphibia nantes with the cartilaginous fishes, made the mistake of reuniting the cetaceans with the fishes; and for the bony fishes he retained the Linnaean divisions Apodes, Jugulares, Thoracici, and Abdominales. His work was useful, however, because of its good drawings and for its little-known historical details.

Goüan, under the overly broad title *Historia piscium,* referred only to genera, which, to be sure, he described in much detail, although in pedantic terms.[39] His classification was Artedi's, based on the consistency of the skeleton and the fin rays, and he subdivided his classes according to the position of the fins, as did Linnaeus, even placing the Chondropterygii with the amphibians, as Linnaeus did in his tenth edition. Nothing was gained in method, but Goüan added three well-defined genera to those his master had established: *Lepadogaster (lépadogaster), Lepidopus (lépidopes),* and *Trachipterus (trachiptères).*

Forster, in his *Enchiridion historiae naturali inserviens,*[40] put the Amphibia nantes back among the fishes, as did Pennant, and took a course opposite that of Goüan for the bony fishes, which he first divided according to the presence or absence of pelvic fins and their position, further subdividing the resulting groups according to whether the fin rays are spiny or soft. He was not even very exact in this last respect, for he regarded the butterfish *(stromatée),* the scabbardfish *(lépidope),* and the silverside *(athérine)* as malacopterygians, and the cuskeel *(ophidium)* and the tarpon *(elops)* as acanthopterygians, which is contrary to the truth. He proposed only two new genera, *Echidna (echidna),* which is a moray eel *(murène),* and *Harpurus (harpurus),* not perceiving that it was the same as *Teuthis (teuthis)* of Linnaeus.

Pallas, a man of genius who at that time readily perceived some of the true correlations of the animals that Linnaeus included indiscriminately under the name Vermes *(vers),* published only some special descriptions of fishes, not to be compared with the combined works of Artedi and Linnaeus.[41] He introduced other fishes in the memoirs of the Academy of Sciences of St. Petersburg, but his principal work on this class, the third volume of his *Zoographia Rosso-Asiatica,* composed toward the end of his life, has not yet been published.[42]

Otherwise the impetus that the influence of Linnaeus gave to research, had he no other merit but that, would suffice to immortalize his name. In simplifying natural history, or at least in making it appear so, he inspired a liking for it generally. Men of the upper class were interested in it; young people full of ardor hurried off in all directions, with the sole intention of completing the system; and at least as regards the species, nature everywhere was made to contribute to the well being of the edifice for which this extraordinary man had drawn the plan.

NOTES

1. Peter Artedi, his father a pastor, was born in the parish of Anunds in Angermanland in 1705. Destined for the church, he was sent in 1716 to the college of Härnösand, and in 1724 to the University of Uppsala, where his taste for alchemy led him to choose the field of medicine. It was there that Linnaeus met him in 1728 and formed a close friendship with him. Artedi left for London in 1734, and in 1735 he came to Leiden to find his friend Linnaeus, who introduced him to [Albertus] Seba as the man most capable of writing the part on fishes in the great description of Seba's cabinet [see n. 17 below]. [In the early hours of 28 September 1735, after an evening of socializing at the house of Seba], Artedi, at the age of thirty, drowned in one of the canals of Amsterdam. [For more on Artedi, see Lönnberg 1905; Merriman 1938, 1941; and Wheeler 1961, 1979, 1987.]

2. [Artedi (1738) added a number of additional] genera in his supplements [to the *Ichthyologia*]: in the ["Appendix" to his] "Genera piscium" [pp. 82–84] he added *Taenia (les cépoles), Silurus, Mustela (blenn. viviparus), Phycis,* and *Sphyraena;* in the "Descriptiones specierum piscium" [pp. 112–18], he added *Cicla (des labres), Hepatus, Capriscus (baliste), Pholis, Citharus, Atherina, Liparis,* and *Chelon (des muges)* [see Gill, 1872, 27–28].

3. Perhaps it is not necessary to dwell on the well-known life of Linnaeus, that great reformer of plant and animal nomenclature, the naturalist who exerted the most incontestable influence on his century, and whose language is spoken in every country where natural history is pursued. We shall confine ourselves to principal dates to refresh our readers' memory. Carolus Linnaeus was born at Roeshult in Smaland on 24 May 1707. He was sent to the college of Vexioe in 1717, went to the University of Lund in 1727, and the next year to Uppsala. He underwent every privation until 1728, when Olof Celsius [1670–1756] and Olof Rudbeck [1660–1740] employed him in their work. It was with Rudbeck that he laid the first foundations of his philosophy of botany. In 1732 he traveled to Lapland, lived for a while at Fahlun, then went to Holland, where for a while he looked after the gardens of a rich businessman named [George] Clifford [1685–1765]. It was in this capacity that he published his works: *Fundamenta botanica* [1736a], *Bibliotheca botanica* [1736b], *Musa Cliffortiana* [1736c], *Critica botanica* [1737a], *Genera plantarum* [1737b], *Methodus sexualis* [1737c], *Flora Lapponica* [1737d], *Hortus Cliffortianus* [1737e], *Classes plantarum* [1738], and what especially interests us, the first edition of his *Systema naturae* (1735) and the *Ichthyologia* (1738) of Artedi. Employed in

1738, through the patronage of the count de Tessin and Baron Carl de Geer, to teach botany at Stockholm, he published there the *Museum Tessinianum* [1753a], the first volume of the *Museum S:ae R:ae M:tis Adolphi Friderici* [1754a], and the *Museum S:ae R:ae M:tis Ludovicae Ulricae reginae* [1764a]. In 1741 he was named professor at Uppsala, and he exercised that responsibility until his death in 1778. It is there that he published his *Philosophia botanica* (1751a), his *Species plantarum* (1753b), his *Mantissa plantarum* (1767, 1771), the numerous dissertations that fill the ten volumes of his *Amoenitates academicae* [the Linnaeus or Stockholm edition was published from 1749 to 1769], and the last four original editions of his *Systema naturae* [see n. 4 below]. In 1773 his memory was already failing, and two apoplectic attacks, in 1774 and 1777, had much altered his health. [He died on 10 January 1778 at Uppsala, where he was buried in the cathedral. For more on Linnaeus and his work, see Frängsmyr 1983.]

4. The original editions of the *Systema naturae* can be reduced to six, the first published at Leiden in 1735, the remaining five all at Stockholm: the second in 1740a, the sixth in 1748a, the eighth in 1753c, the tenth from 1758 to 1759, and the twelfth from 1766 to 1768 [see table 4; Gill 1872, 31–34]. The first four of these consisted of a single volume, the tenth was produced in two volumes, and the twelfth, in three volumes. The third edition, published at Halle in 1740b, is a copy of the first; the fourth, at Paris, 1744, is a copy of the second, supervised by Bernard de Jussieu [1699–1777], who added the names in French. The fifth, published at Halle in 1747a, is also a copy of the second, to which were added the names in German. The seventh edition, produced at Leipzig, 1748b, and the ninth, at Leiden, 1756, are taken from the sixth; however, in the ninth edition the part on fishes is augmented by several genera added by the editor [Laurens Theodorus] Gronovius [1730–77]. The tenth edition was reprinted at Halle from 1760 to 1770 and at Leipzig in 1762; but Linnaeus must not have known about the reprinting at Halle because he counts the one at Leipzig as only the eleventh [the Leipzig edition of 1762 is probably nonexistent; see Hulth 1907, 7; Wheeler 1991b, 9]. The twelfth edition was reprinted at Vienna from 1767 to 1770 as the thirteenth, but this did not prevent [Johann Friedrich] Gmelin [1748–1804] from assigning the number thirteen to his great edition of 1788–93, the last, but which itself was reprinted at Lyons from 1788 to 1796 and in the years following.

5. The Gronovius family, originally from Hamburg and established at Leiden, produced several famous erudites and two naturalists. Johan Frederic Gronovius [1686–1762], second of that name, brother of Abraham [1695–1775], the editor of Aelianus, published several dissertations on fishes, in particular "Pisces Belgii" [1746a] and "Pisces Belgii descripti" [1748] in *Acta Societatis Regiae Scientiarum Upsaliensis* for the years 1741 and 1742, respectively. He treated the same subject in his "Animalium Belgicorum centuria secunda," included in the fourth volume of *Acta Helvetica* ["Animalium in Belgio habitantium" was actually published by Laurens Theodorus Gronovius (see below) in five parts, 1760a, 1760b, 1762a, 1762b, 1762c, *Acta Helvet.* vols. 4 and 5]. He described in particular the loach *(misgurn)*, *Phil. Trans. Roy. Soc. Lond.*, vol. 44 [1747]; and in volumes of the *Acta Soc. Reg. Sci. Upsal.*, the dragonet *(callionyme)* [1744a], the salmon *(bécard)* [1746b], and the mackerel *(maquereau)* and the perch *(perche)* [1751]. It is to him that we owe the

method of preparing fish skins, as one would a herbarium, which he described in the *Phil. Trans. Roy. Soc. Lond.*, vol. 42 [1744b].

Laurens Theodorus Gronovius [1730–77], also the second of that name, son of the preceding, published *Museum ichthyologicum*, two fascicles in folio, at Leiden in 1754 and 1756, with seven plates, in which he described and represented several new fishes. They appear again with others in the first fascicle of his *Zoophylacium*, printed in 1763; the second fascicle, containing insects, appeared in 1764, and the third, which is on worms *(vers)*, did not appear until after his death in 1781. In these works Gronovius included genera that were unknown to Artedi, some of which were adopted by Linnaeus, and some by his successors. [For more on L. T. Gronovius and his fishes, see Wheeler 1956, 1958, 1989.]

6. Mathurin [Jacques] Brisson, born at Fontenay-le-Comte in 1723, assisted [René Antoine Ferchault de] Réaumur [see chap. 11, n. 33] in arranging his cabinets; later a member of the Académie des Sciences and professor of physical science at the Collège de Navarre, he died in Paris in 1806. He began a general zoology under the title *Le Règne animal, divisé en neuf classes*, published at Paris in 1756, one volume in quarto. This first volume, which contains the quadrupeds and the cetaceans, was followed by an "ornithology" in six volumes in quarto, 1760, but Brisson abandoned natural history after the death of Réaumur [1757], and toward the end of his life he had no remembrance of his first works. [For more on Brisson, see Allen 1951, 498–99.]

7. See above [chap. 6, n. 13].

8. The [second] volume [of Catesby's *Natural History of Carolina, Florida and the Bahama Islands*] did not appear until 1743.

9. Patrick Browne [1720?–90], physician at Jamaica, in his civil and natural history of that island, printed in English in folio, at London in 1756, described ninety-three fishes according to Artedi's classification, and much better than Sloane had done.

10. Linnaeus published accounts of his travels in the Swedish provinces as a naturalist, seldom omitting to describe fishes; for example, his visit to Öland and Gotland in 1741 [*Öländska och Gothländska resa*, Stockholm and Uppsala, 1745], Västergötland in 1746 [*Wästgöta-resa*, Stockholm, 1747b], and Skåne in 1749 [*Skånska resa*, Stockholm, 1751b].

11. Linnaeus described a number of collections that contained new fishes: (1) the cabinet that the crown prince Fridrick Adolphus gave to the University of Uppsala: *Museum Adolpho-Fridericianum* published in 1749, *Amoenitates academicae*, vol. 1 [the dissertation of Laurentius Balk; see Linnaeus 1746, 1749]. (2) The cabinet that the same prince, after he became king, assembled at the castle of Ulriksdal: one volume in folio titled *Museum S:ae R:ae M:tis Adolphi Friderici*, Stockholm, 1754a, which includes thirty-six handsome illustrations of fishes; the second part of this work, which includes descriptions of ninety-three fishes, and which Linnaeus also cited in his twelfth edition of the *Systema naturae*, was printed in octavo without illustrations in 1764b. (3) The cabinet that Magnus of Lagerström, director of the Swedish East India Company, received from China: *Chinensia Lagerströmiana*, published in 1759, *Amoenitates academicae*, vol. 4 [the dissertation of Johannes Laurentius Odhelius; see Linnaeus 1754b, 1759]. [The sources of Lin-

naeus's knowledge of fishes are discussed by Wheeler 1979, 1985, 1991a, b; see also Fernholm and Wheeler 1983.]

12. Linnaeus himself provided a list of these young explorers who came before his tenth edition; most of them sent to him specimens of the natural products they were able to gather: [C.] Ternström traveled in Asia, 1745; [Peter] Kalm [1716–79; see Allen 1951, 507–11], in Pennsylvania and Canada, 1747; [Lars] Montin, in Lapland, 1749; [Fredrik] Hasselquist, in Egypt and Palestine, 1749; [Olof] Toren (d. 1753), Malabar and Surat, 1750; [Pehr] Osbeck, Canton and Java, 1750; [Pehr] Löfling, Spain and [South] America, 1751; [M.] Köhler, Italy, 1752; and [Daniel] Rolander, Surinam and St. Eustache, 1755. In the Swedish interior, [Peter Jonas] Bergius went to the island of Gotland in 1752 and [Daniel] Solander to Lapland in 1753 [see Linnaeus 1758–59, 1:ii (of unpaged front matter)].

13. Fredrik Hasselquist [b. 1722], traveled to Egypt and Palestine 1749–52; he died in February 1752. His Voyage was published under the auspices of Linnaeus: *Iter Palaestinum,* Stockholm, 1757, in octavo. Among many objects of natural history, he described thirty-one fishes in great detail. A German translation appeared in 1762, [an English translation in 1766], and a French translation in 1769 by Keralio, which omitted the only useful part, the natural history.

14. Pehr Osbeck [1723–1805] was, like Toren, a ship's chaplain. His *Dagbok öfver en Ostindisk resa,* printed in Swedish at Stockholm in 1757, in octavo, contains the description of sixteen fishes [the pilotfish, *Naucrates ductor,* is figured in pl. 12, fig. 2]. There is a German translation, *Reise nach Ostindien und China,* by Georgi, Rostock, 1765; [and an English translation by Johann Reinhold Forster, London, 1771]. Osbeck also published in *Nova Acta Phys.-Med. Nat. Curio.,* vol. 4, 1770, some "Fragmenta ichthyologiae Hispanicae."

15. Linnaeus cites only letters from Pehr Löfling [1729–56], in the tenth edition. The voyage of this naturalist in Spanish America in 1751 was not printed until 1758 at Stockholm. He described nine fishes. There is a German translation by [Alexander Bernhard] Kölpin, Berlin, 1776.

16. The first fascicle of the *Zoophylacium Gronovianum,* which contained the fishes, was published in 1763 [see n. 5 above].

17. Albertus Seba, born at Etzel, Ostfriesland, in 1665, was a rich pharmacist at Amsterdam who assembled at great cost a very considerable cabinet of natural history, part of which was purchased by Peter the Great and transported to St. Petersburg, and the rest dispersed upon the death of the owner. He had it described and magnificently engraved in four folio volumes, in atlas format, published at Amsterdam in 1734, 1735, 1758, and 1765, the last two volumes appearing posthumously. The text of the third volume contains very good articles on fishes by Artedi, which are rather finer than the rest of the work. The work remained in manuscript long after Seba's death and the dispersal of his cabinet; but Gronovius knew it in that state and used it in his *Museum ichthyologicum* of 1754. Seba died in 1736. [For more on Seba, see Engel 1937, 1961; Holthuis 1969; Boeseman 1970.]

18. [Hieronymus David Gaubius (1705–80), professor of medicine at Leiden University.]

19. Alexander Russell [b. ca. 1715], died in 1768, a Scottish physician who lived at Aleppo, published in 1756, London, in quarto, *The Natural History of*

Aleppo, and Parts Adjacent in which he has good illustrations of fishes of the river Orontes [see Russell 1756, 73–77, pls. 12–13].

20. Jacob Christian Schaeffer was born at Querfurt in 1718, became a pastor at Ratisbon in 1741, and died in 1790. He wrote at length on insects, but he also published *Piscium Bavarico-Ratisbonensium pentas* at Ratisbon, 1761, in quarto, a small book that treats only five species but is remarkable for its accuracy.

21. Joseph Gottlieb Koelreuter [1733–1806], of Karlsruhe, the celebrated producer of plant hybrids *(mulets végétaux),* also interested himself in fishes. He published two memoirs in *Novi Comment. Acad. Sci. Impér. Petro.*, St. Petersburg, vols. 8 and 9, 1763 and 1764, in which he described and illustrated nine species with great accuracy. In vol. 14 of the same journal (1770), he illustrated the navaga [a member of the gadid genus *Eleginus*] *(narwaga),* and in vol. 15 [1771], the lake whitefish *(lavaret);* vols. 16 and 17 [1772, 1773], the anatomy of the sterlet *(sterlet);* vol. 18 [1774], the whitefish *(salmo albula);* vol. 19 [1775], the freshwater burbot *(lote),* and so on. He continued this work in *Nova Acta Acad. Sci. Impér. Petropol.* until vol. 9 (1795), in which he had a final memoir, on the flounder *(flet).*

22. *Tronhiemske Samlinger, udgivne af philaletho* [Memoirs of the Society of Trondheim, a periodical] fairly rich in information on objects of natural history of the North, began to appear at Trondheim in 1761, one volume in duodecimo, in Danish, the second volume in 1762, the third in 1763, and the fourth in 1764, all containing important articles on fishes [see Gunner 1761–64]. The founder and principal collaborator was Johan Ernst Gunner, bishop of Trondheim (1718–73). Hans Strøm, born in 1726 [d. 1797], pastor of the church of Eger, also worked on this series of publications. He produced separately a physical and economic description of the bailiwick of Soendmoer in Norway, printed in Danish at Sorøe, 1762–66, two volumes in quarto, which includes good descriptions of fishes.

23. [Alexander Bernhard Kölpin, 1731–1801; see Kölpin 1770, 1771.]

24. Adam Olearius or Oehlschlaeger, born in 1599 in the area of Anhalt, became secretary to the duke of Holstein, accompanied [Johann Albert] Mandelslo [1616–44] on his travels in Persia, and later [1666] published a description of the cabinet of Gottorp. He died in 1671.

25. James Petiver, an apothecary at London [b. 1663 or 1664], died in 1718. He was the author of several works, the plates of which were collected in two volumes in folio under the title *Jacobi Petiveri opera, historiam naturalem spectantia; or, Gazophylacium . . . ,* London, 1764. There are 306 plates showing in no particular order a multitude of objects of natural history, including some fishes here and there but not very well drawn. [For more on Petiver, see Stearns 1952; Edwards 1981, 29–31.]

26. Alexander Garden, a Scottish physician, born in 1730, lived in South Carolina and died in London in 1791 [see Berkeley and Berkeley 1969]. Letters from Garden are found in the *Correspondence of Linnaeus with Various Scientists* (published in 1821 by Sir James Edward Smith, London, two volumes in octavo), which had been enclosed with objects sent by Garden to the great Swedish naturalist and are often useful in understanding the articles Linnaeus wrote about the objects [see Goode and Bean 1885; Wheeler 1985].

27. *Mantissa plantarum* [Linnaeus 1767, 1771] is a supplement to the sixth

edition of the *Genera plantarum* [Linnaeus 1764c] and to the second edition of the *Species plantarum* [Linnaeus 1762–63]. The appendix [Linnaeus 1771] is an addition to the animal kingdom of the *Systema naturae* and contains references to only three fishes.

28. Johann Daniel Meyer [b. 1713], painter at Nuremberg, published in German a collection of 240 mediocre plates showing illustrations of various animals (including several common fishes), along with their skeletons, Nuremberg, 1748–56, three volumes in folio.

29. John Hill, a pharmacist and later physician in London [b. 1714], died in 1775, is the author of a great quantity of works, among them *A General Natural History*, three volumes in folio, London, 1748–52, in English, the first volume of which is on animals and contains a long chapter on fishes [pp. 201–315], ordered according to Artedi. The illustrations are taken mostly from Willughby [1686]. [On Hill, see Rousseau 1970, 1978; Laundon 1981.]

30. Georg Wolfgang Knorr [1705–61], a painter and engraver at Nuremberg, published several collections of drawings, one of which, titled *Deliciae naturae selectae*, Nuremberg, 1766–67, two volumes in folio, contains fishes; the text is by Philipp Ludwig Statius Müller [1725–76], the German translator of the *Systema naturae* [see Linnaeus 1773–76], a naturalist of little education and a writer of poor taste.

31. [Louis Daniel] Arnault de Nobleville [1701–78] and [François] Salerne [d. 1760], physicians at Orléans, in their *Règne animal*, Paris, 1756–57, six volumes in duodecimo, wrote about fishes in the second volume, but it is only a poor compilation of common fishes.

32. Wilhelm Heinrich Kramer [d. 1765] was a physician at Dresden who lived at Bruck on the Leitha, on the border between Austria and Hungary. In his *Elenchus vegetabilium et animalium per Austriam inferiorem observatorum*, Vienna, 1756, he provides accounts of thirty-eight fishes, which he arranges according to the first method of Linnaeus and describes in his style. [For more on Kramer, see Mearns and Mearns 1988, 217–18.]

33. [Here again (see chap. 6, n. 21), Cuvier indicates the presence in European waters of the strictly New World family Poeciliidae, this time as reported by Kramer (1756, 396): "Umbra Austr. Hundsfisch. Habitat in paludibus, et praesertim in cavernis subterraneis circa Leytaepontum." Certainly not representing a new genus in Cuvier's time, it had been christened *Umbra* some fifty years earlier by Kramer himself (in Scopoli 1777, 450; see Eschmeyer 1990, 417) and the species was later named in his honor, *Umbra krameri* (Walbaum, 1792), the European mudminnow (A. Wheeler, pers. comm., 13 November 1993; see Valenciennes 1847, pl. 590).]

34. Jacob Theodor Klein, born at Danzig in 1685, died in 1759, secretary of that republic, was interested in all areas of zoology. His writings contradict almost all the naturalists of his time. The last three parts of his five-part *Historiae piscium naturalis promovendae*, printed from 1740 to 1749, contain several handsome illustrations, some new species, and some useful ideas. The first part is a description of the otoliths of fishes and is the only specialized treatise on this subject. The number of genera recognized by Klein is sixty-one, exactly as in the twelfth edition of Linnaeus (1766–68), but the divisions are entirely different. His method

is different too; he bases his primary divisions on the general shape of the head and body, and he finishes with the number of dorsal fins, which departs from nature at least as much as Linnaeus's system. [His classification is shown in table 6. For more on Klein, see Gill 1872, 28–31; Adler 1989, 9–10. For a contemporary review of Klein's *Historiae piscium naturalis promovendae*, see Eames 1742.]

35. [Klein's attack on Linnaeus appeared] in a dissertation titled *Summa dubiorum circa classes quadrupedum et amphibiorum in Linnaei systemate naturae*, Danzig, 1743, in quarto.

36. The fourth volume in quarto of Buffon's *Histoire naturelle* [published in forty-four volumes from 1749 to 1804; see below], in which the natural history of the quadrupeds begins, dates from 1753, and the thirteenth dates from 1765; nevertheless, Linnaeus did not cite Buffon in his tenth edition of 1758–59 nor in the twelfth of 1766–68.

[Georges Louis Leclerc, comte de Buffon, French naturalist and philosopher, born at Montbard, Côte d'Or in 1707, died at Paris in 1788, member of the French Academy, perpetual treasurer of the Academy of Science, fellow of the Royal Society of London, and member of most of the learned societies of Europe. At the age of twenty-five he inherited a considerable fortune from his mother, and from this time onward he devoted himself to scientific study. In 1739 he became keeper of the Jardin du Roi and there began to collect materials for his great *Histoire naturelle, générale et particulière,* the first work to present the previously isolated and disconnected facts of natural history in a popular and generally intelligible form. Passing through several editions and translated into various languages, the *Histoire naturelle* was first published in Paris in forty-four quarto volumes between 1749 and 1804. The first fifteen volumes of this edition (1749–67) were prepared with the help of Louis Jean Marie Daubenton and subsequently by P. Guéneau de Montbeillard, the abbé Gabriel Leopold C. A. Bexon, and Charles Nicolas Sigisbert Sonnini de Manoncourt. The following seven volumes (1774–89) form a supplement to the preceding; these were later followed by nine volumes on the birds (1770–83) and five volumes on minerals (1783–88). The remaining eight volumes, which complete this edition, appeared after Buffon's death and include the reptiles, fishes, and cetaceans, completed by Lacepède and published in successive volumes between 1788 and 1804. A second edition was begun in 1774 and completed in 1804, in thirty-six volumes in quarto (*Encyclopaedia Britannica*, 1951, 4:345).]

37. Henri Louis Duhamel du Monceau, an able and hardworking physical scientist and agronomist but a very bad ichthyologist, was born in Paris in 1700 and died in 1782. Among a multitude of works composed with H. L. de La Marre [I cannot find anything on this collaborator of Duhamel], is a *Traité général des pêsches*, which appeared serially from 1769 to 1782, in folio. Here he wrote on the natural history of fishes, but in a most confused manner, which indicates that he had not the least idea of what a natural history should be. Notwithstanding, this work is important to ichthyologists because of the numerous illustrations that adorn it, several being very handsome and quite accurate, although there are also some that are quite faulty, all depending on their sources. It also contains some interesting facts supplied to the author by his correspondents.

38. Thomas Pennant, a Welsh gentleman born at Downing in the county of Flint in 1726, died in 1798, discussed fishes in the third volume of his *British Zoology*, printed in 1768–70 in octavo and reprinted in 1776–77 in quarto. He also wrote about fishes in his *Arctic Zoology* [1784–87; see the "Supplement" of 1787, 99–149], and in a short essay on Indian zoology [the first edition published in 1769 and a much expanded second edition in 1790]. [For more on Pennant, see Allen 1951, 491–94.]

39. Antoine Goüan [1733–1821], professor of botany at Montpellier, was one of the first proponents in France of Linnaeus's methods and nomenclature [Gill 1872, 34–35]. His *Historia piscium*, printed at Strasbourg in quarto, in Latin and in French, in 1770, was probably intended only as an introduction to a true general natural history of these animals, which he never wrote.

40. We shall have more to say about Johann Reinhold Forster as an explorer [see chap. 9, n. 13], but here we mention only his *Enchiridion historiae naturali inserviens*, published at Halle, 1788, in octavo. There is a French translation by Léveillé, Paris, 1799, in octavo.

41. Peter Simon Pallas, the naturalist with perhaps the broadest knowledge and most enlightened spirit in the eighteenth century, was born in Berlin in 1741 and began his scientific career in Holland in 1766 with his *Elenchus zoophytorum* and his *Miscellanea zoologica;* he spent his last years in the Crimea and returned to his native city, where he died in 1811. Fascicles 7 and 8 of his *Spicilegia zoologica*, printed in 1769 and 1770, contain very good descriptions and illustrations of twenty-six foreign fishes with interesting characteristics. His great journey to Siberia lasted from 1769 to 1774. In his *Reise durch verschiedene Provinzen des russischen Reichs* [published from 1771 to 1776], he discussed fishes and described eighteen new species; additional forms are described in the various memoirs of the Imperial Academy of Sciences of St. Petersburg [e.g., see Pallas 1810], but his main work on this class of animals is in the third volume of his *Zoographia Rosso-Asiatica* [1811–14], a posthumous work printed under the supervision of [Wilhelm Gottlieb] Tilesius von Tilenau [see chap. 13, n. 42], which will be discussed below. [For more on Pallas and his fishes, see Rudolphi 1812; Cuvier 1819a; Svetovidov 1976, 1981; Mearns and Mearns 1988, 288–97.]

42. [Difficulty in establishing the date of publication of Pallas's *Zoographia Rosso-Asiatica*, and Cuvier's indication that the third volume of the book had "not yet been published" when he wrote the "Tableau historique des progrès de l'ichtyologie," results from a delay in the production of the plates, which continued for many years after the three volumes of the text had been printed (Svetovidov 1981, 46). Although still often misdated, the publication dates have been established by declaration of the International Commission on Zoological Nomenclature (1954; see also Hemming and Noakes 1958, 1): vols. 1 and 2 date from 1811, vol. 3, from 1814 (see chap. 13, n. 29; see also Sherborn 1934, 1947; Sclatter 1947; Stresemann 1951).]

TABLE 3. The genera of Artedi. [As indicated in the title of his *Genera piscium* (1738), Artedi recognized 242 nominal species in 52 genera, including the cetaceans, which he regarded as constituting an order of fishes called Plagiuri. The cetaceans being subtracted (14 species representing 7 genera), the number is reduced to 228 species in 45 genera (Gill 1872, 28). The number of species recognized for each genus is given below in brackets.]

I. MALACOPTERYGII	II. ACANTHOPTERYGII	III. BRANCHIOSTEGI
Syngnathus [4]	*Blennius* [5]	*Balistes* [6]
Cobitis [3]	*Gobius* [4]	*Ostracion* [22]
Cyprinus [19]	*Xiphias* [1]	*Cyclopterus* [1]
Clupea [4]	*Scomber* [5]	*Lophius* [1]
Argentina [1]	*Mugil* [1]	**IV. CHONDROPTERYGII**
Exocoetus [2]	*Labrus* [9]	*Petromyzon* [3]
Coregonus [4]	*Sparus* [15]	*Acipenser* [2]
Osmerus [2]	*Sciaena* [2]	*Squalus* [14]
Salmo [10]	*Perca* [7]	*Raia* [11]
Esox [3]	*Trachinus* [2]	
Echeneis [1]	*Trigla* [10]	
Coryphaena [3]	*Scorpaena* [2]	
Ammodytes [1]	*Cottus* [5]	
Pleuronectes [10]	*Zeus* [3]	
Stromateus [1]	*Chaetodon* [4]	
Gadus [11]	*Gasterosteus* [3]	
Anarhichas [1]		
Muraena [6]		
Ophidium [2]		
Anableps [1]		
Gymnotus [1]		

TABLE 4. [The editions of Linnaeus's *Systema naturae*

Edition	City	Publisher	Date	Comments
1st	Leiden	Haak	1735	
2d	Stockholm	Kiesewetter	1740	
3d	Halle	—[a]	1740	A reprinting of the 1st, rearranged, and with a new preface, and German translation in parallel columns
4th	Paris	David	1744	A reprinting of the 2nd edited by Bernard de Jussieu, with Swedish names replaced by French equivalents
5th	Halle	—[b]	1747	A reprinting of the 2nd, with Swedish names replaced by German
6th	Stockholm	Kiesewetter	1748	With names of the animals and minerals given in Swedish and in Latin; illustrated with 8 plates
7th	Leipzig	Kiesewetter	1748	A reprinting of the 6th, with Swedish names replaced by German
8th	Stockholm	Salvius	1753	"Regnum vegetabile" only
9th	Leiden	Haak	1756	A reprinting of the 7th, edited by J. F. Gronovius, with German names replaced by French; section on fishes revised presumably by L. T. Gronovius (see Wheeler 1979)
10th	Stockholm	Salvius	1758–59	
Pirated	Halle	Curt	1760–70	A reprinting of the 10th
11th	Leipzig	?	1762	A reprinting of the 10th (probably a ghost edition; see Hulth 1907, 7)
12th	Stockholm	Salvius	1766–68	
—	Vienna	Thomae	1767–70	A reprinting of the 12th
13th	Leipzig	Beer	1788–93	Based on the 12th but edited and greatly expanded by J. F. Gmelin
—	Lyons	Delamollière	1788–96	A reprinting of the 13th

[a]Gedruckt mit Gebauerischen Schriften.
[b]Halae Magdeburgicae.]

TABLE 5. The classification used by Gronovius in his *Museum ichthyologicum* is the same as Artedi's, but he arranged some genera differently and brought the number to fifty-three [see Gronovius 1754–56, pt. 1, "Clavis classium"].

PISCIS

Caudâ horizontali. **PLAGIURI**

Caudâ perpendiculari

- Pinnarum radiis cartilagineis. CHONDROPTERYGII

Callorynchus	*Squalus*	*Petromyzon*	
Acipenser	*Raja*		

- Pinnarum radiis osseis
  - Branchiis ossibus destitutis. BRANCHIOSTEGI

Balistes	*Ostracion*	*Cyclopterus*	*Lophius*

  - Branchiis ossiculatis
    - Pinnis inermibus. MALACOPYERTGII

Syngnathus	*Exocoetus*	*Anarrhichas*	*Gadus*
Cobitis	*Esox*	*Muraena*	*Uranoscopus*
Cyprinus	*Solenostomus*	*Gymnogaster*	*Atherina*
Clupea	*Anableps*	*Coregonus*	*Plecostomus*
Argentina	*Echeneis*	*Osmerus*	*Callichthys*
Silurus	*Ammodytes*	*Salmo*	*Gymnotus*
Aspredo	*Pleuronectes*	*Charax*	

    - Pinnis aculeatis. ACANTHOPTERYGII

Polynemus	*Labrus*	*Perca*	*Cottus*
Blennius	*Sparus*	*Trachinus*	*Zeus*
Scomber	*Sciaena*	*Trigla*	*Chaetodon*
Mystus	*Holocentrus*	*Scorpaena*	*Gasterosteus*
Mugil			

Table 6. In the classification of Klein [as outlined in his *Historiae piscium naturalis promovendae* (pt. 3, 1742, p. 4; pt. 4, 1744, p. 6; pt. 5, 1749, pp. 3, 78)], the number of genera is sixty-one [including 518 nominal species, excluding the cetaceans (Gill 1872, 31)], exactly as in the twelfth edition of Linnaeus [1766–68], but the divisions are entirely different. The method is different, too; he bases his primary divisions on the general shape of the body and the head, and he finishes with the number of dorsal fins, which departs from nature at least as much as Linnaeus's system.

PISCES

PULMONIBUS spirantes sunt PHYSETERES:

Balaena — *Delphaces* — *Delphinus*
Narwhal — *Orca* — *Phocaena*

BRANCHIIS occultis

Spiraculis ad latera
- Pinnata
  - Spiraculis quinque:
    Cynocephalus — *Galeus* — *Cestracion* — *Rhina*
  - Spiraculo unico:
    Batrachus — *Crayracion* — *Capriscus* — *Conger*
- Apenneia
  - Spiraculo unico:
    Muraena
  - Spiraculis septem:
    Petromyzon sive Lampetra

Spiraculis in thorace; spiraculis constanter quinque:
Narcacion — *Rhinobatus* — *Leiobatus* — *Dasybatus*

BRANCHIIS apertis sunt notabiles

Series I. A partibus notabilibus et corpore anguillae formi
- Fasciculus I. Capite et ventre notabilis:
  Silurus
- Fasciculus II. Notabiliter rostrati ore vario
  - A. Ore prono; capite in solidum rostrum exeunte:
    Acipenser
  - B. Ore fisso
    - 1. Rostro retuso, dentibus horridis:
      Latargus
    - 2. Mandibula superiore in rostrum notabile exeunte:
      Xiphias
    - 3. Inferiori mandibula ultra superiorem rostratam producta:
      Mastaccembelus
    - 4. Utraque mandibula aequaliter rostrata:
      Psalisostomus
  - C. Ore in rostri tubulosi extremitate:
    Solenostomus

Continued on next page

Table 6—*continued*

- D. Capite et cauda rostratus:
  - *Amphisilen*
- Fasciculus III. Notabiliter plani et oculati
  - A. In dextro latere oculati:
    - *Solea*, *Passer*
  - B. In sinistro latere oculatus:
    - *Rhombus*
  - C. Utrinque oculati:
    - *Rhombotides*, *Tetragonoptrus*, *Platiglossus*
- Fasciculus IV. Thoracati et notabiliter armati:
  - *Cataphractus*, *Corystion*, *Centriscus*
- Fasciculus V. Vel sterno vel capite notati:
  - *Oncotion*, *Echeneis*
- Fasciculus VI. Corpore teretiusculo:
  - *Enchelyopus*

Series II. Corpore spisso vel lati vel carinati et castigati

- Fasciculus VII. Tripterus:
  - *Callarias barbatus*, *Callarias imberbis*
- Fasciculus VIII. Pseudotripterus:
  - *Pelamys penicillis*, *Pelamys pinnulis*
- Fasciculus IX. Dipterus
  - A. Pinna secunda cutacea:
    - *Trutta dentata*, *Trutta edentula*
  - B. Pinnis ambabus radiatis:
    - *Mullus barbatus*, *Labrax*, *Asperulus*
    - *Mullus imberbis*, *Sphyraena*, *Trichidion*
    - *Cestreus*, *Gobio*
- Fasciculus X. Pseudodipterus
  - A. Pro prima pinna dorsuali aculeis discretis acutissimus et robustis:
    - *Glaucus*
  - B. Praeter pinnam longam, processibus in capite quasi cristatus:
    - *Blennus*
- Fasciculus XI. Monopterus
  - A. Pinna longa
    - 1. Interrupta:
      - *Perca*
    - 2. Sinuosa:
      - *Percis*
    - 3. Coaequata
      - a. Dentibus acutis:
        - *Maenas*, *Cicla*, *Synagris*, *Hippurus*
      - b. Dentibus latis et obtusis:
        - *Sargus*

Continued on next page

Table 6—*continued*

c. Edentulus:
Cyprinus *Prochilus*
B. Pinna brevi
1. Ad medium dorsi
a. Corpore lato et spisso:
Brama
b. Corpore castigato
i. Barbatus:
Mystus
ii. Imberbis:
Leuciscus *Harengus*
2. Caudae proxima:
Lucius
Fasciculus XII. Pseudomonopterus:
Pseudopterus (i.e., *Pterois*)

TABLE 7. In his *Zoophylacium Gronovianum*, Gronovius abandoned the method of subdivision based on spines and used instead the position of the pelvic fins; he moved several groups of bony fishes into the Branchiostegi and increased the number of genera to seventy-eight [see Gronovius 1763–81, pt. 1, pp. 27–28].

PISCES			
Cauda instructi horizontaliter depressa, PLAGIURI			
Physeter	*Delphinus*	*Balaena*	*Monodon*
Cauda instructi perpendiculari			
Gaudentes radiis pinnarum cartilagineis, CHONDROPTERYGII			
Pinnis ventralibus praesentibus			
Acipenser	*Callorynchus*	*Squalus*	*Raja*
Pinnis ventralibus nullis			
Petromyzon			
Gaudentes radiis pinnarum osseis			
Branchiarum aperturis foramine exiguo tantum apertis, BRANCHIOSTEGI			
Pinnis ventralibus nullis			
Muraena	*Gymnotus*	*Syngnathus*	*Ostracion*
Pinnis ventralibus spuriis			
Balistes	*Cyclopterus*	*Cyclogaster*	
Pinnis ventralibus veris praesentibus			
Gonorynchus	*Cobitis*	*Uranoscopus*	*Lophius*
Branchiarum aperturis subtus atque in lateribus laxe apertis, BRANCHIALES			
I. Pinnis ventralibus in pectore sub pectoralibus			
Pinna dorsi solitaria			
Sciaena	*Coracinus*	*Callyodon*	*Blennius*
Cynaedus	*Scarus*	*Pleuronectes*	*Enchelyopus*
Sparus	*Chaetodon*	*Echeneis*	*Pholis*
Holocentrus	*Labrus*		
Pinnis dorsi unâ pluribus			
Cottus	*Gobius*	*Mullus*	*Zeus*
Amia	*Eleotris*	*Perca*	*Gadus*
Trachinus	*Trigla*	*Scomber*	
II. Pinnis ventralibus inter pinnas pectorales et analem sitis			
Pinna dorsi solitaria			
Clarias	*Clupea*	*Erythrinus*	*Anableps*
Silurus	*Argentina*	*Umbra*	*Esox*
Aspredo	*Synodus*	*Cataphractus*	*Solenostomus*
Cyprinus	*Hepatus*	*Exocoetus*	*Belone*
Pinnis dorsalibus duabus, posteriore spuria seu adiposa			
Salmo	*Anostomus*	*Charax*	*Mystus*
Pinnis dorsalibus duabus veris seu radiatis			
Callichthys	*Centriscus*	*Polynemus*	
Plecostomus	*Mugil*	*Atherina*	
III. Pinnis ventralibus veris nullis			
Anarrhichas	*Ammodytes*	*Gasteropelecus*	*Gymnogaster*
Ophidion	*Gasterosteus*	*Xiphias*	
Mastacembelus	*Channa*	*Leptocephalus*	

TABLE 8. Here is the list of genera of Linnaeus as it appears in the twelfth edition of his *Systema naturae* [see Linnaeus 1766–68, vol. 1, pt. 1, pp. 349, 423–24]. [A total of sixty-one genera and 477 species were recognized; new genera added to the fifty-seven found in the tenth edition (Linnaeus 1758–59, vol. 1) include *Amia, Elops, Cepola,* and *Teuthis* (Gill 1872, 34).]

Classis III. AMPHIBIA NANTES			
Spiracula composita			
Petromyzon	*Raja*	*Squalus*	*Chimaera*
Spiracula solitaria			
Lophius	*Balistes*	*Diodon*	*Syngnathus*
Acipenser	*Ostracion*	*Centriscus*	*Pegasus*
Cyclopterus	*Tetrodon*		
Classis IV. PISCES			
I. APODES			
Muraena	*Trichiurus*	*Ammodytes*	*Stromateus*
Gymnotus	*Anarhichas*	*Ophidium*	*Xiphias*
II. JUGULARES			
Callionymus	*Trachinus*	*Blennius*	
Uranoscopus	*Gadus*		
III. THORACICI			
Cepola	*Scorpaena*	*Sparus*	*Gasterosteus*
Echeneis	*Zeus*	*Labrus*	*Scomber*
Coryphaena	*Pleuronectes*	*Sciaena*	*Mullus*
Gobius	*Chaetodon*	*Perca*	*Trigla*
Cottus			
IV. ABDOMINALES			
Cobitis	*Salmo*	*Argentina*	*Exocoetus*
Amia	*Fistularia*	*Atherina*	*Polynemus*
Silurus	*Esox*	*Mugil*	*Clupea*
Teuthis	*Elops*	*Mormyrus*	*Cyprinus*
Loricaria			

9 Great Maritime Expeditions and Regional Contributions

A royal competition, of which King George III of England had the honor of being the first example,[1] and which at the time caused reigning princes to order up great maritime expeditions for the sole purpose of extending our knowledge of the globe, offered naturalists every means of exerting their ardor fruitfully; and they hastened to profit from these opportunities to extend their discoveries.

Commerson[2] embarked with Bougainville [fig. 29],[3] and on the return voyage he was left at Mauritius to explore its productions and from there made an excursion to Madagascar. Indefatigable in work, full of ardor and knowledge, he made immense collections in the three kingdoms and, on the subject of ichthyology particularly, left a series of descriptions more exact and more detailed than those of any of his predecessors. They included fishes of the Atlantic, the Brazilian coast, the entire [East] Indian archipelago, and especially Mauritius and Madagascar, amounting to more than 160 species, more than two-thirds of which were new at the time. He established several good genera, which have been maintained to this day. Drawings by Sonnerat or Commerson himself, or by a painter named Jossigny, accompanied the text; so that the accuracy of the drawings could be readily judged, Commerson kept the fishes themselves, dried according to the methods of Gronovius. Unfortunately his works met the same fate as those of Plumier, though his were much superior. His papers and the collections they were based on, sent after his death to the Ministry of the Navy, were given to Buffon, who inserted fragments of them in his natural history of birds but ignored the rest. A part of Commerson's work on fishes has since been used by Lacepède, who also had some of his drawings engraved; but since he had only rough drafts of the descriptions, which were not in good order and not always possible to match with the figures, the use he made of them is

FIG. 29. Louis Antoine, comte de Bougainville (1729–1811). Courtesy of M.-L. Bauchot, M. Ducreux, and the Bibliothèque Centrale, Muséum National d'Histoire Naturelle, Paris.

not free of error and confusion.[4] As our good fortune would have it, several years ago Duméril recovered the dried fishes, which had remained crated in the attics of the Muséum National d'Histoire Naturelle, Paris, since the time of Buffon; and several months ago [early 1828] two manuscripts, fair copies in Commerson's own hand, on the animals of Mauritius and Madagascar, containing precise references to the drawings, were discovered in the library of the late Johann Hermann in Strasbourg,[5] thus permitting us at last to do justice to that excellent observer and put his work in ichthyology to better use.

We also call attention to the ichthyological collections of Sonnerat,[6] one of Commerson's collaborators who stayed in India and became established at Pondicherry. He returned to France in 1814 and sent us the fishes he had gathered on that coast, dried according to the method of Commerson and Gronovius; but his description of them, which we have seen, has remained in the hands of his heirs, and we do not know what has become of it.

Events no less singular have given us the advantage of being the first to profit from a large part of the collections made at about the same time by Banks[7] and Solander,[8] and a short time afterward by the two Forsters. Banks, voluntarily accompanying Captain Cook[9] on his first voyage round the world, took with him Solander, one of Linnaeus's best students. They collected many fishes on the very productive beaches of the [East] Indian archipelago and South Pacific and had several of them drawn by Parkinson;[10] but except for ten species that Broussonet[11] introduced in his first and only *décade ichtyologique [Ichthyologia sistems piscium descriptiones et icones, decas I],* the fishes and the drawings remained in Banks's collection. Fortunately, some samples of fishes that he had given to Broussonet for continuing his work, and that had remained at Montpellier until now, have just been transmitted to us, owing to the generosity of the Faculty of Medicine of that city, and anything new that can be extracted from Parkinson's drawings has been put at our disposal by Mr. Brown.[12]

The same was true about the drawings made by the two Forsters.[13] These German naturalist-scientists were, as we know, appointed by the English government to accompany Cook on his second voyage of 1772–75, and fishes were not forgotten in their observations. But having quarreled with the Admiralty upon their return, the elder Forster found himself obliged to leave his drawings in the hands of his creditors; and thence they passed to Banks's collection, where they remain today. The manuscript of his descriptions was purchased after his death for the Royal Library of Berlin, where Schneider took excerpts from it that he included in Bloch's posthumous *Systema ichthyologiae,* published in 1801.[14]

The ease with which we have been able to consult the drawings,[15] and thereby complete whatever the descriptions left vague and uncertain about the characters of the species, has given us a means of clarifying many obscure points in this part of ichthyology, of comparing several of these species with Commerson's, and thus of eliminating a quantity of those double usages so injurious to the true progress of the science.

Would to God we had had the same good fortune relative to another observer at that time, no less zealous or less capable, who also described many of the same fishes. We refer to Forsskål,[16] sent to Arabia by King Frederick V of Denmark, a generous protector of all the sciences. He was particularly interested in studying the many beautiful fishes that populate the Red Sea [fig. 30]. His descriptions were published after his death by the efforts of his friend Niebuhr,[17] but without illustrations. When he wrote the descriptions, he had no other guide but the tenth edition of Linnaeus's *Systema naturae,* and he was often confused about the correct classification, to the extent that he mistook the electric catfish *(silure electrique)* for the electric ray *(torpille), Centriscus scolopax (centriscus scolopax)* for a catfish *(silure),* the ten-pounder *(elops)* for an argentine *(argentine),* and so on. Unfortunately his successors did not perceive his

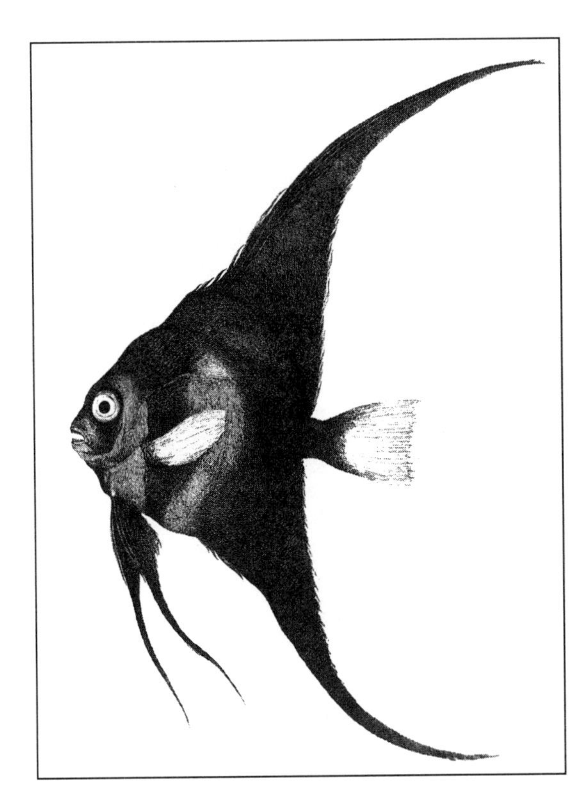

FIG. 30.
A batfish, *Platax teira,* from Peter Forsskål's *Icones rerum naturalium,* 1776, pl. 22. Drawing by G. W. Baurenfeind. Courtesy of J. Nielsen and the Zoological Museum, University of Copenhagen.

mistakes, causing them to list many incorrect species in their systems. This part of Forsskål's work is nonetheless among the most valuable ichthyological productions of the era. In it he described, quite as well as any of the other students of the Linnaean school, 121 species or varieties, and the genera *Scarus (scares)* and *Siganus (sidjans)* appear for the first time [fig. 31].

While the naturalists of France and England traveled the seas and in the face of great danger painstakingly prepared works that were to remain unnoticed in their own countries, Russia had its own people conduct a general exploration of its wide territory and made certain the results were more useful to the public. In thus reforming itself, Russia gave an example to other states worthy of being followed. Its first voyagers, too, had been quite neglected. Messerschmidt,[18] who traveled throughout Siberia from 1720 to 1726 at the behest of Peter the Great and made large collections, died of disappointment and in poverty in 1735. His papers remained in the archives of the Academy of Sciences, St. Petersburg, which took no steps to publish them. An expedition sent by the empress Ann, Peter's granddaughter, composed of several scien-

FIG. 31. Pages from Peter Forsskål's "fish herbarium" (Herbarium Ichthyologicum Forsskåli). Forsskål adopted a technique of preparing dried specimens, described earlier (1744b) by Johan Frederic Gronovius (see chap. 8, n. 5), in which only the skin and parts of the skeleton were retained, the remains then pressed flat on paper like a plant (see Klausewitz and Nielsen 1965). Courtesy of J. Nielsen, T. Wolff, and the Zoological Museum, University of Copenhagen.

tists,[19] explored the same country between 1733 and 1743, with much more care. Only the botanist Johann Georg Gmelin[20] succeeded in publishing his work. The zoologist Steller,[21] one of the men who were most knowledgeable about marine animals, did not live to see the publication of his research on seals and manatees; after his death, which was hastened by treachery and the malice of others, all his memoirs on ichthyology save one, which contains only some general remarks on the class of fishes, were buried and forgotten, along with those of Messerschmidt. It was not until a few years ago that Pallas and Tilesius brought to light some fragments.

Catherine II, on the advice of Count Vladimir Orlof, saw to it that the third exploration, which she sponsored in 1768, was performed with more care and regularity and that science profited as early as possible from the efforts of the men engaged in it.[22] To that end she ordered that the observations be written down during each winter quarter and sent immediately to St. Petersburg along with the collections made during the year; a precaution all the more desirable since three of the naturalists,

Falck,[23] Gmelin,[24] and Güldenstädt,[25] lost their lives either during the journey or before being able to put their writings into final form. But their colleagues, especially Pallas, finished them, and nothing was lost to science.

From these expeditions, ichthyology gained the knowledge of several fishes of the Siberian rivers, Lake Baikal, and the Caspian Sea; and after these first results, others soon followed. Communications established by the explorers brought the arrival at St. Petersburg of species from the eastern seas. In general, it is in the memoirs of the Academy of Sciences of St. Petersburg that this part of natural history has been treated with the most consistency. Pallas and other members of this body have continued to provide the academy with interesting fishes up to the moment of this writing.[26] All this research and all these descriptions have been done methodically in the style and spirit of Linnaeus.

During this same time, isolated naturalists studied the fishes of the northern seas and described them with equal exactitude. Johann Christian Fabricius,[27] the celebrated entomologist of Norway, Otto Fabricius[28] working on the icy coasts of Greenland, and Olafsen and Povelsen[29] on those of Iceland, all bent their efforts to applying the nomenclature of Linnaeus to the products of those forbidding climates. They were not always successful, but their descriptions, especially those of Otto Fabricius, make up for the small errors they were led into by lack of help from the literature. Ascanius[30] published colored drawings of some species from the German sea [North Sea]; Müller[31] introduced species in his Danish zoology from all the shores then belonging to the Crown of Denmark [which at the time included Norway and Greenland]. Others published special memoirs in the collections of the northern academies on the fishes of their country and included some exotic species as well.[32] Thunberg[33] included descriptions of fishes he brought back from Japan,[34] and the collections he made there also served as materials for Houttuyn's[35] memoirs, published by the Society of Sciences of Haarlem in 1782.

Memoirs on fishes also appeared in the *Philosophical Transactions of the Royal Society of London,* but in smaller quantity.[36] Except for Broussonet,[37] few Frenchmen studied these animals in a scientific way. It was a Dane, Brünnich,[38] who was the first since the sixteenth-century ichthyologists to study the fishes of Marseilles and the Adriatic, and he made an effort to arrange them according to the Linnaean system. Cetti[39] published a sketchy indication of those of Sardinia, in a manner to be expected from the state natural history was in at that time in the south of Europe, where the works of Linnaeus had been penetrating only slowly. Toward the end of that period, a work analogous to Cetti's, but more detailed, was written on the fishes of Galicia by Cornide de Saavedra,[40]

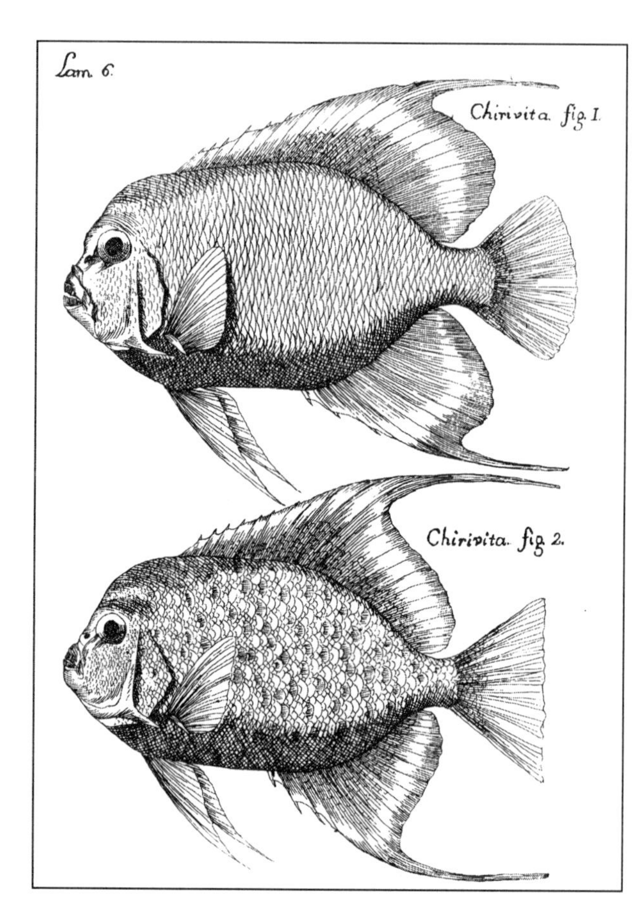

FIG. 32. Angelfishes from Antonio Parra's *Descripcion de diferentes piezas de historia natural las mas del ramo maritimo,* 1787, pl. 6. Courtesy of Melinda K. Hayes and the Allan Hancock Library of Biology and Oceanography, University of Southern California, Los Angeles.

and another Spaniard, Antonio Parra, wrote an account of the fishes of Cuba that was infinitely more valuable for its illustrations [fig. 32].[41]

The Germans, as usual, were more industrious and better informed in the science. The publications of their societies—particularly those of the society of naturalists of Berlin,[42] the periodical titled *Naturforscher,*[43] and others—received a great number of writings on the fishes of Germany. Wulff[44] produced a catalog, using the Linnaean system, of the fishes of Prussia; Fischer,[45] those of Livonia; Birkholz,[46] those of Brandenburg; Sander,[47] those of the Rhine; and Seetzen,[48] those of Westphalia. Leske[49] described the minnows *(cyprins)* of the waters of Leipzig; Meidinger[50] published handsome illustrations of the fishes of Austria; and Paula Schrank[51] described some fishes of Bavaria.

For more distant countries, Pennant,[52] in his *Indian Zoology,* described a shark *(squale)* and a wrasse *(labre),* drawn in Ceylon by Loten, [Dutch] governor of that island. The natural history of Sumatra by Marsden[53] and that of Chile by Molina[54] included more fishes but with less precision. Forster[55] published one on America, especially Hudson Bay; Schoepff,[56] on the United States; and Pennant,[57] on the whole Northern Hemisphere.

Some new genera appeared in these publications. In this way Houttuyn[58] created the genus *Centrogaster,* which is the same as *Buro* of Commerson and *Amphacanthus* of Bloch; Hermann[59] described the genus *Sternoptyx,* which is still valid; Scopoli[60] tried to distinguish *Cottus japonicus* of Pallas under the name *Percis,* and *Coryphaena velifera* or *Pteraclis* of Gronovius under *Pteridium;* Sevastianof reallocated the species of wrasses *(girelles)* with long snouts *[Gomphosus]* to the genus *Acarauna.*[61] All these genera have reappeared under other names in the works of subsequent writers.

NOTES

1. The first expeditions mounted in this spirit were those of [John] Byron [1723–86], [Samuel] Wallis [1728–95], and [Philip] Carteret [d. 1796], described by [John] Hawkesworth [1715?–73], along with that of the first voyage of [James] Cook [see n. 9 below], London, 1773, three volumes in quarto.

2. Philibert Commerson, born at Châtillon-les-Dombes, Ain, in 1727, was passionately devoted to natural history as early as his medical studies at Montpellier. The claim has been made that, at the invitation of Linnaeus, he collected Mediterranean fishes for the queen of Sweden, which surprises us because Linnaeus never mentioned it. Sailing with Bougainville in 1766, he visited the coast of Brazil, Montevideo, Buenos Aires, the Falkland Islands, Tierra del Fuego, Tahiti, some islands near New Guinea, and Java, stayed on at Mauritius, and died there [of pleurisy] in 1773. [For more on Commerson's life and work, see Oliver 1909.]

3. Louis Antoine, comte de Bougainville, famous for his bravery on land and sea, was born in Paris in 1729 and died there in 1811. [Son of a notary, he was called to the bar in Paris. While *aide-major* in the Picardie regiment (1753), then secretary at the London embassy (1754), he pursued his interest in mathematics and by 1754 had published a treatise on integral calculus.] In 1763 he [joined the navy, with the rank of captain, and] supervised the establishment of a trading post in the Falkland Islands that was later described by Father [A. J.] Pernety [1769]. From 1766 to 1769 [in command of the *Boudeuse* and *Etoile*] he circumnavigated the globe [becoming the first French commander to do so] and published an account of his voyage, Paris, 1771, in quarto; another edition appeared in 1772, two volumes in octavo. [For more on Bougainville, see Hammand 1970.]

4. It very often happened that Lacepède [in his *Histoire naturelle des poissons* (1798–1803)] created three or four different species from the description, the drawings, and the notes written on the back of the drawings, and he even placed these imaginary fishes in different genera. We shall see many examples of this.

5. [The manuscripts of Commerson are at present housed in the Bibliothèque Centrale du Muséum National d'Histoire Naturelle, Paris (see Laissus 1978).]

6. Pierre Sonnerat, born at Lyons [in 1749], was the nephew of the famous [Pierre] Poivre [born at Lyons in 1719, died there in 1786] who was [general commissioner and later] intendant of Mauritius. He died in Paris in 1814, the day the city was taken by the Allies [the allied nations of Europe that forced the abdication of Bonaparte on 11 April]. [Traveling to Mauritius in 1767, he explored Réunion and Madagascar with Philibert Commerson from 1768 to 1771. In 1771 he sailed on the naval transport vessel *Isle de France* as clerk and naturalist, participating in the last expedition to New Guinea and returning with a large collection of plants and animals for the king's cabinet.] He is well known to the public for his two voyages: the first to New Guinea in 1769 [a description of which was] published in 1776 in quarto; the second to India and China, 1774–81, printed in 1782, two volumes in quarto. He provides many illustrations of quadrupeds and birds but does not discuss fishes, which he reserved for another work. [For more on Sonnerat, see Rookmaaker 1989, 37–38.]

7. Joseph Banks, privy councillor, knight of the bath, and president of the Royal Society of London, was a man to be commended for having used his fortune to further the cause of science and his credit to protect scientists. Born in London in 1743 [he studied at Eton, Harrow, and Oxford]. [Upon the death in 1761 of his father, a rich landowner in Lincolnshire, he inherited a fortune that permitted him to devote himself to natural history. He studied the works of Linnaeus and Buffon, set up an exhaustive herbarium of the flora of Great Britain, and assembled one of the largest and most famous libraries of natural history in Europe. In 1766 he traveled to Newfoundland and Labrador, and from 1768 to 1771 he participated in the first voyage of Captain Cook, contributing money to it and bringing with him three artists, Sidney Parkinson (1745?–71), Alexander Buchan (d. 1769), and Herman Diedrich Spöring (1733?–71) (on the artists of the *Endeavour* voyage, see Wheeler 1986). In 1772 he organized at his own expense a scientific expedition to the Hebrides and to Iceland. He was elected president of the Royal Society of London in 1778, was knighted in 1781, and was elected commander of the Order of the Bath and member of the Privy Council in 1797. Originator of numerous British exploratory missions, he extended his patronage to naturalists of all countries, even those at war with Great Britain. His botanical collections formed the original basis of the Botanical Department of the British Museum.] He died [at Spring Cove, Isleworth, Middlesex] in 1820. [For book-length biographies of Banks, see Carter 1988 and O'Brian 1993; for biographic and bibliographic sources relative to Banks, see Carter 1987.]

8. Daniel Solander, student of Linnaeus, born at Pitea, Sweden, in 1736. [From 1753 to 1756 he traveled to Russia, Lapland, and the Canary Islands. Sent by Linnaeus to London in 1760 to expound the Linnaean system, he took up residence there. By 1763 he was employed in classifying and cataloging natural history collections at the British Museum, where he became assistant librarian in 1765 and chief curator in 1773. In 1764 he was elected to the Royal Society of London, and] from 1768 to 1771 he was Joseph Banks's companion on Cook's first voyage. He died [in London] in 1782. [For more on Solander, see Diment and Wheeler 1984; Jonsell 1984; Marshall 1984; Stearn 1984; Tingbrand 1984; Wheeler, 1984a, b, 1986.]

9. James Cook [1728–79] is another of those men whose biography we need not rehearse. We shall confine ourselves to mentioning only the dates of his three great voyages, which were so rich in discoveries and which added so much to our knowledge of natural history [see Whitehead 1969a, b, 1978b]. The first voyage, accompanied by Banks and Solander, lasted from 1768 to 1771; it was described by Hawkesworth in 1773. The second voyage, on which he took with him the two Forsters, from 1772 to 1775, was described by himself, London, 1777, two volumes in quarto, and by the younger Forster in 1778, two volumes, translated into German and printed in Berlin. In the same year, the elder Forster published his observations separately in one volume in quarto, London, 1778. On the third voyage, begun in 1776, during which Cook lost his life [1779], he did not wish to take any naturalists; it was completed in 1780, under the leadership of [Charles] Clerke [1741–79] and [John] Gore [d. 1790] and described by [Captain James] King, London, 1784, three volumes in quarto. All three voyages have been translated into French: the first in 1774; the second in 1778, with observations by the two Forsters, five volumes in octavo; the third in 1785, four volumes in octavo.

10. Sidney Parkinson [born at Edinburgh in 1745 and died at sea in 1771], was an English painter employed on Cook's first voyage [1768–71], about which he wrote an account published at London, 1773, in quarto. [For more on Parkinson, see Carr 1983.]

11. Pierre Marie Auguste Broussonet, born at Montpellier in 1761, secretary of the Agricultural Society of Paris, then consul to Morocco, and professor at Montpellier at his death in 1807, was very interested in fishes. It is even said that he prepared a general natural history of fishes in which he described twelve hundred species; but he published only a fragment containing ten species, printed in London in 1782. [In addition, he published a number of smaller papers]: an article on sharks [1784, 1785a] in which he described twenty-seven species, nine of which were new to science; memoirs on the electric catfish *(silure électrique)* [1785b, c]; the wolffish *(anarrhique)* [1788a]; the sailfish *(voilier)* [1788d]; and research on the scales of fishes [1787a], the spermatic vessels [1788b], and the regeneration of fins [1788e]. [For more on Broussonet, see Bauchot 1969.]

12. [Robert Brown, botanist, born at Montrose, Scotland, 1773, and died in London in 1858, was curator-librarian for Joseph Banks and later underlibrarian for the custody of the Banksian Collection at the British Museum (see Edwards 1976; Mabberley 1985; Wheeler 1993).]

13. Johann Reinhold Forster, born at Dirschau [Tczew] in Polish Prussia in 1729, Protestant minister near Danzig, emigrated to Russia, then to England, seems not to have had a very conciliatory disposition; he quarreled with Cook and was harshly treated by the Admiralty on his return. He decided then to enter the service of Prussia and was professor at Halle from 1780 to 1798, when he died there. We cite among his numerous works his *Indische Zoologie,* Halle, 1781, reprinted at London, 1790, and at Halle, 1795.

Johann George Adam Forster, son of Johann Reinhold, born in 1754, companion and aide to his father during the voyage round the world, professor at Kassel in 1778, at Vilna in 1784, then at Mayence, died on the revolutionary scaffold in Paris in 1794. He concurred in the statements on physical science and natural history made by his father and found in the French edition of the *Voyage* [1778]. [On the two Forsters, see Allen 1951, 501–3; Whitehead 1978a; Mearns and Mearns 1988, 154–59; Rookmaaker 1989, 43–59.]

14. See the preface of *Systema ichthyologiae* [Bloch and Schneider 1801], p. xiv.

15. Mrs. Bowdich, well known for her courage in accompanying her husband on perilous expeditions and for the distinguished talents she has devoted to an amiable science, has been willing—with the concurrence of the present depository, the great botanist Mr. Robert Brown [see n. 12 above]—to make us copies of all these drawings. We consider it one of our primary duties to express here our gratitude to her.

[Sarah Bowdich (née Wallis), born at Bristol in 1791, married the traveler and writer Thomas Edward Bowdich and accompanied him to the coast of Africa, first in 1814–18 and again in 1822–24. Her husband born in 1791, became associated with Cuvier during the two years he spent at Paris studying mathematics and the natural sciences; he died at Bathurst in Gambia in 1824. She edited and illustrated her husband's work as well as making copies for Cuvier of numerous drawings by her husband and those by Forster and Parkinson. She also sent Cuvier fishes from England and Scotland.]

16. Peter Forsskål, born in Finland in 1732, was chosen by the king of Denmark on the recommendation of Linnaeus to participate as naturalist on a scientific expedition to Arabia in 1761; he died in that country in 1763. Niebuhr [see n. 17 below] collected his papers and drew from them the *Descriptiones animalium*, Copenhagen, 1775, in quarto; *Flora Aegyptiaco-Arabica*, Copenhagen, 1775; and *Icones rerum naturalium*, Copenhagen, 1776. [For more on Forsskål and his collection of fishes, see Klausewitz and Nielsen 1965; Wolff 1967, 15–47.]

17. Karsten Niebuhr, astronomer and cartographer, was born at Ludingsworth in Lauenburg in 1733 and died in 1815. A simple peasant who became an engineer, he was employed as such on the expedition to Arabia [with Peter Forsskål; see Wolff 1967, 16–47], from which he alone returned in 1767. In 1772 he published a description of Arabia, and in 1774–78, in two volumes in quarto, an account of his journey [a third volume was published in 1837].

18. Daniel Théophile Messerschmidt, from Danzig, was born in 1685 and died in 1735. His researches seem to have been immense. To him we are obliged for the first [description of a] fossil elephant cranium.

19. [Joseph Nicolas] Delisle de la Croyère [1688–1768], astronomer; [?] Müller and [?] Fischer, historians; [Aleksei] Chirikof and [Vitus Jonassen] Bering [1681–1741], mariners; [Johann Georg] Gmelin, botanist; [Georg Wilhelm] Steller, zoologist, and others.

20. Johann Georg Gmelin, born at Tübingen in 1709, followed his compatriots [George Bernard] Bülfinger [professor of theology at Tübingen, 1693–1750] and [Johann Georg] Duvernoy [physician and botanist, native of Tübingen; 1691–1759] to St. Petersburg, and there he occupied the chairs of botany and chemistry. After his return from [the expedition to] Siberia [1733–43], he published the first two volumes of the flora of that country, St. Petersburg, 1747–49; his nephew [Samuel Gottlieb Gmelin (see n. 24 below)] published the remaining two volumes in 1768–69. He returned to Tübingen in 1749 and died there in 1755. He published an

account of his journey at Göttingen, 1751–52, in German, four volumes in octavo. An abridgment in French in two volumes in duodecimo was produced by [Louis Félix Guinement de] Keralio, and another is in the eighteenth volume in quarto of the *Histoire générale des voyages* [see Prévost 1768]. It contains little on fishes.

21. Georg Wilhelm Steller, one of the most courageous and most able naturalists Russia has had in its service, and whom it has treated with the greatest ingratitude, was born in 1709 at Winsheim in Franconia. At several German universities he studied theology, medicine, and natural history, and he was made physician in the Russian army that besieged Danzig in 1734. The baron of Korff, president of the Academy of Sciences of St. Petersburg, sent him in 1738 to join the expedition that had departed in 1734. [Vitus Jonassen] Bering, who was to explore the islands between Siberia and America, invited him in 1741 to accompany him. Steller suffered horribly on this journey and in the end found himself misled by all the promises of that captain. He went to St. Petersburg to seek justice, but orders were forwarded to him to return to Irkutsk to defend himself against who knows what imputation. He was about to return when he received a second order of the same nature, in 1746, and this time the guard sent to escort him left him to freeze on the high road. His description of Kamchatka was published in German in 1774, under the supervision of J. B. Scherer, an employee of the French foreign affairs office. There is an excellent memoir by Steller on seals and manatees in *Novi Comment. Acad. Sci. Impér. Petro.* (St. Petersburg), vol. 2 [1751], and in the third volume of the same periodical some general observations on fishes [1752], from which it is obvious that he had studied them with care. He composed [but left unpublished] an "Ichthyology of Siberia" from which Pallas and Tilesius have provided interesting excerpts in their own works. [For an English translation of Steller's 1741–42 voyage with Bering, see Steller 1988; for a full biographical treatment, see Stejneger 1936; see also Mearns and Mearns 1988, 347–55.]

22. In addition to astronomers and surveyors, there were five naturalists and some students on the expedition of 1768: Pallas turned toward the Jaïk and the Caspian Sea, examined mines in the Urals and the Altais in the district of Kolywan, crossed Lake Baikal, and approached the frontier of Chinese Tartary; he returned by way of the Caucasus. Gmelin went to the south, saw the settlements of the Cossacks on the Don and Astrakhan, and made two excursions into Persia. Falck explored the province of Orenburg and adjacent lands up to the Ob. Georgi was assistant first to Falck and then to Pallas. Güldenstädt concentrated on the Caucasus. Lepechin visited the Urals, Astrakhan, and the shores of the White Sea.

23. Johan Peter Falck, born in Sweden in 1725, was a student of Linnaeus and then professor of botany at the Apothecaries' Garden in St. Petersburg. Afflicted with hypochondria and sufferings of every sort, he killed himself at Kazan in March 1774. A description of his voyage was published, under the supervision of Georgi, in three volumes in quarto, 1785–86.

24. Samuel Gottlieb Gmelin, born at Tübingen in 1745, was the nephew of Johann Georg Gmelin, a member of the Siberian expedition of 1733–43 [see n. 20 above]. The first three volumes of his travels appeared from 1770 to 1774. But he (the nephew) died in 1774, a prisoner of the khan of Khaïtakes, and the editing of the fourth volume was entrusted to Güldenstädt and, after his death, to Pallas,

who issued it in 1784. There are some descriptions and three drawings of fishes, and many details on the fisheries.

25. Johann Anton von Güldenstädt, born at Riga in 1745, studied at Berlin. The favor of the czar of Georgia procured him much assistance in his exploration of the Caucasus, but he contracted diseases in that country that enfeebled him. Nevertheless he returned to St. Petersburg, where he died at age thirty-six of a wasting fever rampant in that city. A description of his voyage, printed under the direction of Pallas, in two volumes in quarto, 1787–91, includes accounts of some fishes. He also described and illustrated several fishes in *Novi Comment. Acad. Sci. Impér. Petro.* (St. Petersburg), vols. 16 [1772], 17 [1773], and 19 [1775]. [For more on Güldenstädt, see Mearns and Mearns 1988, 179–81.]

Johann Gottlieb Georgi, born in Pomerania in 1738, sent to join Falck in 1770, published a description of his voyage, two volumes in quarto, 1775, in which he describes a few fishes. The seventh part of his description of Russia, in eight parts in octavo, 1797–1802, contains a natural history of fishes of that empire, although incomplete.

Ivan [Ivanovich] Lepechin (ca. 1750–1802) studied at St. Petersburg and Strasbourg and in 1783 was secretary of the Russian Academy of Sciences. He published [the description of] his voyage in Russian, three volumes in quarto, 1771–80 [a fourth volume was published in 1805], in which he describes several species of fishes. There is a German translation by [Christian Heinrich] Hase, Altenburg, 1774–83.

Nikolai [Petrovich] Rychkov [1746–84], one of the students attached to this expedition, published [a description of] his voyage in Russian [1770–72], and Hase translated it into German, Riga, 1774, in octavo.

26. In addition to the numerous memoirs of Koelreuter mentioned above [see chap. 8, n. 21], which appear in various volumes of the *Nova Acta* of the Imperial Academy of Sciences, St. Petersburg, this periodical contains a number of additional articles on fishes: a knifefish *(carape)* described by Basil Zuiew, vol. 5, 1789; a sturgeon *(esturgeon)* by Lepechin, vol. 9, 1795; and the natural history of the salmon in the Frozen Sea by [Nikolai Yakovlevich] Ozeretskovsky, vol. 12, 1801.

27. Johann Christian Fabricius, Danish entomologist, born at Tondern in the duchy of Schleswig in 1745, professor [of natural history] at Kiel, died in 1808. His immense work on insects does not pertain to our subject; we shall cite only his travels in Norway, published in German, Hamburg, 1779, in octavo, in which he discusses fourteen species of fishes. This work has been translated into French by Millin [and published in 1802 in octavo].

28. Otto Fabricius [1744–1822], clergyman, was employed in the Danish colony of Greenland, and then in Norway and in Denmark. He published a description of the fauna of Greenland, Copenhagen and Leipzig, 1780, in octavo, one of the best works of this sort, in which he gives precise descriptions of forty-four species of fishes, and for several of them gives very interesting details on their natural history. However, one must occasionally be cautious with his nomenclature [see Wolff 1967, 48–50].

29. The voyage to Iceland written by Eggert Olafsen, Icelandic naturalist (1726–68) and by Bjorne Povelsen, chief physician on that island (d. 1778), was

published in Danish at Sorøe in 1772, and in German at Copenhagen in 1774–75, two volumes in quarto. It contains descriptions and drawings of fishes, although they are somewhat crudely done. There is a French translation by Gauthier de la Peyronie, Paris, 1802, five volumes in octavo, with an atlas in which the nomenclature of natural history is often mutilated.

30. Peder Ascanius [1723–1803], inspector of mines in the north of Norway, published several colored drawings of fishes, some of which represent new species, in his *Icones rerum naturalium,* or drawings in color of the natural history of the North [a work initiated in oblong folio in 1767], Copenhagen and Geneva [and continued in folio, 1772–1806, Copenhagen].

31. Otto Fredrik Müller, naturalist, born in Copenhagen in 1730, died in 1784; one of the most careful and diligent observers of the eighteenth century, who was made famous by his discoveries with the microscope, illustrated some fishes in *Zoologiae Danicae,* 1777–89, in folio, and his example was followed by his successors, [Peder Christian] Abildgaard [1740–1801], Viborg [probably Eric Nissen Viborg, 1759–1822], and [Martin Heinrich] Rathke [1793–1860]. He included a general catalog of Danish fishes in his *Zoologiae Danicae prodromus,* Copenhagen, 1776, in octavo [see Wolff 1967, 48–50].

32. A large number of memoirs are to be found in the collections of the Academy of Sciences of Stockholm, the Royal Society of Copenhagen, and the Society of Sciences of Norway. The most notable authors are [Hans] Strøm [1726–97], [Bengt Anders] Euphrasén, [Morten Thrane] Brünnich [1737–1827; see n. 38 below], Strupenfeld, [Nils] Gissler [1715–71], [Théodore] Ankarcrona [1687–1750], [Henrik] Tonning [1732–96], [Martin Hendriksen] Vahl [1749–1804], [Clas Fredrik] Hornstedt [1758–1809], [Theodore] Holm, [Anders Jahan] Retzius [1742–1821], [Lars] Montin [1722–85], and others.

33. Carl Peter Thunberg [a Swedish doctor, botanist, and traveler], born in 1743 [d. 1828], was a student of Linnaeus and professor at Uppsala. [In the 1770s, on his way to Japan under the employ of the Dutch East India Company, he stopped off at the Cape of Good Hope, where from 1772 to 1775 he made large collections of plants and animals. He then continued on to Japan, Java, and Ceylon before returning to the Cape in 1778 and finally Sweden in 1779. All of his specimens were presented to the University of Uppsala, where he was appointed director of the museum and professor of natural history. For more on Thunberg, see Rookmaaker 1989, 148–62.]

34. [Thunberg] *Kongl. Svenska Vetensk. Acad. Nya Handling.* (Stockholm), 1790, 1792, and 1793. There is also a dissertation by him [see Thunberg 1789] on the moray eel *(murène)* and the snake eel *(ophichte).*

35. Martinus Houttuyn [1720–98] was a diligent but poorly educated naturalist who translated and paraphrased in Dutch the *Systema naturae* [of Linnaeus], Amsterdam, 1761–85 [see Houttuyn 1761–85].

36. [Memoirs on fishes published in the *Philosophical Transactions of the Royal Society of London*]: Johann Friedrich Gronovius, on the loach *(misgurn),* vol. 44 [1747]; [James] Parsons, on the anglerfish *(baudroie),* vol. 46 [1752]; Cromwell Mortimer, on the opah *(zeus luna),* vol. 46 [1752]; Farrington, on the char *(truite des Alpes),* vol. 49 [1756]; [James] Ferguson, on the anglerfish *(baudroie),* vol. 53 [1764];

[John Albert] Schlosser, on *Chaetodon rostratus,* vol. 54 [1765, 1767]; [Pieter Simon] Pallas, on the archerfish *(toxotes),* vol. 56 [1767]; [Michael] Tyson, on a perch *(perche)* of the South Pacific, vol. 61 [1772]; Daines Barrington, on the trout *(truite),* vol. 64 [1774]; [Thomas] Brown, on the flyingfish *(exocet),* vol. 68 [1779]; [William] Watson, on the blue shark *(squale glauque),* vol. 68 [1779]; [William] Bell, on *Chaetodon nodosus,* vol. 83 [1793].

37. Broussonet [see n. 11 above], a memoir on the sailfish *(voilier),* 1788d, and another on the different species of dogfish sharks *(chien de mer),* 1784, both published in the *Hist. Mém. Acad. Roy. Sci.* (Paris); the latter article was reprinted in the *J. Physique,* vol. 26 [Paris, 1785a].

38. Morten Thrane Brünnich [1737–1827], professor at Copenhagen, author of the *Ichthyologia Massiliensis,* Copenhagen and Leipzig, 1768, in octavo, in which he describes fairly accurately 101 species, some of which were new. His nomenclature cannot always be trusted; his *Perca pusilla,* for example, is but *Zeus [Capros]* aper. This work contains an appendix titled "Spolia Maris Adriatici," in which he indicates another 13 species, but some of these are the same as those included in the main body of the work. [For more on Brünnich, see Mearns and Mearns 1988, 93–95.]

39. Francesco Cetti [1726–78], ex-Jesuit, author of *Storia naturale di Sardegna,* four volumes in octavo, Sassari, 1774–78. He discusses fishes in the third volume, but all except the tuna *(thon)* are mentioned only briefly. [For more on Cetti, see Mearns and Mearns 1988, 113–15.]

40. Don José Andrés Cornide de Saavedra [b. 1734], governor of Santiago, author of an essay on the natural history of fishes and other marine organisms of the coast of Galicia, based on the system of Linnaeus, in Spanish, 1788, in duodecimo.

41. *Descripcion de diferentes piezas de historia natural las mas del ramo maritimo* by Don Antonio Parra [fl. 1763–99], in Spanish, Havana, 1787, small in quarto, with seventy-five plates. It is one of the most useful works for information on the fishes of the Gulf of Mexico, not so much for the text as for the very exact illustrations of these fishes. [On Parra and his fishes, see Poey y Aloy 1863; González 1989.]

42. The *Gesellschaft Naturforschender Freunde zu Berlin* began to publish its works in 1775 in octavo in German [see Kronick 1976, 105]. The first four volumes were titled *Beschäftigungen der Berlinischen Gesellschaft Naturforschender Freunde;* the next eleven, from 1780 to 1794, *Schriften der Berlinischen Gesellschaft Naturforschender Freunde;* the series was continued in quarto from 1795 to 1803 under the title *Neue Schriften, Gesellschaft Naturforschender Freunde zu Berlin.* The earlier, octavo collection contains several memoirs on ichthyology by [Marcus] Bloch [1779, 1780, 1785, 1788, 1792a, b], [Bernhard] Wartmann [1777, 1783a, b, c], [Heinrich] Sander [1779], [Petrus Camper, 1787], [Johann David] Schoepff [1788], [Johann Julius] Walbaum [1783, 1784a, b], [Franz von Paula] Schrank [1780, 1781, 1783], and [Petrus Christian] Abildgaard [1792].

43. *Naturforscher,* an interesting collection, printed at Halle from 1774 to 1804 in thirty parts. It contains, among other things, ichthyological memoirs by [Johann] Hermann [1781, 1782] and [Heinrich] Sander [1781], and by [Johann David] Schoepff on the American perch [1784a] and the pike [1784b].

44. Joannes Christophorus Wulff [d. 1767], physician at Königsberg, published *Ichthyologia, cum amphibiis regni Borussici methodo Linneana disposita,* Regiomonti, 1765. This is a catalog of fifty-three species, occasionally ill named. For example, he confuses a whitefish *(marène)* with a minnow *(cyprin).*

45. [Jacob Benjamin] Fischer [1730–93], an essay on the natural history of Livonia, in German, Leipzig, in octavo, 1778, reprinted [at Königsberg] in 1791; he discusses forty species of fishes.

46. Johann Christoph Birkholz published before Bloch, in German, an economic description of fishes found in the waters of the Electoral March of Brandenburg, Berlin, 1770, in octavo.

47. Heinrich Sander [1754–82] published some materials for a natural history of fishes of the Rhine in *Naturforscher,* vol. 15 [1781], and in the same periodical, vol. 25 [1791], some remarks on this memoir were published by Bernard Sebastian Nau [1766–1845].

48. [Ulrich Jasper] Seetzen [1767–1811] published a catalog of the fishes of the principality of Jever in Westphalia in the first volume [1794] of the *Zoologische Annalen* (Weimar), edited by [Friedrich Albrecht Anton] Meyer.

49. Nathanael Gottfried Leske [1751–86], professor at Leipzig, published *Ichthyologiae Lipsiensis specimen,* Leipzig, 1774, in octavo, in which there are detailed descriptions of seventeen species of minnows *(cyprins).*

50. Baron Carl von Meidinger [1750–1820], secretary to the emperors Joseph II and Leopold II, was the author of a collection of handsome colored illustrations titled *Icones piscium Austriae indigenorum,* in five decades, in folio, Vienna, 1785–94, in which are represented several interesting fishes of the Danube and its tributaries.

51. Franz von Paula Schrank, professor at Ingolstadt, born in 1747 [d. 1835]. In [his description of] a journey through Bavaria, Munich, 1786, he describes one or two trouts *(truites).*

52. [Pennant, see chap. 8, n. 38], *Indian Zoology,* London [1769], in folio, with twelve plates; a [much expanded] second edition appeared in 1790 in quarto. [His "tiger shark" and "Ceylon wrasse" are shown in pl. 16.]

53. William Marsden [1754–1836], *History of Sumatra,* London, 1783. There is a French translation, Paris, 1788, two volumes in octavo. A third edition appeared in 1811 in quarto.

54. [Giovanni] Ignazio Molina [1740–1829], ex-Jesuit, wrote from memory in Italy his *Saggio sulla storia naturale del Chili,* Bologna, 1782, in octavo, translated into French by Gruvel, Paris, 1789, in octavo. The second Italian edition was published at Bologna, 1810, in quarto. This book contains several descriptions that need to be verified.

55. Johann Reinhold Forster [see n. 13 above], *Catalogue of the Animals of North America,* London, 1771, in octavo; [see also] "An Account of Some Curious Fishes, Sent from Hudson's Bay," published in the *Phil. Trans. Roy. Soc. Lond.,* vol. 63 [1773].

56. [Johann David Schoepff (1752–1800), "Beschreibungen einiger nordamerikanischer Fische," 1788], a remarkable memoir published in *Schr. Berl. Ges. Naturf. Fr.,* vol. 8, the best thing we had until the works of Mitchill [see chap. 13, n. 30]. [For biographical information on Schoepff, see Adler 1989, 14–15.]

57. [Pennant, see chap. 8, n. 38], in the ["Supplement" or] third volume of his *Arctic Zoology* [1787].

58. Martinus Houttuyn, in addition to his memoir on the fishes of Japan cited above [Houttuyn 1782], published one on the eggs of sharks [1764] and another on exotic fishes [1765].

59. Johann Hermann, professor of natural history at Strasbourg (1738–1800), author of several memoirs included in German periodicals, of special interest being his description of the genus *Sternoptyx* [1781]. He gives his views on the relationships of fishes in his *Tabula affinitatum animalium*, 1783, one volume in quarto, and describes some new species in his *Observationes zoologicae*, 1804, one volume in quarto.

60. Giovanni Antonio Scopoli, born in the See of Trent in 1725 [d. 1788], professor [of mineralogy] at Schemnitz and then [professor of chemistry and botany] at Pavia, wrote about fishes in his *Deliciae florae et faunae insubricae* [Ticini, 1786–88, three parts in folio].

61. [A. Sevastianof], *Nova Acta Acad. Sci. Impér. Petro.* (St. Petersburg), vol. 13 [1802].

10 Bloch and His Contemporaries

Bloch[1] was regularly writing special memoirs on fishes by 1780[2] as a prelude to his great and magnificent work with which he enriched ichthyology, and which made him peerless and considered even today one of the primary authors on the natural history of fishes. We therefore owe our readers an extensive analysis of his work, such as we provided for the works of Willughby, Artedi, and Linnaeus.

It is composed of two essentially distinct parts: the economic natural history of the fishes of Germany, and the natural history of foreign fishes [fig. 33].[3] The first, resulting primarily from the observations of the author and illustrated under his direction from fresh samples, contains good descriptions, faithful pictures, and interesting and accurate observations. In it he discusses 115 species, of which some minnows *(cyprins)* and salmon *(saumons)* were not well known or very clear before him; but he does not include fishes of the Mediterranean, even though, because of the coast of Austrian Istria, they would belong to Germany as well. As a matter of fact, Bloch had very little knowledge of Mediterranean fishes, which is not surprising when one considers that he lived in an unfavorable location, in the midst of the beaches of Brandenburg. What is more extraordinary, and what we ourselves have difficulty believing, is that there are very common fishes in the ocean that he did not know well: the silverside *(athérine),* for example, which he represented very poorly [his pl. 393, fig. 3]; the sardine *(sardine),* for which he substituted in his plate 29 [fig. 2] another small species from the Baltic; and the allis shad *(alose),* in the place of which he shows in plate 30 a drawing of the twaite shad *(feinte).*

In his fishes of Germany, Bloch includes drawings of foreign fishes borrowed from the manuscripts of Plumier and Marcgrave, which are much less authentic or less correct than those he had had made under his

D. MARCUS ELIESER BLOCH'S,

ausübenden Arztes zu Berlin: der Utrechter, Frankfurter, Göttinger, Harlemmer, Vlieſſinger, Böhmiſchen und Coppenhagner Geſellſchaften der Wiſſenſchaften; der Berliner, der Römiſch-Kaiſerlichen, der Danziger, Halliſchen und Zürcher Naturforſchenden Geſellſchaften; der St. Petersburger, Leipziger, Baierſchen und Zelliſchen ökonomiſchen Geſellſchaften; der Ackerbau-Geſellſchaft und des Muſée zu Paris, Mitglieds oder Correſpondenten,

NATURGESCHICHTE

DER AUSLÄNDISCHEN FISCHE.

MIT SECHS UND DREISSIG AUSGEMALTEN KUPFERN NACH ORIGINALEN UND EINEM TITELKUPFER.

DRITTER THEIL.

BERLIN, 1787.

Auf Koſten des Verfaſſers, und in Commiſſion in der Buchhandlung der Realſchule.

FIG. 33. Title page of Marcus Elieser Bloch's *Naturgeschichte der ausländischen Fische*, 1785–95, vol. 3. Courtesy of Leslie Overstreet and the Special Collections Department, Smithsonian Institution Libraries, Washington, D.C.

own supervision [fig. 34]. He used them even more frequently in the second part, in which, as a specialist, he presents foreign fishes, and which is composed of very different elements. The species the author possessed in a natural state, either dried or in preservative, are often well drawn and well described, except for the colors, which are almost always misleading in subtle ways because there is no art that can preserve colors after death. Their natural history is fairly exact when details have been given him along with the fishes by known explorers such as the Mission-

FIG. 34. Marcus Elieser Bloch's striped mojarra, *Diapterus plumieri*, from his *Naturgeschichte der ausländischen Fische*, 1785–95, vol. 4, pl. 247 (copied by Bloch from a manuscript of Charles Plumier; see fig. 24.) Courtesy of M. Ducreux and the Bibliothèque Centrale, Muséum National d'Histoire Naturelle, Paris.

ary John,[4] who was one of the most useful to him. But as for the species he bought from vendors or merchants, it is often only haphazardly that he indicates their origin and behavior, depending on his having been more or less fortunate in finding them in the works of other authors, which he was not equal to consulting as an enlightened critic. In fact, it happens more than once that he took one species for another, that he confused one from India with one from America, that he claimed as identical some species that are distinct but close relatives, and so on. Sometimes he even took the liberty of altering the drawings to make them square with his opinions, and in other instances the artists he used were so negligent that we could not have recognized his species had we not had the opportunity of examining the original artwork.

The drawings he borrowed from the manuscripts of Prince Maurits and Plumier are the least dependable of all. Not only did he preserve most of the faults that are in these drawings, executed at a time when there was little exact knowledge of the anatomy of fishes, but also when he tried to correct these faults, he did so in an inappropriate manner, exchanging them for other mistakes. It is only by conjecture that he gave counts for the numbers of fin rays, which the artists had never thought of showing.[5]

Bloch was not very well versed in the anatomy of fishes and hardly ever rose to a philosophical consideration of their relationships and distribution; nonetheless, he established some genera[6] founded on good characters and real analogies. But he also recognized some that are purely artificial[7] and some that can only be regarded as simple subdivisions, more or less well done, of the natural genera of Artedi and Linnaeus.[8]

In this great work, Bloch followed the system of Linnaeus as modified by Pennant; that is, the Amphibia nantes were returned to the class of fishes and divided, following Artedi, into two orders, the Branchiostegi and Chondropterygii. In his fishes of Germany, however, he reversed Linnaeus's order and began with the Abdominales, because among them are the most species that can be produced profitably. But by the end of his life, Bloch had prepared a general system[9] in which he placed not only the species described in his great work, but also all those for which authors had furnished him sufficient descriptions. To classify them he conceived a method based solely on the number of fins, as the sexual system of Linnaeus is based on the number of stamens, then subdividing according to the relative position of the pelvic and pectoral fins, the same characteristic Linnaeus had used in his first division [see table 9].

He could not have done better had he had the express intention of exposing artificial methods to ridicule and showing the absurd relationships they can lead to. In fact, there has never been a more bizarre classification: the silverside *(athérine)* is placed next to the snipefish *(centrisque)*; the armored catfish *(loricaire)* is near the shark *(squale)*; the ray *(raie)* is far from the shark *(squale)* but near the catfish *(silure)* and the pike *(brochet)*; freshwater eels *(anguilles)* and pufferfishes *(tétrodons)* are in the same class; and on and on. Entire genera are based on groupings no less bizarre. For example, in his *Grammistes* he has included species from eighteen natural genera, only because their bodies are marked by longitudinal stripes; in his *Cichla (cichla)*, species from seven natural genera, and so forth. There are some genera, however, that are good, some he was the first to establish—his *Synanceia (synanceia)*, for example, which had been confused with *Scorpaena (scorpènes)*—and he was correct in adopting several genera of Gronovius, Brünnich, and others of his predecessors.

The greatest utility of this curious production consists in the author's having included several new species that he received after his great work was finished. His editor, Schneider,[10] made important additions himself, taken from the papers of Forster and more recent authors. He also included critical remarks worthy of attention and some anatomical observations, and so it became an almost complete survey of known fishes at the beginning of the present century. The number of genera is 113; of species, 1,519. But in this number, there are at least a hundred that are doubtful or repeated two or three times.

Having been given the opportunity, by the courtesy of the naturalists in Berlin, to go back to the sources used by Bloch, to examine the very fishes he possessed in his cabinet, and to match all the duplicate uses with the proper species, we shall indicate in the present work the correct genus of the species he has thus misplaced, and we shall show only too often proof of the negligence with which he worked. In order to consider the complete body of Bloch's works, we have taken our natural history through his posthumous system. Now let us go back and consider other ichthyological works published while he was working on his own. The appearance of the first volumes of his great natural history of fishes seems to have been a signal for recommencing works in general on this class of animals.

Haüy[11] did not yet know of Bloch's existence when he was editing the ichthyological portion of the *Dictionnaire de l'encyclopédie méthodique* under the name of Daubenton, although he did not publish it until 1787; but this section of the dictionary, written by a man who had no knowledge of fishes, consists of hardly anything but excerpts from Willughby and from other authors cited in the twelfth edition of Linnaeus, from Klein's *Historiae piscium naturalis promovendae,* and from the *Spicilegia* of Pallas [fig. 35]. In it there are no observations or views of the author himself.

Bonnaterre,[12] in charge of gathering drawings for the same enterprise, felt obliged to produce a text more at the level of current science. In the ichthyological plates, published in 1788, he copied everything of Bloch's that had appeared and completed his task with drawings from the works of Pallas, Koelreuter, Gronovius, Broussonet, the museum of Frederick Adolphus, and Pennant; and when these authors failed him, he used those by Catesby and Willughby, and even by Rondelet and Marcgrave. Thus he collected more than four hundred drawings; but this collection, although useful to people who did not have the originals, should be consulted with caution. No more than Haüy's contribution to the *Encyclopédie méthodique* is it based on a solid knowledge of the subject. When Bloch, Pallas, and Broussonet do not guide the author, he follows citations given by Linnaeus and goes so far astray as to figure the grayling *(ombre, Salmo thymallus)* for *Sciaena umbra* and the pilotfish *(Scomber ductor)* for *Coryphaena pentadactyla,* which is a razorfish *(rason),* and so forth.

One should perhaps judge even more severely Walbaum's[13] *Artedius renovatus,* the publication of which also began in 1788. This is Artedi's text supplemented by notes taken from all subsequent authors—Linnaeus, Forsskål, Pallas, Gronovius, Bloch, and so on—amassed without com-

FIG. 35. Three views of a scorpionfish, *Inimicus didactylus,* from Peter Simon Pallas's *Spicilegia zoologica,* 1769, vol. 7, pl. 4. Courtesy of Eveline Nave Overmiller and the Library of Congress, Washington, D.C.

parison, without criticism, and in the authors' own terms. Schneider proved that Walbaum knew little about fishes, and this compilation shows he had as little taste as judgment.[14]

Nonetheless, these sorts of books are necessary. Although one cannot rely on their authority alone, they give information on earlier authorities and thus save time for anyone who wishes to investigate a particular branch of the science. Unfortunately, books of this kind cannot be done well except by people who have already investigated the science and, what is even rarer, who do not consider themselves above such a task; it is thus that the responsibility seems almost always to fall into hands that are less capable.

This is what happened at the time, in a most vexing way, to the *Systema naturae.* A second Linnaeus would have been needed to produce a new edition and include in it the riches acquired in thirty years, but instead it was a mediocre chemist, practically a stranger to natural history, Gmelin,[15] who took charge of this vast enterprise, which could have been so worthwhile.

I believe Gmelin had not seen a single one of the animals he was to classify in the work; perhaps he did not even read the books from which he took excerpts for the work. But as too often happens in Germany, the task was carried out as in a factory, a certain number of young people taking charge of the excerpts, and the editor confining himself to assembling and classifying.

Thus they accumulated under the genera of Linnaeus the species indicated or described by Pallas, Brünnich, Klein, Olafsen, Soujew, Strøm, Forsskåll, Fabricius, Molina, Hermann, Houttuyn, Pennant, Meidinger, Broussonet, and especially those that Bloch dealt with at the time in his "fishes of Germany" and the first two volumes of his "foreign fishes." Also included were species collected by the various explorers whose writings we have cited and, whenever possible, citations of the more ancient authors whom Linnaeus did not use.

As a guide for going back to the sources, this large collection of citations is certainly very valuable; one could not succeed in bringing together another, with so many citations, except with a great deal of trouble. But anyone who trusts the results expressed in the book will often be led into error. Gmelin ranked the species as did the authors from whom he took them. All the drums *(sciènes)* and the perch *(perca)* of Forsskåll, are drums *(sciènes)* and perch *(perca)* for him as well; like that traveler, he placed the snipefish *(centrisque)* among the catfishes *(silures);* the grenadier or rattail *(macroure),* which Gunner took to be a dolphinfish *(coryphène),* is a dolphinfish *(coryphène)* for Gmelin as well. He followed Houttuyn just as blindly, and since these various observers had different conceptions of genera, and several had none of the correct genera of

Linnaeus, it often happens that the species are very much out of place; and more often than that, they are duplicated, sometimes two or three times.

On the other hand, different species are confounded as if they were one species; but overall, the apparent number of species, which is 826, should be decreased: there are at least 50 too many. Only five genera, *Sternoptyx (sternoptyx), Leptocephalus (leptocéphales), Kurtus (kurtus), Scarus (scares),* and *Centrogaster (centrogastres),* were added to the sixty-one genera of Linnaeus, so the total is only sixty-six. The species that make up the other new genera described by other authors are distributed among the old genera, often haphazardly; as to the distribution of the orders, Gmelin submitted to general opinion and restored the cartilaginous forms to the class of fishes, placing them at the end, as Artedi did, under the names Branchiostegi and Chondropterygii.

To bring to a close the natural history of ichthyology before Lacepède, it only remains for us to run through the anatomist's works on fishes in the eighteenth century, as we have already done for the sixteenth and seventeenth centuries.

NOTES

1. Marcus Elieser Bloch, Jewish surgeon in Berlin, born at Ansbach in 1723 of very poor parents, did not try to correct his lack of education until late, and then only imperfectly, as is easily seen in his writings. It was not until he was fifty-six that he began to write about fishes, and prodigious perseverance and industry were needed to bring to a successful conclusion the considerable enterprise that was his great *Ichthyologie.* He died in 1799 when he was sixty-nine. [For more on Bloch see Karrer 1978, 1980; Wells 1981.]

2. For example, Bloch's earliest publication on fishes, "Naturgeschichte der Maräne," appeared in 1779 *(Beschäft. Berl. Ges. Naturf. Fr.,* vol. 4). In 1780 he published an economic natural history of the fishes of the Prussian states, especially of the Marches and Pomerania *(Schr. Berl. Ges. Naturf. Fr.,* vol. 1). Later, in 1785, he wrote about two species of flounders *(pleuronectes) (Schr. Berl. Ges. Naturf. Fr.,* vol. 6); in 1789, two species of scorpionfishes *(scorpènes) (Kongl. Svenska Vetensk. Acad. Nya Handling.* [Stockholm], vol. 10); and two articles in 1792, one on two species of perches *(perches),* the other a description of two new fishes and remarks on Abildgaard's memoir on the hagfish *(myxine) (Schr. Berl. Ges. Naturf. Fr.,* vol. 10).

3. *Oeconomische Naturgeschichte der Fische Deutschlands* appeared in German in three volumes in quarto, with 108 plates in folio, Berlin, 1782–84, and in octavo, 1783–85; and in French in folio, 1785–86. *Naturgeschichte der ausländischen Fische,* in nine volumes in quarto, with 324 plates in folio, 1785–95, and in octavo, 1786–87 [two volumes only]; in French in folio, 1787–97. The two were combined, in French, in twelve volumes in folio, with 432 plates, under the name *Ichthyologie,*

ou Histoire naturelle générale et particulière des poissons [Gill 1872, 35–36], a title that promises far too much, for the author had no intention and made no claim of treating all known fishes, but only those for which he was able to show original drawings. There is also an edition in octavo. [René Richard] Castel republished the text, arranging it according to the Linnaean system but omitting the synonyms and other scholarly citations, printed following the Buffon of Déterville, in ten small volumes in duodecimo, with figures much reduced in size.

4. [Missionary John], a Danish missionary priest at Tranquebar on the coast of Coromandel [see Wells 1981, 9].

5. See Schneider, in the preface to *Systema ichthyologiae* [Bloch and Schneider 1801, xv], and the memoirs of Lichtenstein (1818, 1819, 1822a, b, 1829) in the volumes of the Royal Academy of Sciences of Berlin.

6. [For example], his *Batrachus (batrachus),* in which he fittingly unites certain codfishes *(gades)* and certain sculpins *(cottes)* of Linnaeus.

7. For example, his *Lutjanus (lutjans),* in which he assembles perches *(perches),* drums *(sciènes),* and wrasses *(labres),* only because their preopercle is denticulated.

8. Thus he separated from the perches *(perches)* his *Epinephelus, Anthias, Holocentrus, Bodianus,* and *Gymnocephalus.*

9. *M. E. Blochii Systema ichthyologiae iconibus CX illustratum. Post obitum auctoris opus inchoatum absolvit, correxit, interpolavit Jo. Gottlob Schneider,* Berlin, 1801, one volume in octavo, with 110 plates [see Bloch and Schneider 1801; see also Gill 1872, 39-40].

10. [Johann Gottlob Theaenus Schneider, German philologist and naturalist, was born in 1750 at Kollmen, near Oscharz, just southeast of Leipzig, and died at Breslau in 1822. He studied philology, especially the Greek classics, and natural history at universities in Leipzig (1769), Göttingen (1772), and finally Strasbourg, where he received his Doctor of Philosophy degree (1774). In 1776 he was appointed professor of philology at the University of Frankfurt (an der Oder), where he produced a large number of translations and commentaries on classical works, including a complete edition of Oppian (1776), Aelianus's *De natura animalium* (1784), Aristotle's *Historia animalium* (1811), and Nicander's *Theriaca* (1816). Schneider's studies in natural history were secondary to his literary work, but he made several significant contributions to zoology, the most important in ichthyology being his edition of Bloch's *Systema ichthyologiae* of 1801. For more on Schneider, see Adler 1989, 13.]

11. René Just Haüy (1743–1822) became quite famous for his excellent contributions to mineralogy [1801] and his discoveries in crystallography [1822]. He spent his youth in the obscure functions of regent of the lower classes of a college but later became professor at the Muséum National d'Histoire Naturelle, Paris. He took courses under [Louis Jean Marie] Daubenton [1716–1800], who inspired in him a taste for natural history and induced him to work on the *Encyclopédie méthodique* [see below] under his direction and under his name. As a matter of fact, it was not until after Daubenton's death that Lacepède cited Haüy as an author of this dictionary. [On Haüy, see Walsh 1906–17, 1:169–92; for comments on Haüy's knowledge of fishes as demonstrated in the *Encyclopédie méthodique,* see Gill 1872, 36–37.]

[*Encyclopédie méthodique (Dictionnaire encyclopédie méthodique)*, ou par ordre de matières; par une Société de Gens de Lettres, de savans et d'artistes . . . précédé d'un vocabulaire universel, 196 vols. (in 186), in quarto, Paris and Liège, Panckoucke et Plomteux, 1782–1832. For contents and dates of publication of the various parts of the zoological portion, see Sherborn and Woodward 1893, 1899, 1906.]

12. [Pierre Joseph] Bonnaterre [b. ca. 1752], died 1804, priest in the Rouergue and professor at the central school of Rodez (Rhodès), had been assigned by the publisher Panckoucke to take charge of the section of plates in the *Encyclopédie méthodique* dealing with mammals, birds, reptiles, fishes, and insects, as was [Jean Guillaume] Bruguière [1749–98] for worms *(vers)* and [Jean Baptiste Pierre Antoine de Monet de] Lamarck [1744–1829] for plants. Bonnaterre composed an extensive text for the classes he was to present [the part on fishes, "a poor compilation, arranged according to the Linnaean classification, by an individual who . . . availed himself of the works of most of the authors preceding, and collected illustrations of more than 400 species" (Gill 1872, 37)]. His successor in mammals was [Anselme Gaëtan] Desmarest [see chap. 13, n. 26], who was more able than he to fulfill such a mission. For worms *(vers)*, Lamarck succeeded Bruguière, and [Jean Vincent Félix] Lamouroux [born at Agen in 1779, died at Caen in 1825] succeeded Lamarck.

13. Johann Julius Walbaum, physician at Lubeck (1724–1800). The "Bibliotheca" and the "Philosophia" of his *Petri Artedi renovatus* appeared in 1788–89; the "Genera piscium," in 1792; the last two parts ["Synonymia nominum piscium" and "Descriptiones specierum piscium"] in 1793 ["a poor compilation," according to Gill 1872, 38]. Under each genus he places the species described by writers who came after Artedi, and at the end of the volume, the new genera and their species.

14. See the preface of Bloch and Schneider [1801], p. xxi.

15. Johann Friedrich Gmelin, born at Tübingen in 1748, of the same family as the explorers of Siberia, professor of chemistry at Göttingen, died in 1804, author of a multitude of works, gave his name to the thirteenth edition of the *Systema naturae;* but one need only consider that he issued the first seven parts of that work, covering the whole animal kingdom and containing more than four thousand pages, in the space of five years, from 1788 to 1792, to guess that, despite the imperfection of his compilation, he did not work on it alone. [For more on Gmelin and the thirteenth edition of the *Systema naturae*, see Gill 1872, 37–38; Kohn 1992, 39–40.]

TABLE 9. Disposition of the genera in the posthumous system of Bloch [see Bloch and Schneider 1801, vol. 1, *Conspectus operis vel index classium et generum*, xxiii–lx, a compilation, according to Gill (1872, 39), "in which the various species described by authors are collected together, and referred with very little judgment to the genera admitted."]

Classis I. HENDECAPTERYGII
1. *Lepadogaster*

Classis II. DECAPTERYGII
Ordo I. Jugulares
2. *Gadus*
Ordo II. Thoracici
3. *Trigla*
Ordo III. Abdominales
4. *Polynemus*

Classis III. ENNEAPTERYGII
5. *Scomber*

Classis IV. OCTOPTERYGII
Ordo I. Jugulares

6. *Callionymus*	8. *Uranoscopus*	10. *Trachinus*
7. *Batrachus*	9. *Enchelyopus*	11. *Phycis*

Ordo II. Thoracici

12. *Platycephalus*	18. *Mullus*	24. *Monocentris*
13. *Cottus*	19. *Sciaena*	25. *Lonchurus*
14. *Periophtalmus*	20. *Perca*	26. *Macrurus*
15. *Eleotris*	21. *Xiphias*	27. *Agonus*
16. *Gobius*	22. *Zeus*	28. *Eques*
17. *Johnius*	23. *Brama*	

Ordo III. Abdominales

29. *Cataphractus*	32. *Centriscus*	35. *Gasterosteus*
30. *Sphyraena*	33. *Fistularia*	36. *Loricaria*
31. *Atherina*	34. *Mugil*	37. *Squalus*

Classis V. HEPTAPTERYGII
Ordo I. Jugulares

38. *Lophius*	41. *Kyrtus*	44. *Blennius*
39. *Pteraclis*	42. *Trichogaster*	45. *Percis*
40. *Pleuronectes*	43. *Centronotus*	46. *Trichonotus*

Ordo II. Thoracici

47. *Monoceros*	56. *Alphestes*	65. *Epinephelus*
48. *Grammistes*	57. *Ophicephalus*	66. *Anthias*
49. *Scorpaena*	58. *Lepidopus*	67. *Cephalopholis*
50. *Synanceia*	59. *Echeneis*	68. *Calliodon*
51. *Cyclopterus*	60. *Cepola*	69. *Holocentrus*
52. *Amphiprion*	61. *Labrus*	70. *Lutjanus*
53. *Amphacanthus*	62. *Sparus*	71. *Bodianus*
54. *Acanthurus*	63. *Scarus*	72. *Cichla*
55. *Chaetodon*	64. *Coryphaena*	73. *Gymnocephalus*

Continued on next page

Table 9—*continued*

Ordo III. Abdominales		
74. *Acipenser*	83. *Acanthonotus*	92. *Cobitis*
75. *Chimaera*	84. *Esox*	93. *Cyprinus*
76. *Pristis*	85. *Synodus*	94. *Amia*
77. *Rhina*	86. *Salmo*	95. *Poecilia*
78. *Rhinobatus*	87. *Clupea*	96. *Pegasus*
79. *Raja*	88. *Exocoetus*	97. *Mormyrus*
80. *Platystacus*	89. *Chauliodus*	98. *Polyodon*
81. *Silurus*	90. *Elops*	99. *Argentina*
82. *Anableps*	91. *Albula*	
Classis VI. HEXAPTERYGII		
Ordo I. Apodes, i.e., pinnis ventralibus carentes		
100. *Balistes*	101. *Rhynchobdella*	
Ordo II. Pinna anali carentes		
102. *Trachypterus*	103. *Gymnetrus*	
Classis VII. PENTAPTERYGII		
Ordo I. Apodes		
104. *Ophidium*	107. *Ammodytes*	112. *Ostracion*
104a. *Pomatias*	108. *Sternoptyx*	113. *Tetrodon*
104b. *Gnathobolus*	109. *Anarrhichas*	114. *Orthagoriscus*
105. *Muraena*	110. *Channa*	115. *Diodon*
106. *Stromateus*	111. *Sternarchus*	116. *Syngnathus*
Classis VIII. TETRAPTERYGII, APODES		
117. *Trichiurus*	118a. *Taenioides*	
118. *Bogmarus*	119. *Stylephorus*	
Classis IX. TRIPTERYGII		
Ordo I. Apodes		
120. *Gymnonotus*		
Ordo II. Achiri et Apodes		
121. *Synbranchus*	122. *Gymnothorax*	
Classis X. DIPTERYGII		
Ordo I. Apodes		
123. *Ovum*		
Ordo II. Apodes et Achiri		
124. *Petromyzon*	125. *Leptocephalus*	
Classis XI. MONOPTERYGII, Apodes et achiri		
126. *Gastrobranchus*	127a. *Fluta, Addendorum (Monoptère)*	
127. *Sphagebranchus*	128. *Typhlobranchus*	

11 Anatomical and Physiological Contributions of the Eighteenth Century

The zeal for comparative anatomy had diminished at the beginning of the eighteenth century, when physicians rightly recognized that man was best studied by studying man himself, and that in every detail of the structure of a species, the anatomy of another species could be a deceptive guide. Nevertheless, there remained imitators of Duverney, who were making comparative observations on the different organs and who sometimes used those of fishes. Thus Petit, in his research on eyes, noted the proportions of the eyeball of fishes and the almost spherical form of its lens.[1]

Various authors on human anatomy occasionally published illustrations of animal skeletons or their parts. For fishes in particular, Cheselden,[2] in the plates of a treatise on bones, illustrated the skeleton of the ray *(raie)* and the jaws and teeth of the pike *(brochet)*, the parrotfish *(scare)*, and the bonefish *(glossodonte)*.

In addition, there appeared illustrations of skeletons and other interior parts of fishes in some books, such as those by Meyer[3] and by Duhamel and La Marre,[4] which we have already mentioned and that were essentially devoted only to their natural history.

But toward the middle of the century Haller [fig. 36] gave comparative anatomy a new luster with his important applications of it to general physiology.[5] Almost at the same time, Buffon and Daubenton showed that comparative anatomy has no less importance for simple natural history and for distinguishing among animals. Following in their footsteps, such people as Monro, Camper, Hunter, Vicq-d'Azyr, and Scarpa studied the subject from these new points of view and made discoveries that benefited the class of fishes as well as all the other classes, although the ichthyologists of that time, confined within the narrow limits of Linnaean systems, gave little consideration to them.

FIG. 36. Albrecht von Haller (1708–77) from his *Elementa physiologiae corporis humani*, 1757, vol. 1, frontispiece. Courtesy of Colleen M. Weum and the Health Sciences Library and Information Center; photograph by Patricia L. McGiffert, Health Sciences Photography, University of Washington, Seattle.

Thus Haller himself produced excellent descriptions of the eye[6] and the brain[7] of several fishes; above all, he made known the various modes of suspension of the lens and sought to determine the correspondence of the various parts of their encephalon with our own. Pieter Camper,[8] at about the same time, described completely the ear of fishes and gave interesting observations, although incomplete, on the brain of the codfish *(morue)*, the ray *(raie)*, the anglerfish *(baudroie)*, and others, as well.[9]

And yet the strict sectarians of the Linnaean school, fixed on exterior characters only, paid no attention to these discoveries. Who would believe, for example, that, in his anatomical work in 1770 Goüan would still seriously maintain that the brains of these animals had only three lobes and that they possessed neither an internal nor an external ear?[10]

It was not until some years later that Vicq-d'Azyr[11] began to include more fish anatomy in natural history. He introduced their brain and their ear into the comparisons he made of these two organs in vertebrate animals. He also made this class in general the object of a comparative examination, but the division he made of the class, into cartilaginous forms, eellike forms *(anguilliformes),* and bony forms, which he called spiny, proves that he still had only a rather slight knowledge of the class. His illustrations are even better proof. On the other hand, his memoirs contain several interesting observations that had not been made before.

But the primary author on this subject is Alexander Monro the Younger.[12] In his essay on the nervous system,[13] he provided drawings of the brain and part of the nerves of the codfish *(morue).* In his essay on the anatomy and physiology of fishes,[14] he showed the internal organs of these animals—especially the intestines, circulatory system, nervous system, sense organs, mucous ducts—in large, handsome plates. Finally, in his essay on the ear, he gave a complete representation of that of the ray *(raie).*[15]

After these authors, who described all or several parts of the anatomy of fishes, we ought also to mention those who concentrated on one or another of their organs in particular. Their organ of hearing occupied physical scientists no less than it did the anatomists. Klein, as early as 1740, described the otoliths in their ear.[16] Nollet performed experiments in 1743 that proved that fishes can hear underwater.[17] In 1748 Arderon experimented directly on the ability of fishes to hear.[18] In 1755, Geoffroy described the bony labyrinth of the ray *(raie).*[19] Independent of the discoveries of Camper and Monro mentioned above, on the membranous labyrinth of divers fishes, which were subsequent to Geoffroy's memoir, there appeared on this subject in 1783 a memoir by Hunter,[20] in which he claimed to have known about this organ since before 1760 and described for the first time the exterior orifice of the ear of chondropterygian fishes.

In 1789 Scarpa brought out his fine essay[21] on smelling and hearing, representing these organs in fishes in quite handsome plates. Comparetti in the same year published a work[22] on hearing in which he also very carefully described their ear, but his drawings were not very well executed. In 1788 Ebel, in his "neurological observations,"[23] described the brains of several fish species.

Also published were some observations on the teeth of fishes. Hérissant, in 1753, described those of the shark *(squale);*[24] more recently, Andre described those of the wolffish *(anarique)* and the butterflyfish *(chétodon).*[25]

Broussonet wrote in 1785 a memoir on their respiration;[26] Spallanzani commenced important experiments on respiration[27] that were completed by Silvestre,[28] and at the end of this period (in 1795) Fischer drew attention to the relation their swim bladder might have to that function.[29] Before then, in 1776, Erxleben conducted research on the use of this singular organ and the source of the air in it.[30]

Some anatomical descriptions of the species increased our knowledge of their viscera, especially the abdominal viscera.[31] For Hewson, their lymphatic vessels became the object of continuous, painstaking research.[32] Réaumur [fig. 37] revealed the substance that gives color to fish scales, which is used in making artificial pearls.[33] Baster described the scales of some fishes,[34] and Broussonet wrote a special memoir on the subject.[35] The organ found in the snout of certain sharks *(squales)* that secretes abundant mucus was described by Lamorier.[36]

During that century, anatomists and physical scientists frequently studied the electric fishes and the organs by which they exert their singular faculty. In 1714 Réaumur gave some idea of the structure of these organs in the electric ray *(torpille),* but with an entirely erroneous explanation of their effects.[37] The strength of this faculty in the knifefish *(gymnote)* made it possible to form more exact notions about it. Richer tested it as early as 1677 at Cayenne,[38] but Allamand, in 1755, aroused naturalists' interest in the subject when he announced that it resulted from the same cause as the phenomenon of the Leiden jar, which he had just discovered.[39] Adanson advanced the same idea in the catfish *(silure)* in 1757.[40] Lott[41] and Bancroft[42] made Allamand's conjecture seem all the more probable, and Walsh demonstrated it in 1774 in precise experiments performed not only on the knifefish *(gymnote)* but also on the electric ray *(torpille).*[43] At about the same time, Hunter published accurate descriptions of the anatomy of the electric organs of these two fishes.[44] In 1786 Paterson added a species of pufferfish *(tétrodon)* to the list of fishes that have this capacity.[45]

Interested observers were writing about the natural[46] and artificial[47] fecundity of useful species, how long they live,[48] how to feed them[49] and transport them,[50] the harm caused by some fishes,[51] their diseases,[52] and even the castration of fishes.[53] Broussonet made observations on the spermatic vessels.[54] Bloch tried to prove that the curious appendages attached to the pelvic fins of male sharks *(squales)* and rays *(raies)* are not intromittant organs.[55]

The reproduction of the freshwater eel *(anguille)* was a special problem for which a solution was long sought, and which perhaps has not yet been solved. Allen[56] and Dale[57] studied the problem as early as the eighteenth century; in this century, the nineteenth, Vallisneri,[58] Marsigli,[59] Geer,[60] Monti,[61] Mondini,[62] Spallanzani,[63] and several others made it the object of their research.

FIG. 37. René Antoine Ferchault de Réaumur (1683–1757). Engraving by P. Simonneau from a painting by A. S. Belle. Courtesy of Jacqueline Lanson and the Bibliothèque Nationale, Paris.

Cavolini,[64] in his observations on reproduction in fishes, confirmed among other curious facts the habitual hermaphroditism of *Serranus (serran)*, reported by Aristotle. Incidental hermaphroditism was observed in several other species.[65]

NOTES

1. François Pourfour du Petit, born in Paris in 1664, a longtime army physician, member of the Academy in 1722, died in 1741. He published, in *Hist. Mém. Acad. Roy. Sci.* (Paris), 1728, an article in which he described several discoveries made on the eyes of man, quadrupeds, birds, and fishes; and in that same periodical, 1732, a memoir on the crystalline lens in the eye of man, four-footed animals, birds, and fishes. [Duverney's anatomical studies (see chap. 5, n. 19) described in *Hist. Mém. Acad. Roy. Sci.*, Paris, for the year 1701, were not actually published unitl 1743; thus, though his work certainly predated that of Petit (1728, 1732) its actual publication did not.]

2. William Cheselden, famous English surgeon (1688–1752), author of an *Osteography* embellished with handsome plates, London, 1733, large in folio.

3. Meyer [see chap. 8, n. 28] included illustrations of the skeletons of several common fishes [in a collection of 240 plates of various animals, Nuremberg, 1748–56, three volumes in folio].

4. Duhamel du Monceau [and L. H. de La Marre, in their *Traité général des pêsches* published from 1769 to 1782 (see chap. 8, n. 37)] illustrated the skeletons of the common carp *(carpe)* [vol. 2, pt. 2, sec. 1, pl. 3]; the ray *(raie)* [vol. 3, pt. 2, sec. 9, pl. 7, fig. 3]; the electric ray *(torpille)* [vol. 3, pt. 2, sec. 9, pl. 13, figs. 5–6]; and the plaice *(carrelet)* [vol. 3, pt. 2, sec. 9, pl. 12].

5. Albrecht von Haller, poet, botanist, anatomist, almost universal scientist, but famous mainly for his physiological works, born at Bern of a patrician family in 1708, professor at Göttingen from 1736 to 1753, became one of the magistrates of his country, where he died in 1777. The list of his works is immense, but only the ones I [Cuvier] have noted in the text concern our subject.

6. [Haller], in a memoir sent to the Royal Academy of Sciences of Paris in 1762 [but not published until 1764] and in more detail in a memoir addressed to the Royal Society of Göttingen in 1766, reprinted in his *Operum anatomici argumenti minorum,* vol. 3 [1768c].

7. [Haller], in the fourth volume of his *Elementa physiologiae corporis humani* [1762], and in a memoir published by the Dutch Academy of Sciences of Haarlem (1768a) and reprinted in his *Operum anatomici argumenti minorum,* vol. 3 [1768b].

8. Petrus Camper, anatomist, full of genius, and perhaps the man most responsible for making the study of comparative anatomy interesting, on account of his provocative discoveries, was born at Leiden in 1722, professor at Franeker in 1749, at Amsterdam in 1755, at Gröningen in 1763, member of the state council of federated provinces of the Netherlands in 1787, died of pleurisy at The Hague in 1789. He published no great work, but we have a multitude of memoirs by him, included among those of the principal academies. After his death his son, Adriaan [Gilles] Camper [1759–1820], published his anatomical descriptions of the elephant [1802; see Bruggen and Pieters 1990] and the whale [1820], based on his notes and sketches.

9. [Petrus Camper], in a memoir originally published in *Verh. Holl. Maatsch. Wetensch. Haarlem,* vol. 7, 1762, but later reprinted in vol. 1, pt. 2, of a German translation of Camper's works by [Johann Frederik Mauritz] Herbell [1784–90]; and in another memoir sent to the Royal Academy of Sciences, Paris, and printed in 1774. This latter work can also be found in vol. 2, pt. 2, of Herbell's translation [1787a], but neither one is present in the French collection [of Camper's works] published by [Hendrik J.] Jansen [1803].

10. Goüan [see chap. 8, n. 39], in his *Historia piscium* [1770], 2, 79. We should note, however, that the anatomy he gives for this class includes a myology rather new for the time, but the osteology is only roughly sketched.

11. Félix Vicq-d'Azyr, famous physician and anatomist, and brilliant writer, born at Valogne in 1748, secretary of the Royal Society of Medicine in 1773, member of the Royal Academy of Sciences, Paris, in 1774 and of the French Academy in 1788, professor at the veterinary school, died in 1794. He published several memoirs on the brain, including a large work with magnificent plates,

and for the *Encyclopédie méthodique* he began a series of special anatomical descriptions of species taken from various authors [see Vicq-d'Azyr 1792]. Hippolyte Cloquet [1787–1840] continues that work [with contributions appearing in 1823 and 1830]. The writings we refer to in the text are (1) two memoirs for "use in the anatomical natural history of fishes" (*Mém. Math. Phys. Acad. Sci.* [Paris], vol. 7, 1776a, b); and (2) a memoir on the structure of the brain in animals, compared with the brain of man (*Hist. Mém. Acad. Roy. Sci.* [Paris], 1786). All three articles are reprinted in the collection of Vicq-d'Azyr's works [published in 1805] by [Jacques Louis] Moreau de la Sarthe [1771–1826] (5:165 ff.); the first volume of this collection features a biography and a selected bibliography of Vicq-d'Azyr [pp. 1–88], drawn up with care by the editor.

12. Alexander Monro the Elder, born in London in 1697, professor at Edinburgh, died in 1767, left a short essay on comparative anatomy published [in 1744; a revised edition appeared] after his death [1783a]. However, the man we are speaking of in the text is his son (1733–1817), also named Alexander and professor at Edinburgh.

13. [Monro], *Observations on the Structure and Functions of the Nervous System*, Edinburgh, 1783b, in folio.

14. [Monro], *The Structure and Physiology of Fishes Explained, and Compared with Those of Man and Other Animals*, Edinburgh, 1785, in folio. There is a German translation by Schneider [see Monro 1787].

15. [Monro], "Observations on the Organ of Hearing in Man and other Animals" [the third of three treatises, on the brain, the eye, and the ear, published together], Edinburgh, 1797, in quarto.

16. [Jacob Theodor Klein, "De lapillis eorumque numero in craniis piscium"], in the first part of his *Historiae piscium naturalis promovendae* [1740, 9–23, pls. 1–3], mentioned above [chap. 8, n. 34].

17. [Jean Antoine Nollet, 1700–1770; his experiments were apparently performed in 1743 but not published until 1746 by the] Royal Academy of Sciences, Paris.

18. [William Arderon (1703–67), extracted from a letter to Mr. Henry Baker (1698–1774), concerning the hearing of fishes; Arderon's experiments were apparently conducted in 1748, but the results were not published until 1750], *Phil. Trans. Roy. Soc. Lond.*, vol. 45. [On Arderon and Baker, see Whalley 1971.]

19. [Etienne Louis Geoffroy, 1727–1810] in a memoir on the ear of reptiles published by the Royal Academy of Sciences, Paris [1755].

20. [John Hunter (1728–93), Scottish surgeon and anatomist, brother of the famous anatomist and physician William Hunter (1718–83), in a memoir published in the] *Phil. Trans. Roy. Soc. Lond.*, vol. 72 [1783], and reprinted in his *Observations on Certain Parts of the Animal Oeconomy* [1786]; a second edition of the later work appeared in [1792]. [On Hunter, see Paget 1897; Dobson 1969.]

21. *Anatomicae disquisitiones de auditu et olfactu*, Ticini, 1789, in folio, by Antonio Scarpa [1747–1832], professor at Pavia and one of the most capable anatomists of recent times.

22. *Observationes anatomicae de aure interna comparata*, Padua, 1789, in quarto, by Andrea Comparetti [1745–1801], professor at Padua.

23. *Observationes neurologicae ex anatome comparata* by [Johann Gottfried] Ebel [1764–1830], Frankfurt an der Oder, 1788, in octavo, reprinted in *Scriptores neurologici minores,* vol. 3 [1791–95], of [Christian Friedrich] Ludwig [see Ebel 1793].

24. François David Hérissant, capable anatomist and member of the Royal Academy of Sciences of Paris, was born at Rouen in 1724 and died in 1773. [His research on shark's teeth was published in 1753.]

25. [William Andre], *Phil. Trans. Roy. Soc. Lond.*, vol. 74 [1784].

26. [Broussonet, see chap. 9, n. 11] *J. Physique* [Paris], vol. 31 [apparently submitted in 1785 but not published until 1787b]. It was later reprinted in *Hist. Mém. Acad. Roy. Sci.* (Paris) [1788c].

27. [Lazzaro Spallanzani, 1729–99], in his works on respiration [which were not published until after his death (1803, 1807)]. [On Spallanzani, see Walsh 1906–17, 3:115–46.]

28. [Augustin François de Silvestre, 1762–1851], *Bull. Sci. Soc. Philomat.* (Paris), vol. 1 [1791].

29. [Gotthelf Friedrich Fischer von Waldheim, a German zoologist born at Waldheim, Saxony, in 1771, died at Moscow in 1853, received his doctor of medicine degree in 1798 and became professor of natural history and librarian at the Centralschule at Mainz. From 1804 on, he was professor of natural history at the University of Moscow and director of its natural history museum. He later became president of the Moscow Imperial Society of Naturalists. His numerous zoological publications included works on the insects of Russia, fossil mollusks, brachiopods, fossil fishes, the swim bladder of fishes, and comparative anatomy. For more on Fischer von Waldheim, see Mearns and Mearns 1988, 151–53; Gould 1993.] His *Versuch über die Schwimmblase der Fische* was published at Leipzig, 1795, in octavo.

30. [Johann Christian Polycarp Erxleben, 1744–77], in his *Physikalisch-Chemische Abhandlungen* [1776].

31. Koelreuter [see chap. 8, n. 21], whose ichthyological memoirs included many observations on splanchnology, printed from 1763 to 1795 in various volumes of the Academy of Sciences of St. Petersburg, published separately in *Novi Comment. Acad. Sci. Impér. Petro.* (St. Petersburg), his observations on the viscera of the beluga sturgeon *(hausen),* vol. 16 [1772], and the sterlet *(sterlet),* vol. 17 [1773]. Steller [see chap. 9, n. 21] carefully dried the fishes he collected, and descriptions excerpted from his papers, either in the volumes of the Academy of Sciences of St. Petersburg [1751, 1752] or in Pallas's *Zoographia Rosso-Asiatica,* vol. 3 [1814], offer very good observations on fish splanchnology.

32. William Hewson, London surgeon [b. 1739], died in 1774. [His work on the lymphatic system of fishes was first published in English], *Phil. Trans. Roy. Soc. Lond.,* vol. 59 [1770]; and [later in French] in the *J. Physique* [Paris], vol. 1 [1772].

33. René Antoine Ferchault de Réaumur, intendant of the Order of Saint-Louis, member of the Royal Academy of Sciences, Paris, a scholar in all fields, but famous especially for his admirable memoirs on insects, was born at La Rochelle in 1683 and died in Paris in 1757 [see Allen 1951, 498–99]. [His work on fish scales was presented to the] Royal Academy of Sciences, Paris, in 1716 [but not published until 1718.]

34. [Job Baster, 1711–75] in [a paper titled "De squammis piscium," part of] his *Opuscula subseciva* [vol. 1, pt. 3, pp. 129–37, pl. 15, 1759–65]; [the work was later reprinted in] *Verh. Holl. Maatsch. Wetensch. Haarlem*, vol. 6 [1762].

35. [Broussonet, see chap. 9, n. 11], *J. Physique* (Paris), vol. 31 [1787a].

36. [Louis Lamorier, French surgeon and naturalist born at Montpellier in 1696, died 1777, was a member of the Royal Society of Sciences in Montpellier and an associate member of the Royal Academy of Surgeons in Paris. His work on a mucus-secreting gland of the snout of sharks was presented to the] Royal Academy of Sciences of Paris in 1742 [but not published until 1745].

37. [Réaumur, see n. 33 above] on the electric ray *(torpille)*, [presented to the] *Hist. Mém. Acad. Roy. Sci.* (Paris), for the year 1714 [but not published until 1717].

38. [Jean Richer, 1630–96, published his work on the electric organs of the knifefish in 1679, at Paris, in folio.]

39. Jean Nicolas Sébastien Allamand, professor of physical science and natural history at Leiden (1713–87), well known, apart from his discoveries in electricity, for the supplements he provided for the animals of Buffon's *Histoire naturelle* [see chap. 8, n. 36]. [For more on Allamand and his additions to Buffon's *Histoire naturelle*, see Rookmaaker 1992.]

40. [Michel Adanson, on the *poisson trembleur*] in his *Histoire naturelle du Sénégal* [1757], 134–35. [Adanson, a philosopher and voyager-naturalist, born in Aix-en-Provence in 1727, lived from 1749 to 1753 in Senegal, where he gathered plants and animals as well as abundant information on Senegalese meteorology, cartography, and linguistics. On his return to France, living at first near the Jardin des Plantes and later, until 1772, at the Grand Trianon, where he was named king's botanist, he published numerous works. As resident member of the Royal Academy of Sciences, Paris, he was known and appreciated all over Europe. In 1760 a chair in natural history was proposed for him at the University of Louvain; in 1766 the empress of Russia invited him to teach at the Academy of Sciences of St. Petersburg; and in 1779 he traveled for six months, collecting plants in the south of France, Spain, Italy, and Switzerland. Married in 1770, he was separated from his wife in 1785, who left with her daughter for England during the Revolution. He had other disappointments, for example, not replacing Buffon at the Jardin des Plantes. He died in 1806. For more on Adanson, see Chevalier 1934; Bertin 1950.]

41. [Frans van der Lott], *Verh. Holl. Maatsch. Wetensch. Haarlem*, vol. 6 [1762].

42. [Edward Bancroft, 1744–1821] in his *Naturgeschichte von Guiana in Süd-Amerika* [Frankfurt and Leipzig, 1769, and a London edition of the same year, both in octavo].

43. [John Walsh (1725?–95), "Of the Electric Property of the Torpedo," in a letter to Benjamin Franklin that appeared in the] *Phil. Trans. Roy. Soc. Lond.*, vol. 63 [1774].

44. [Hunter (see n. 20 above), in two articles published separately, 1774 and 1775], *Phil. Trans. Roy. Soc. Lond.*, vols. 63 and 65.

45. [William Paterson (born 1755, died at sea in 1810), "An Account of a New Electrical Fish," in a letter to Sir Joseph Banks published in] *Phil. Trans. Roy. Soc. Lond.*, vol. 76 [1786]; and [later reprinted in] *J. Physique* (Paris), vol. 30 [1787]. [On Paterson and his zoological contributions, see Rookmaaker, 1989, 163–76.]

46. Anders Hellant, *Kongl. Svenska Vetensk. Acad. Handling.* (Stockholm), vol. 6, 1745; [W.] Grant, *Königl. Schwed. Akad. Wissenschaft. Abhandl.*, vol. 14, 1752; and Ferris, *J. Physique* (Paris), vol. 20 [1782], [all writing] on the reproduction of salmon *(saumons).* [Abraham] Argillander, on the fecundity of the pike *(brochet), Kongl. Svenska Vetensk. Acad. Handling.* (Stockholm), vol. 14, 1753. [Martinus] Houttuyn [b. 1720], on the reproduction of sharks *(squales), Uitgez. Verhandl. Soc. Wetensch. Europa* (Amsterdam), vol. 9 [1764]. [Giovanni Antonio] Battarra [1714–89], on the reproduction of rays *(raies), Atti Accad. Sci. Siena Fisio-Crit.*, vol. 4 [1771]. Thomas Harmer [1715–88], on the fecundity of fishes, *Phil. Trans. Roy. Soc. Lond.*, vol. 57 [1768].

47. [Johann Gottlieb] Gleditsch [1714–86], on the artificial fecundation of trout and salmon, published by the Royal Academy of Sciences, Berlin, 1766.

48. [Friederich Heinrich Wilhelm] Martini [1729–78], on the age of fishes [1776]; Hans Hederström [b. 1710], also on the age of fishes, 1759; [Ernst Gottfried] Baldinger [1738–1804], on the age of a pike *(brochet)* [1802] (it was said to have lived 267 years).

49. [Johann Reinhold] Forster [see chap. 9, n. 13], [in a letter to Daines Barrington (1727–1800), vice president of the Royal Society of London], on the method of raising carp in Polish Prussia, *Phil. Trans. Roy. Soc. Lond.*, vol. 61 [1772].

50. [Herrn von Marwitz] on transporting fishes, *Beschäft. Berl. Ges. Naturf. Fr.*, vol. 4 [1779].

51. Martini [see n. 48 above], [on the harm caused by some fishes], *Berl. Samml.*, vol. 7 [1775]; [William] Anderson (1750–78), on poisonous fishes, *Phil. Trans. Roy. Soc. Lond.*, vol. 66 [1777].

52. Anton Rolandsson Martin [1729–86], on diseases of the skin *(gale)* of fishes, *Kongl. Svenska Vetensk. Acad. Handling.* (Stockholm), vol. 21, 1760; and on worms in fishes, *Kongl. Svenska Vetensk. Acad. Handling.* (Stockholm), vol. 31, 1771. [Joseph] Beckmann, on tapeworms in fishes, *Hannov. Mag.*, 1769. [Cuvier says that Beckmann's paper describes *la fic des poissons*, or the "warts" of fishes, but Beckmann's subject is clearly *Fiecks* (or *Fieks) in Fischen*, tapeworms in fishes; see Bloch 1782.]

53. [Samuel] Tull, on the method of castrating fishes, [originally published] in *Hist. Mém. Acad. Roy. Sci.* (Paris), 1745 [but later reprinted] in *Phil. Trans. Roy. Soc. Lond.*, vol. 48 [1755].

54. [Broussonet (see chap. 9, n. 11), in a paper presented to the] Royal Academy of Sciences of Paris in 1785 [but not published until 1788b].

55. [Bloch, see chap. 10, n. 1], *Schr. Berl. Ges. Naturf. Fr.*, vols. 6 and 8 [1785, 1788a].

56. [Benjamin Allen (1663–1738), on "the manner of the generation of eels"], *Phil. Trans. Roy. Soc. Lond.*, vol. 19, 1698.

57. [Samuel Dale (1659–1739), "An Account of a Very Large Eel"], *Phil. Trans. Roy. Soc. Lond.*, vol. 20, 1699.

58. [Vallisneri (see chap. 5, n. 21), "Dissertationem de ovario anguillarum," originally published in 1712] in *Misc. Curio. Medico-Physica*, centur. 1–2 [but later reprinted] in his collected works, 2:89 [1733].

59. [Marsigli, see chap. 6, n. 20], *Gior. Letterati d'Italia* (Venice), vol. 29 [1718]; reprinted in *Act. Vratisl.* 5:1690 [not found as cited; date unknown].

60. [Carl de Geer, 1720–78], *Kongl. Svenska Vetensk. Acad. Handling.* (Stockholm), vol. 11, 1750.

61. [Gaetano Lorenzo (or Cajetani) Monti, professor of botany at Bologna University, 1712–97; see Heniger 1986, 106–8, 123], *Comment. Bonon. Sci. Inst. Acad.* (Bologna), vol. 6 [1783].

62. [Caroli Mondini, "De anguillae ovariis," reprinted in] *Comment. Bonon. Sci. Inst. Acad.* (Bologna), vol. 6 [1783].

63. [Spallanzani, see n. 27, above] in vol. 6 of the French edition [1800] of his *Travels in the Two Sicilies [and Some Parts of the Apennines,* originally published in Italian, 1792–97; there is also an English translation of 1798].

64. Filippo Cavolini [1756–1810], *Memoria sulla generazione dei pesci e dei granchi,* Naples, 1787, in quarto; translated into German by [E. A. W.] Zimmermann, Berlin, 1792, in octavo.

65. In the common carp *(carpe),* by Alischer, *Samml. Nat. Med.* (Leipzig), 1725 [this article, "Von einigen merckwürdigen Fischen Curland. IV. Die unvermuthete Karpffen," was actually written by Samuel Joannes Rhanaeus]; and by [Franz Ernst] Brückmann [1697–1753], *Commerc. Litter. Med. Sci. Nat.* (Nuremburg), 1734. In the codfish *(morue)* by [apparently Heinrich Friedrich] Link [1767–1851], *Act. Vratisl.* 18:617 [not found as cited; date unknown].

Part Five
COLLECTION BUILDING

12 Lacepède and His Immediate Successors

Such was the progress ichthyology had made by the end of the eighteenth century, at which time Lacepède began to make it the object of his studies.[1] During the 120 years after Willughby, ways had been discovered of firmly establishing the nomenclature of fishes; a considerable number of fish species had become established by detailed descriptions and accurate drawings; highly diverse methods of distribution had been attempted for classifying them; almost all their organ systems had been studied by capable anatomists; and close observations had been made of their habits and economy. It is not to be doubted that this eloquent writer—who had conceived the plan of his book in a grand and elevated manner and who had a talent for finding the appealing side of the story of these organisms that seem to touch us so little and do not arouse our imagination—it is not to be doubted, I say, that he would have built an impressive monument had he found himself in more favorable circumstances. But writing his book during the stormiest years of the Revolution, when France was separated from neighboring states by a cruel war, he could not profit from the wealth of material in foreign works. Even the great ichthyology of Bloch, that capital work that was finished by the time Lacepède began to publish his own work,[2] was not yet available to him in its entirety, and it was not until the fourth of his volumes that he began to cite the last six volumes of the ichthyologist from Berlin. Likewise Bloch himself, while composing his *Systema ichthyologiae,* which was published after his death, and even his editor Schneider, had knowledge of only the first two volumes of Lacepède's work. These circumstances must be borne in mind when comparing the works of these two famous ichthyologists.

Another difficulty no less great, at a time when we had lost all our colonies, and none of our ships ventured across the seas, was that of procuring fishes from distant waters and examining them in a state of nature.

The French naturalist thus found himself obliged to accept as foundation for his work the lists of fishes drawn up by Gmelin and Bonnaterre, and it is from these that he took the characters of his divisions and of most of his genera, in the meantime adding species from various sources. The king's cabinet furnished him some; he found some in the cabinet of the stadholder, which was brought to Paris in 1795.[3] A few were given him by Le Blond, a physician at Cayenne.[4] Bosc, a learned naturalist who had been consul at New York for some time, sent drawings made in that country.[5] Other individuals, especially Noël[6] and Mesaize of Rouen,[7] sent drawings and notices on fishes that had come to them by chance and that seemed remarkable to them; but his most abundant materials came from the manuscripts of Commerson and the drawings made under the supervision of that observer, to which he added those that Aubriet had copied from Plumier's manuscripts for the vellum collection. Unfortunately, as mentioned above, he was not able to profit from the fishes themselves that Commerson had sent with his drawings and that had remained unknown after Buffon's death.

These materials were not all of the same value. The men who provided him information were by no means all professional ichthyologists. The copist Aubriet had altered the originals in more than one place, and the originals themselves had often omitted essential characters. Commerson's drawings were not always verified against his descriptions, and Lacepède often made one species from the description and another from the drawing; and it is difficult to believe, but it also happened more than once that he made yet another species from the descriptive phrase written on the drawing in question. These strange aberrations can be explained only by the fact that he composed his articles in the countryside where the Terror had banished him, far from the papers he had consulted and with only notes that he had made of them, and also by the fact that he named the fishes engraved on his plates according to what he believed he recognized and not from what was written on the original drawing, which he no longer had in front of him.

Like many other naturalists, Lacepède was liable not to recognize certain species that earlier authors had already described, either because the organisms had lost their color or their shape, or because the descriptions themselves had been based on altered organisms, or simply because he had not sufficiently taken into consideration the descriptive terms.

The result was that to the duplications that already existed in the published classifications of various authors, which he consulted with too much confidence, he added a large number of others, and the total number of his species (1,463) should be reduced by more than 200; but his multiplication of genera has contributed even more mightily to the confusion in his work.

The primary source of these multiplications was again in not sufficiently comparing the figures and the descriptions in the Commerson manuscripts. One genus based on a note from that traveler next reappeared under another name according to his drawings, and often it was reproduced a third time according to some other naturalist.[8]

The implicit confidence he had in all these predecessors was another source of these imaginary genera. Whenever Brünnich, Houttuyn, Forskål, or Gmelin placed a fish in one of their genera, Lacepède, believing they could not be mistaken, supposed the fish had all the characteristics common to that genus,[9] and then finding in its particular description some trait that seems useful for distinguishing it, from his first supposition and the species character he composed his new genus character. It sometimes happened that he created new genera from fishes he observed in nature without noticing that they were already in his book, taken from other authors and represented under other names.[10]

Moreover, the motives that made him remove some species from a genus, or leave them in, were subject to strange variations. For example, he left the anchovies *(anchois)* in the genus *Harengus (harengs)*, although they have none of the characteristics he assigned to this genus, and he distinguished *Clupanodon (clupanodons)*, which are very little different from the other herrings *(harengs)*.

The general classification established by Lacepède is that of Pennant, into cartilaginous and bony fishes, with the subdivisions of Linnaeus based on the position of the pelvic fins applied to both; but between these two partitions he intercalates another one, based on the presence or absence of opercula and branchiostegal rays. Even had the facts been consistently followed, this intercalated division would still have the disadvantage of being artificial, because it would, for example, separate the moray eels *(murènes)* and the swamp eels *(synbranches)* from the freshwater eels *(anguilles)*; but what is more objectionable is that the characters assigned to the classes are not always to be found in the fishes placed in them. Thus the anglerfishes *(baudroies)*, the triggerfishes *(balistes)*, and the elephantfishes *(mormyres)* have opercula, although Lacepède supposed the contrary; and there are opercula and branchiostegals in the moray eels *(murènes)*, the swamp eels *(synbranches)* and the other genera separate from freshwater eels *(anguilles)*, although he denies it [see table 10].

The natural history of fishes that Sonnini published after his edition of Buffon, and to which he gave his own name and even his portrait, is almost a word-for-word copy of Lacepède's, with preliminary articles taken from Artedi on the authors of ichthyology and on terminology, and with memoirs by Duverney and Broussonet on various parts of their organization.[11]

Philippe Loos has begun a translation into German of the work of Lacepède, Berlin, in octavo.[12]

Such as it is, this natural history of fishes by Lacepède forms an epoch also in ichthyology, and it has served, along with the great work by Bloch, as the main basis for what has been written on this science up to the present time.

That cannot be said in all truth about the natural history of the fishes of Vizagapatam by Russell.[13] It was published in 1803, but because it was written in India several years before, it still followed Pennant's classification, the only thing added being a genus taken from Bloch. But despite some rash placements of species, it is incontestably the most important work we have on fishes in the oriental seas, and the English East India Company, in ordering its publication, has acquired a right to the gratitude of naturalists. It contains two hundred species, accurately drawn by an artist of the country and carefully engraved in England. There is hardly a drawing that does not sufficiently present all the characters needed for determining the species, and even for placing it in its genus or subgenus. In the text, the author added a description of principal colors and interesting facts of life history.

The influence of Lacepède is more noticeable in the part on fishes in the *General Zoology* by Shaw [fig. 38].[14] This is hardly anything more than a development of Gmelin's system augmented by species taken from Bloch and Lacepède along with a few of their genera, the others being based on those of Gmelin. Moreover, there is no criticism of either the duplications or the incorrect placement of the species. A large number of the new species are distributed among the old genera in a fashion that may be qualified as absurd. For example, within the genus *Sparus (spares),* Shaw placed some species of *Serranus (serrans), Crenilabrus (crénilabres), Girellus (girelles),* and *Sciaena (sciènes);* within *Labrus (labres)* he included some species of *Sciaena (sciènes),* perches *(perches),* and so on. Most of the figures are copied from Lacepède and Bloch, except for perhaps five or six taken from fishes that Shaw examined in the British Museum and of which he had custody [fig. 39]. One of these fishes forms a new genus, *Stylephorus,* but it would be difficult to determine its affinities from what the author says about this genus and the drawing he gives, which is of a deteriorated individual.[15]

The two general works written at this time by Duméril for student use, his *Traité élémentaire d'histoire naturelle* of 1804[16] and his *Zoologie analytique* of 1806, worked together to make Lacepède's work more popular in facilitating the determination of his genera. The latter work in particular shows the genera in synoptic tables and distributes them

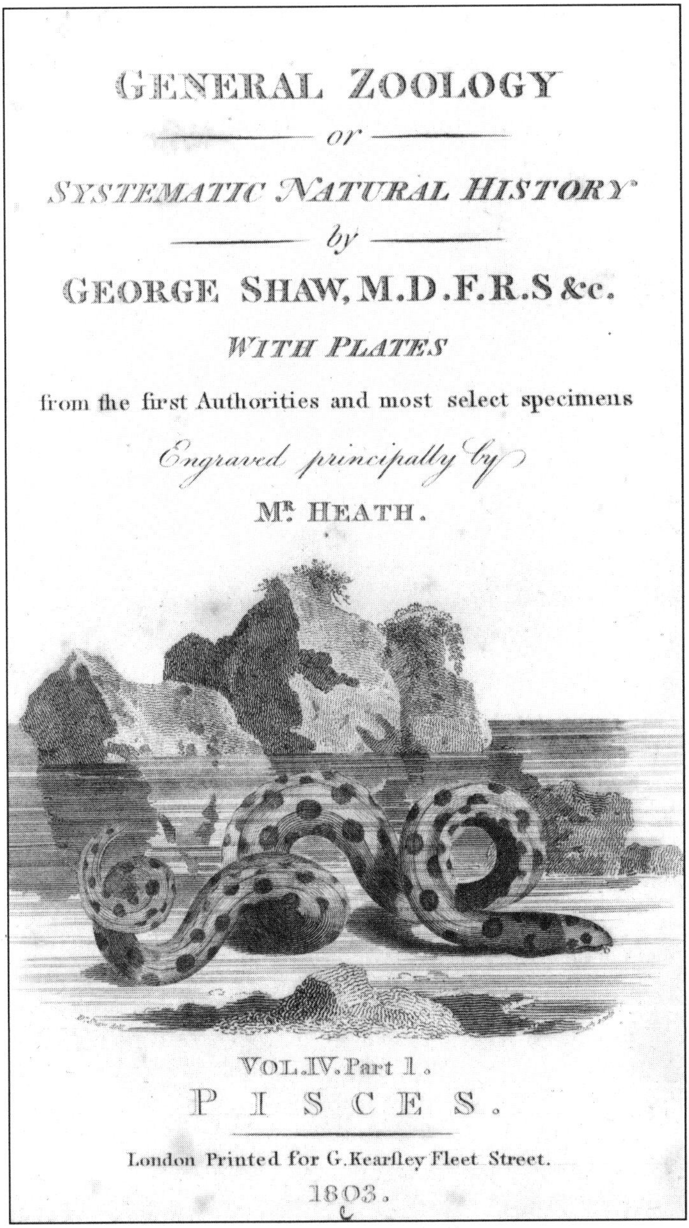

GENERAL ZOOLOGY

or

SYSTEMATIC NATURAL HISTORY

by

GEORGE SHAW, M.D.F.R.S &c.

WITH PLATES

from the first Authorities and most select specimens

Engraved principally by

M^R^. HEATH.

VOL.IV.Part 1.

PISCES.

London Printed for G.Kearsley Fleet Street.

1803.

FIG. 38. Title page of George Shaw's *General Zoology*, 1800–1826, vol. 4, pt. 1. T. W. Pietsch collection.

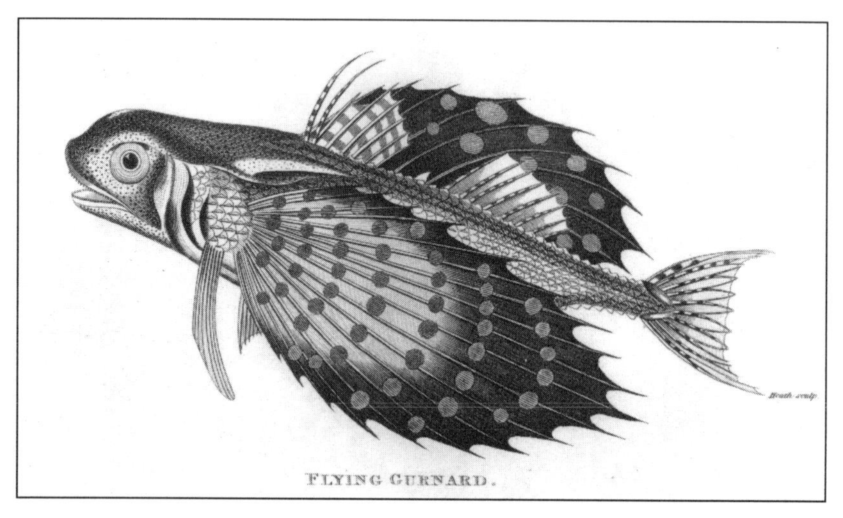

FIG. 39. George Shaw's flying gurnard, *Dactylopterus volitans,* from his *General Zoology,* 1800–1826, vol. 4, pt. 2, pl. 91 (copied by Shaw from Bloch's *Naturgeschichte der ausländischen Fische,* 1785–95, vol. 7, pl. 351). T. W. Pietsch collection.

among orders and families with precisely fixed distinctions [fig. 40]; however, he too used for a foundation characters based on the alleged absence of opercula and [branchiostegal] rays which Lacepède had proposed [see table 11].

From that time the French ichthyologist's work was used by the authors of several special treatises as the basis of their work. For example, Delaroche,[17] a young naturalist taken away from science too early, collected many fishes at Ibize, Majorca, and Bayonne and in 1809 published a catalog of 105 species according to Lacepède's system, with notes on their habits and utility. He added detailed descriptions of 32 species and drawings of 18 that were either new or incompletely determined at that time.[18] This work, which included his observations on the swim bladder, was an important addition to ichthyology not only because of its great accuracy but also because the author deposited his specimens in the king's cabinet, where their characters can still be verified.

The first edition of the *Ichthyologie de Nice* by Risso, published in 1810, was also arranged, overall and in detail, according to Lacepède's classification, even in its errors;[19] for the moray eels *(murènes)* are again stated to lack opercula and branchial membranes. But this work was interesting because it contained information on a great number of Mediterranean fishes almost forgotten since Rondelet, or even unknown to that earlier naturalist, and also because of the new and precise details it gave of their habits. It describes 317 species according to observations from nature

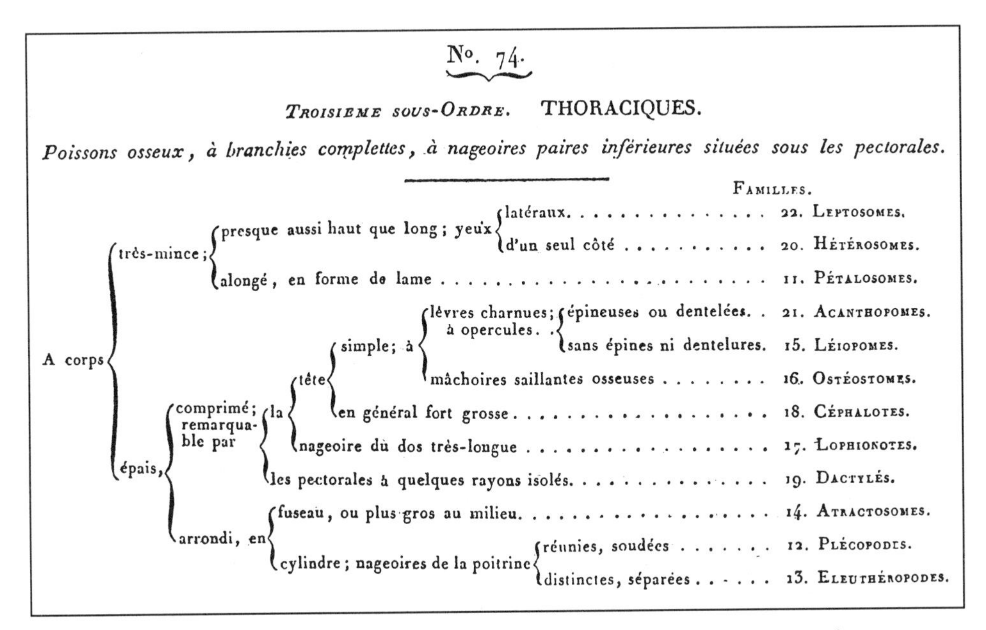

No. 74.

TROISIEME SOUS-ORDRE. THORACIQUES.

Poissons osseux, à branchies complettes, à nageoires paires inférieures situées sous les pectorales.

Key	Familles.
A corps très-mince; presque aussi haut que long; yeux latéraux	22. LEPTOSOMES.
A corps très-mince; presque aussi haut que long; yeux d'un seul côté	20. HÉTÉROSOMES.
A corps très-mince; alongé, en forme de lame	11. PÉTALOSOMES.
A corps épais, comprimé; remarquable par la tête simple; à lèvres charnues; à opercules épineuses ou dentelées	21. ACANTHOPOMES.
A corps épais, comprimé; remarquable par la tête simple; à lèvres charnues; à opercules sans épines ni dentelures	15. LÉIOPOMES.
A corps épais, comprimé; remarquable par la tête simple; à mâchoires saillantes osseuses	16. OSTÉOSTOMES.
A corps épais, comprimé; remarquable par la tête en général fort grosse	18. CÉPHALOTES.
A corps épais, comprimé; remarquable par la nageoire du dos très-longue	17. LOPHIONOTES.
A corps épais, comprimé; remarquable par les pectorales à quelques rayons isolés	19. DACTYLÉS.
A corps épais, arrondi, en fuseau, ou plus gros au milieu	14. ATRACTOSOMES.
A corps épais, arrondi, en cylindre; nageoires de la poitrine réunies, soudées	12. PLÉCOPODES.
A corps épais, arrondi, en cylindre; nageoires de la poitrine distinctes, séparées	13. ELEUTHÉROPODES.

FIG. 40. A dichotomous branching of families of the "Thoraciques," one of forty-one synoptic tables used by André Marie Constant Duméril to illustrate his classification of fishes, as presented in his *Zoologie analytique, ou Méthode naturelle de classification des animaux*, 1806, pp. 97–153.

made by the author, who deposited the most interesting specimens at the Muséum National d'Histoire Naturelle, Paris. Several are new, of which 40 are represented, but the nomenclature of the established species is not always free of error, for it was in fact difficult to verify the nomenclature using such a work as that of Lacepède.

Risso presented to the Institut de France in 1820 a supplement that contained additional interesting species[20] and described others in the memoirs of the Academy of Sciences of Turin.[21] He then worked on a new edition of his ichthyology in which he sought to profit from the progress of the science, and he added several more new species; it is included in volume 3 of a general work that he published under the title *Histoire naturelle de l'Europe méridionale*.[22] We shall discuss elsewhere the method he used in arranging it.

NOTES

1. Bernard Germain Etienne de la Ville, the comte de Lacepède, born at Agen in 1756, curator of the king's cabinet in 1785, professor at the Muséum National d'Histoire Naturelle in 1795, member of the Institut de France in 1796, of the

senate in 1800, grand chancellor of the Legion of Honor in 1802, died in 1826. Lacepède was a writer of great eloquence and man of great benevolence who published works on music, general physical science, electricity, and the natural history of oviparous quadrupeds and snakes [1788–89], fishes [1798–1803], and whales [1804b] as a sequel to Buffon's great natural history of viviparous quadrupeds [1753–65] and birds [1770–83], the materials of which were assembled partly by [Louis Jean Marie] Daubenton, [P.] Guéneau de Montbeillard, and [G. L. C. A.] Bexon. Lacepède's *Histoire naturelle des poissons* was printed in five volumes in quarto [1798–1803] and in fourteen volumes in duodecimo [1799–1804a]. [On Lacepède, see Cuvier 1827a; Swainson 1827; Gill 1872, 38–39; Appel 1973; Adler 1989, 14; Bornbusch 1989.]

2. The twelfth volume of the French edition of Bloch's *Histoire naturelle, générale et particulière des poissons* was published in 1797. The first volume, in quarto, of the *Histoire naturelle des poissons* by Lacepède was published in 1798; the second in 1800; the third in 1801, the fourth in 1802; and the fifth in 1803.

3. [In January 1795, after the French revolutionary army, under the command of General Charles Pichegru (1761–1804), invaded the Netherlands in the abnormally cold winter of 1794–95 (advancing over the frozen rivers, which then did not present the usual formidable barriers against invasion), the stadholder and his family fled to England (see Edmundson 1922, 342–43). The French army then confiscated the natural history collection of the prince and transported it to Paris to enrich the Muséum National d'Histoire Naturelle, where the first shipment, composing about half of the collection and packed in ninety-five crates, arrived in October 1795 (Boyer 1971, 397; Pieters 1980, 541–42). In 1815, shortly after the fall of Napoléon Bonaparte (1769–1821), professor of natural history Sebald Justinus Brugmans (1763–1819) of Leiden University was sent to Paris to retrieve the stolen material, but he met with strong opposition, especially from Jean Baptiste de Lamarck (1744–1829), who at the time was writing his *Histoire naturelle des animaux sans vertèbres* (published between 1815 and 1822) and did not want to release the invertebrates. Many other objects, particularly the stuffed giraffe, were highly valued by the French, and they argued strongly for their retention. Moreover, the prince's collection had been integrated into the Paris museum in such a way that most of the specimens were no longer recognizable as having belonged to the stadholder (Pieters 1980, 542). Finally, after long negotiations, Alexander von Humboldt (see chap. 15, n. 58) intervened and suggested a compromise: in exchange for leaving a large part of the Dutch collection in Paris, Brugmans was given some ten thousand duplicate items from the Muséum National d'Histoire Naturelle (Hamy 1906, 23). This compensatory material, along with that part of the stadholder's collection that had been retrieved earlier, was presented by his son King William I to Leiden University, and from there it ultimately made its way into the Rijksmuseum van Natuurlijke Historie at Leiden after its foundation in 1820 (see Gijzen 1938, 5–21, 26–27, 46; Boeseman 1970, 184–87; Pieters 1980, 543).]

4. [Jean Baptiste Le Blond, born at Toulongeon, Saône-et-Loire, in 1747, traveled to Martinique in 1766, ascended the Orinoco, and continued on to Peru. Returning to France in 1785, and appointed physician-botanist in 1786, he left

again for the New World, where he remained at Cayenne until 1802. He died in 1815.]

5. [Louis Augustin Guillaume Bosc d'Antic, born at Paris in 1759, sailed as French consul to Carolina, where for two years he collected an immense amount of natural history material. Returning to France, he was named inspector of the gardens and nurseries of Versailles in 1803 and of those belonging to the ministry of the interior in 1806. He died at Paris in 1828.]

6. [Noël is probably Simon Barthélémy Joseph Noël de la Morinière, 1765–1822; see chap. 16, n. 67.]

7. [Pierre François Mesaize, a medical officer at Rouen, was a resident member of the Rouen Emulation Society in 1799. He sent several fishes from the area around Rouen, along with observations that Lacepède used in his natural history of fishes.]

8. For example, a drawing by Commerson is reproduced [by Lacepède] under the name *Synode renard* [or *Synodus vulpes;* 1798–1803, 5:321, pl. 8, fig. 2]; a note inscribed on this drawing gives reason for establishing the genus *Butirin* and the species *Butirin banane* [see Lacepède 1798–1803, 5:46]. Another drawing of the same species found in the manuscripts of Plumier appears under the name *Clupée macrocéphale* [or *Clupea macrocephala;* Lacepède 1798–1803, 5:458, pl. 14, fig. 1]. This drawing by Plumier reappears in Schneider's edition of Bloch under the name *Albula plumieri* [see Bloch and Schneider 1801, 432–33, pl. 86]; and neither Schneider nor Lacepède perceived that this fish is the same as *Argentina glossodonta* that they adopted from Forsskål [1775a, 68].

9. Thus it was that *Centriscus scolopax* of Linnaeus [1766–68, 1:415], mistaken by Forsskål [1775a, xiii, 66] for a catfish *(silure)* and named *Silurus cornutus,* became under Lacepède [1798–1803, 5:136] the genus *Macroramphosus (macroramphose).*

10. Such a case is his genus *Pogonias* [Lacepède 1798–1803, 3:137], which does not differ, not even on the species level, from *Pogonathus (pogonathe),* which he [Lacepède 1798–1803, 5:120] presents according to Commerson.

11. [Charles Nicolas Sigisbert Sonnini de Manoncourt, 1751–1812], *Histoire naturelle, générale et particulière des poissons,* Paris, 1803–4, thirteen volumes in octavo.

12. [Philippe Werner Loos (1754–1819), *Naturgeschichte der Fische, als eine Fortsetzung von Buffons Naturgeschichte*]. Pts. 1 and 2 of the first volume were published in 1799; pt. 1 of vol. 2 in 1803; pt. 2 of vol. 2 in 1804. I do not know whether more volumes have appeared [see Lacepède 1799–1804b].

13. *Descriptions and Figures of Two Hundred Fishes; Collected at Vizagapatam on the Coast of Coromandel,* by Patrick Russell [1726–1805], M.D., London, 1803, two volumes in folio. [On Russell, see Adler 1989, 16–17.]

14. George Shaw, born in 1751 at Bierton, Buckingham, died in 1813, curator of the zoological collection at the British Museum, author of several memoirs among those of the Linnaean Society of London, of a collection of plates titled *The Naturalist's Miscellany* [with figures by Frederick Polydore Nodder (d. 1800?) and afterward by Elizabeth and Richard P. Nodder, published from 1789 to 1813]; a *Zoology of New Holland* [with figures by James Sowerby (1757–1822), published in 1794]; and in particular, a compilation titled *General Zoology, or Systematic Natural*

History, thirteen volumes published from 1800 to 1826. Vols. 4 and 5 of the last, containing the fishes [a total of 1,230 nominal species; see Gill 1872, 40–41], each divided into two parts, were published in 1803 and 1804, immediately after the last of Lacepède's [1798–1803] volumes. [For more on Shaw, see Adler 1989, 17.]

15. [See Shaw 1803, pt. 1, pp. 86–89, pl. 11.] [Henri de] Blainville's [1818] drawing and description [of Shaw's *Stylephorus*], published in the *J. Physique* (Paris), vol. 87, are much to be preferred.

16. [André Marie Constant Duméril, born at Amiens in 1774, was appointed professor of anatomy in Paris in 1801, and from 1803 he substituted for Lacepède in the chair of ichthyology and herpetology before becoming his successor in 1825. He published works on the classification of fishes but devoted himself especially to herpetology (see Adler 1989, 31–32), leaving to Cuvier the task of putting the fish collections in order. He died in 1860. A much expanded second edition of his *Traité élémentaire d'histoire naturelle* appeared in 1807.]

17. François Etienne Delaroche [also spelled De La Roche and de Laroche], born at Geneva in 1789, died at Paris in 1812 [a doctor of medicine who served as a naturalist to the Commission of Spain, charged with pursuing the measurement of the meridian]. [He was the brother-in-law of André Marie Constant Duméril (see n. 16 above).]

18. [Delaroche], in *Ann. Mus. Hist. Nat.* (Paris), vol. 13 [1809a].

19. [Joseph Antoine] Risso, pharmacist and professor at Nice [born there in 1777, he became professor of physical sciences and natural history at the Lycée Impérial in 1813, professor of botany at the medical-surgical school of Nice in 1814, and professor of mineralogical chemistry and botany at the preparatory school of medicine and pharmacology, which he established at Nice in 1832. In ichthyology his two major works are] *Ichthyologie de Nice, ou Histoire naturelle des poissons du département des Alpes Maritimes*, Paris, 1810, one volume in octavo [and *Histoire naturelle des principales productions de l'Europe méridionale et particulièrement de celles des environs de Nice et des Alpes Maritimes*, Paris, 1827a, five volumes in octavo, vol. 3 containing the fishes. He died at Nice in 1845]. [For more on Risso's life and work, see Monod and Hureau 1978; Vayrolatti et al. 1978.]

20. [See Risso 1820a.]

21. [Risso] on four lanternfishes *(scopèles)* [from off Nice], *Mem. R. Accad. Sci. Torino*, vol. 25 [1820b]; and on a new genus that he calls *Alepocephalus (alépocephale)* [from those same waters], *Mem. R. Accad. Sci. Torino*, vol. 25 [1820c].

22. [Risso, *Histoire naturelle de l'Europe méridionale*], Paris, 1827a, five volumes in octavo.

TABLE 10. Here is the classification of Lacepède [see Lacepède 1798–1803, vol. 1, "Première Table méthodique de l'Histoire naturelle des poissons: Tableau général de la classe, des sous-classes, des divisions et des ordres"]. ["A work by an able man and eloquent writer (even prone to add rhetoric by the aid of imagination, in absence of desirable facts), but which . . . is entirely unreliable" (Gill 1872, 39), Lacepède's (1798–1803) four-volume compendium includes descriptions of 1,463 nominal species.]

Classe. POISSONS

Sous-classe. POISSONS CARTILAGINEUX [cartilaginous fishes]

I. Division. Opercula and branchial membranes absent

1. Ordre. Apodes
   Pétromyzons | *Gastrobranches*
2. Ordre. Jugulaires
   —
3. Ordre. Thoracins
   —
4. Ordre. Abdominaux
   Raies | *Squales* | *Aodons*

II. Division. Opercula absent, branchial membranes present

5. Ordre. Apodes
   —
6. Ordre. Jugulaires
   Lophies
7. Ordre. Thoracins
   Balistes
8. Ordre. Abdominaux
   Chimères

III. Division. Opercula present, branchial membranes absent

9. Ordre. Apodes
   —
10. Ordre. Jugulaires
   —
11. Ordre. Thoracins
   —
12. Ordre. Abdominaux
   Polyodons | *Acipensères*

IV. Division. Opercula and branchial membranes present

13. Ordre. Apodes
   Ostracions | *Ovoïdes* | *Sphéroïdes*
   Tétrodons | *Diodons* | *Syngnathes*
14. Ordre. Jugulaires
   —
15. Ordre. Thoracins
   Cycloptères | *Lépadogastères*
16. Ordre. Abdominaux
   Macrorhynques | *Pégases* | *Centrisques*

Continued on next page

Table 10—*continued*

Sous-classe. POISSONS OSSEUX [bony fishes]

I. Division. Opercula and branchial membranes present

17. Ordre. Apodes

Cécilie	*Ophisure*	*Murène*	*Makaira*
Monoptère	*Triure*	*Ammodyte*	*Anarhique*
Leptocéphale	*Aptéronote*	*Ophidie*	*Coméphore*
Gymnote	*Régalec*	*Macrognathe*	*Stromatée*
Trichiure	*Odontognathe*	*Xiphias*	*Rhombe*
Notoptère			

18. Ordre. Jugulaires

Murénoïde	*Uranoscope*	*Batrachoïde*	*Kurte*
Callionyme	*Trachine*	*Blennie*	*Chrysostrome*
Calliomore	*Gade*	*Oligopode*	

19. Ordre. Thoracins

Lépidope	*Bostryche*	*Apogon*	*Léiostome*
Hiatule	*Bostrychoïde*	*Lonchure*	*Centrolophe*
Cépole	*Echenéis*	*Macropode*	*Chevalier*
Taenioïde	*Macroure*	*Labre*	*Léiognathe*
Gobie	*Coryphène*	*Chéiline*	*Chaetodon*
Gobioïde	*Hémiptéronote*	*Chéilodiptère*	*Acanthinion*
Gobiomore	*Coryphénoïde*	*Ophicéphale*	*Chaetodiptère*
Gobiomoroïde	*Aspidophore*	*Hologymnose*	*Pomacentre*
Gobiésoce	*Aspidophoroïde*	*Scare*	*Pomadasys*
Scombre	*Cotte*	*Ostorhynque*	*Pomacanthe*
Scombéroïde	*Scorpène*	*Spare*	*Holocanthe*
Caranx	*Scombéromore*	*Diptérodon*	*Enoplose*
Trachinote	*Gastérostée*	*Lutjan*	*Glyphisodon*
Caranxomore	*Centropode*	*Centropome*	*Acanthure*
Caesio	*Centrogastère*	*Bodian*	*Aspisure*
Caesiomore	*Centronote*	*Taenianote*	*Acanthopode*
Coris	*Lepisacanthe*	*Sciène*	*Sélène*
Gomphose	*Céphalacanthe*	*Microptère*	*Argyréiose*
Nason	*Dactyloptère*	*Holocentre*	*Zée*
Kyphose	*Prionote*	*Persèque*	*Gal*
Osphromène	*Trigle*	*Harpé*	*Chrysotose*
Trichopode	*Péristédion*	*Piméleptère*	*Capros*
Monodactyle	*Istiophore*	*Chéilion*	*Pleuronecte*
Plectorhinque	*Gymnètre*	*Pomatome*	*Achyre*
Pogonias	*Mulle*		

Continued on next page

Table 10—*continued*

20. Ordre. Abdominaux			
Cirrhite	*Pogonathe*	*Mégalope*	*Mugiloïde*
Chéilodactyle	*Cataphracte*	*Notacanthe*	*Chanos*
Cobite	*Plotose*	*Esoce*	*Mugilomore*
Misgurne	*Agenéiose*	*Synode*	*Exocet*
Anableps	*Macroramphose*	*Sphyrène*	*Polynème*
Fondule	*Centranodon*	*Lépisostée*	*Polydactyle*
Colubrine	*Loricaire*	*Polyptère*	*Buro*
Amie	*Hypostome*	*Scombrésoce*	*Clupée*
Butyrin	*Corydoras*	*Fistulaire*	*Myste*
Triptéronote	*Tachysure*	*Aulostome*	*Clupanodon*
Ompok	*Salmone*	*Solénostome*	*Serpe*
Silure	*Osmère*	*Argentine*	*Méné*
Macroptéronote	*Corégone*	*Athérine*	*Dorsuaire*
Malaptérure	*Characin*	*Hydrargyre*	*Xystère*
Pimélode	*Serrasalme*	*Stoléphore*	*8Cyprinodon*
Doras	*Elope*	*Muge*	*Cyprin*
II. Division. Opercula present, branchial membranes absent			
21. Ordre. Apodes			
Sternoptix			
22. Ordre. Jugulaires			
—			
23. Ordre. Thoracins			
—			
24. Ordre. Abdominaux			
—			
III. Division. Opercula absent, branchial membranes present			
25. Ordre. Apodes			
Styléphore			
26. Ordre. Jugulaires			
—			
27. Ordre. Thoracins			
—			
28. Ordre. Abdominaux			
Mormyre			
IV. Division. Opercula and branchial membranes absent			
29. Ordre. Apodes			
Murène	*Murénoblenne*	*Unibranchaperture*	
Gymnomurène	*Sphagebranche*		
30. Ordre. Jugulaires			
—			
31. Ordre. Thoracins			
—			
32. Ordre. Abdominaux			
—			

TABLE 11. This is the classification of the families and genera of Duméril [which he has shown diagrammatically by charts (numbered 58–98) in his *Zoologie analytique, ou Méthode naturelle de classification des animaux* (1806, 97–153)]. To show it to its full advantage, it would have been necessary to present the divisions and subdivisions by which he goes from families to genera, but lack of space does not allow this.

Classe. POISSONS
Sous-classe. CARTILAGINEUX
I. Division. Opercula and branchial membranes absent
- 1. Ordre. TRÉMATOPNÉS
  - Ventral fins absent, mouth circular
    - 1. Famille. CYCLOSTOMES
      - *Lamproie* *Gastrobranche*
  - Ventral fins present, mouth transverse
    - 2. Famille. PLAGIOSTOMES
      - *Torpille* *Rhinobate* *Squale*
      - *Raie* *Squatine* *Aodon*

II. Division. Opercula absent, branchial membranes present
- 2. Ordre. CHISMOPNÉS
  - 3. Famille. CHISMOPNÉS
    - *Baudroie* *Lophie* *Baliste* *Chimère*

III. Division. Opercula present, branchial membranes absent
- 3. Ordre. ELEUTHÉROPOMES
  - 4. Famille. ELEUTHÉROPOMES
    - *Polyodon* *Pégase* *Acipensère*

IV. Division. Opercula and branchial membranes present
- 4. Ordre. TÉLÉBRANCHES
  - Pelvic fins present, abdominal
    - 5. Famille. APHYOSTOMES
      - *Macrorhinque* *Solénostome* *Centrisque*
  - Pelvic fins present, thoracic
    - 6. Famille. PLÉGOPTÈRES
      - *Cycloptère* *Lépadogastère*
  - Pelvic fins absent
    - 7. Famille. OSTÉODERMES
      - *Ostracion* *Ovoïde* *Sphéroïde*
      - *Tétraodon* *Diodon* *Syngnathe*

Sous-classe. OSSEUX
I. Division. Opercula and branchial membranes present
- 5. Ordre. HOLOBRANCHES
  - Pelvic fins absent
    - 1. Sous-ordre. APODES
      - Remaining fins absent
        - 8. Famille. PÉROPTÈRES
          - *Caecilie* *Gymnote* *Ophisure*
          - *Monoptère* *Trichiure* *Aptéronote*
          - *Leptocéphale* *Notoptère* *Régalec*

Continued on next page

Table 11—*continued*

Remaining fins present

9. Famille. PANTOPTÈRE

Murène	*Macrognathe*	*Coméphore*
Ammodyte	*Xiphias*	*Stromatée*
Ophidie	*Anarrhique*	*Rhombe*

Pelvic fins present, jugular in position.

2. Sous-ordre. JUGULAIRES

10. Famille. AUCHÉNOPTÈRES

Murénoïde	*Uranoscope*	*Batrachoïde*	*Kurte*
Callionyme	*Vive*	*Blennie*	*Chrystostrome*
Calliomore	*Gade*	*Oligopode*	

Pelvic fins present, thoracic in position.

3. Sous-ordre. THORACIQUES

Body highly elongate

11. Famille. PÉTALOSOMES

Lépidope	*Taenioïde*	*Bostrichoïde*
Cépole	*Bostrichte*	*Gymnètre*

Round-bodied, cylindrical, pelvic fins united

12. Famille. PLÉCOPODES

Gobie	*Gobioïde*

Round-bodied, cylindrical, pelvic fins separate

13. Famille. ELEUTHÉROPODES

Gobiomore	*Gobiomoroïde*	*Echénéïde*

Spindle-shaped

14. Famille. ATRACTOSOMES

Scombre	*Caranxomore*	*Gastérostée*	*Céphalacanthe*
Scombéroïde	*Caesion*	*Centropode*	*Istiophore*
Caranx	*Caesiomore*	*Centronote*	*Pomatome*
Trachinote	*Scombéromore*	*Lépisacanthe*	

Thick-bodied, laterally compressed, head ordinary, lips thick, opercle without denticulated spines

15. Famille. LÉIOPOMES

Hiatule	*Monodactyle*	*Chéiline*	*Spare*
Coris	*Plectorhinque*	*Chéilodiptère*	*Dipterodon*
Gomphose	*Pogonias*	*Ophicéphale*	*Chéilion*
Osphronème	*Labre*	*Hologymnose*	*Mulet*
Trichopode			

Jaws bony

16. Famille. OSTÉOSTOMES

Scare	*Ostorhinque*	*Léiognathe*

Dorsal fin highly elongate

17. Famille. LOPHIONOTES

Coryphène	*Coryphénoïde*	*Centrolophe*
Hémiptéronote	*Taenianote*	*Chevalier*

Head extremely large

18. Famille. CÉPHALOTES

Gobiésoce	*Aspidophoroïde*	*Scorpène*
Aspidophore	*Cotte*	

Continued on next page

Table 11—*continued*

Pectoral fin with isolated rays			
19. Famille. DACTYLÉS			
Dactyloptère	*Prionote*	*Trigle*	*Péristédion*
Body flat, oval, eyes on one side			
20. Famille. HÉTÉROSOMES			
Pleuronecte	*Achire*		
Opercula spiny or denticulate			
21. Famille. ACANTHOPOMES			
Lutjan	*Bodian*	*Sciène*	*Holocentre*
Centropome	*Taenianote*	*Microptère*	*Persèque*
Eyes on both sides			
22. Famille. LEPTOSOMES			
Chétodon	*Pomacanthe*	*Aspisure*	*Zée*
Acanthinion	*Holacanthe*	*Acanthopode*	*Gal*
Chétodiptère	*Enoplose*	*Sélène*	*Chrysostose*
Pomacentre	*Glyphisodon*	*Argyréiose*	*Capros*
Pomadasys	*Acanthure*		
Pelvic fins present, abdominal in position			
IV. Sub-ordre. ABDOMINAUX			
Body cylindrical, mouth at end of long snout			
23. Famille. SIPHONOSTOMES			
Fistulaire	*Aulostome*	*Solénostome*	
Body cylindrical, mouth not extended			
24. Famille. CYLINDROSOMES			
Cobite	*Fondule*	*Butyrin*	
Misgurne	*Colubrine*	*Triptéronote*	
Anableps	*Amie*	*Ompolk*	
Body conical or compressed, pectoral-fin rays free or distinct, one rigid			
25. Famille. OPLOPHORES			
Silure	*Doras*	*Agénéiose*	*Hypostome*
Macroptéronote	*Pogonate*	*Macroramphose*	*Corydoras*
Malaptérure	*Cataphracte*	*Centranodon*	*Tachysure*
Pimélode	*Plotose*	*Loricaire*	
Several flexible			
26. Famille. DIMÉRÈDES			
Cirrhite	*Chéilodactyle*	*Polynème*	*Polydactyle*
Pectoral fin without distinct rays, opercula scaly, mouth toothless			
27. Famille. LÉPIDOPOMES			
Muge	*Chanos*	*Exocet*	
Mugiloïde	*Mugilomore*		
Dorsal fins with rays			
28. Famille. GYMNOPOMES			
Argentine	*Buro*	*Clupanodon*	*Dorsuaire*
Athérine	*Clupée*	*Serpe*	*Xystère*
Hydrargyre	*Myste*	*Mené*	*Cyprin*
Stoléphore			

Continued on next page

Table 11—*continued*

Jaw simple, adipose fin present			
29. Famille. DERMOPTÈRES			
Salmone	*Corrégone*	*Serrasalme*	
Osmère	*Characin*		
Opercula smooth, jaw highly developed, punctured.			
30. Famille. SIAGONOTES			
Elope	*Esoce*	*Sphyrène*	*Polyptère*
Mégalope	*Synodon*	*Lépisostée*	*Scombrésoce*
Opercle present, branchial membranes absent			
VI. Ordre. STERNOPTYGES			
31. Famille. STERNOPTYGES			
Sternoptyx			
Opercle absent, branchial membranes present			
VII. Ordre. CRYPTOBRANCHES			
32. Famille. CRYPTOBRANCHES			
Styléphore	*Mormyre*		
Opercle and branchial membranes absent			
VIII. Ordre. OPHICHTHYCTES			
33. Famille. OPHICHTHYCTES			
Murénophis			
Gymnomurène			
Murénoblenne			
Unibranchaperture			
Spagebranche			

13 Rafinesque, Regional Contributions, and Maritime Expeditions of the Early Nineteenth Century

In 1810, the same year Risso's ichthyology appeared, a naturalist of French origin, living then in Sicily, Rafinesque-Schmaltz [fig. 41], published two small books that are also important to the natural history of Mediterranean fishes: *Caratteri di alcuni nuovi generi e nuove specie di animali e piante della Sicilia*[1] and *Indice d'ittiologia Siciliana*.[2] The latter, which is the more recent, brings the number of known species to 390; in the two books combined, about 180 species are described as new and 73 are illustrated. A large number are in fact new, but there are not nearly so many new ones as are given as such, even omitting from consideration the work of Risso. The author does not seem to have had at his disposal all the works of his predecessors, especially memoirs scattered among those of the academies, which prevented him from recognizing that several of his fishes had already been described. In addition, he listed in his catalog, without scrutiny, all the species described by Lacepède and Linnaeus as being Mediterranean, which caused him to include several that are purely imaginary, and that includes even some of his genera. Thus his genus *Aodon (aodon)*, taken from Lacepède, is the manta ray *(raie céphaloptère)*; his *Macroramphosus (macroramphose)*, also taken from Lacepède, is *Centriscus (centrisque)*. He greatly multiplied the number of genera, sometimes on slight characters, so that, without counting those that are foreign to the Mediterranean, he recognized 139. Despite his readiness to divide them, he did not do it in circumstances where the laws of the system would make it imperative to do so. For example, he left the anchovies *(anchois)* in the genus of herrings *(harengs)* and the plaice *(plies)* in the genus of soles *(soles)*, and of the one genus of sharks *(squales)* recognized by Linnaeus he created sixteen.

The general classification in the first work is that of Lacepède; in the second, the author altered the system only in intercalating the carti-

FIG. 41.
Constantine Samuel Rafinesque-Schmaltz (1783–1840).
Courtesy of S. L. Jewett and L. F. Palmer, Division of Fishes, National Museum of Natural History, and D. Burgevin, Photographic Services, Smithsonian Institution, Washington, D.C.

laginous fishes with the others, assigning them places according to what Lacepède said about their opercula and gills; for in this regard Rafinesque deferred entirely to the French naturalist and believed that the anglerfishes *(baudroies)* and triggerfishes *(balistes)* have no opercula and that the moray eels *(murènes)* have neither opercula nor branchial membranes. The genera in each division were distributed among certain orders, numbering seventy-one, but without regard for natural relations: *Trachurus (trachures)* and *Labrus (labres)* were placed in the order of the sparoids *(spares)*; mullets *(muges)* are in that of the minnows *(cyprins)*; the swordfish *(xiphias)* is far removed from the spearfish *(tétraptures)*, and so on.

The two works of Rafinesque are nonetheless quite worthy of attention for the original ideas in them and on account of some fishes the descriptions and illustrations of which are not found elsewhere, as well as for the care the author took to give us the Sicilian names of most of his species [see table 12].

It seems from his citations that he took some of his material from the work that Cupani had prepared under the title *Panphyton Siculum*, and which therefore must have contained something other than plants,[3] but it is a book we do not know.

One ought also to count among the writings that contributed to extending our knowledge of Mediterranean fishes the lists of common names and the particular descriptions given in various collections by Italian naturalists or by travelers in Italy: Viviani,[4] Spinola,[5] Giorna,[6] Bonelli,[7] Otto,[8] Ranzani,[9] and Valenciennes [fig. 42].[10] I might also place in that number the monographs I have included in the memoirs of the Muséum National d'Histoire Naturelle, Paris [fig. 43].[11]

Fishes of the Adriatic Gulf have been studied with remarkable care by Naccari[12] and Nardo,[13] and according to the prospectus the latter has just put out, one may expect from him a handsome publication in which these fishes are considered in every aspect.[14]

Low's work on the fauna of the Orkneys, published in 1813 and edited with a preface by Leach,[15] added very interesting details to the natural history of fishes of the North Sea, but the number of species presented is only fifty-two. The late George Montagu has left in the memoirs of the Wernerian Society descriptions of several rare fishes of the south coast of Great Britain.[16] A fine posthumous memoir by Jurine on the fishes of Lake Geneva has just been published by the Physical Science Society of that city.[17]

Among the particular studies made in this period on fishes in distant climes, one may place in the first rank those of Etienne Geoffroy Saint-Hilaire on the fishes of the Nile and the Red Sea, to be found either in the annals of the Muséum National d'Histoire Naturelle, Paris,[18] or in the great work on Egypt.[19] His studies have informed us of a multitude of unusual catfishes *(silures)* and the quite extraordinary genus *Polypterus (polyptère)* and have given us a better notion of many species incompletely described by Hasselquist and Forsskål. The value of these studies is increased by the beautiful drawings made from nature by Redouté the Younger (1812).[20] Moreover, these studies have led the author to important works on the osteology of this class, which we shall discuss below. Isidore Geoffroy Saint-Hilaire, the son, has just published a general edition of these descriptions that presents them with order and clarity.[21] Lacepède himself described, in studies published separately from his great *Histoire naturelle des poissons,* some species sent to him from the Indian Ocean by Péron [fig. 44].[22]

The various dictionaries of natural history published in France[23] and abroad also contain important articles on fishes, which should not be neglected when one is studying this class. Among the naturalists who have published original observations in these dictionaries, we cite Bosc,[24] Bory de Saint-Vincent,[25] Desmarest,[26] and Cloquet.[27]

The members of the Academy of Sciences of St. Petersburg have continued to describe fishes from the sea of Kamchatka, and Tilesius especially has discussed some very remarkable ones.[28]

The third volume of *Zoographia Rosso-Asiatica* by Pallas,[29] printed

FIG. 42. The wreckfish, *Polyprion americanus,* described from the Mediterranean Sea by Achille Valenciennes in his "Description du cernié," *Mém. Mus. Hist. Nat.* (Paris), 1824, 11, pl. 17. Photograph by Patricia L. McGiffert, Health Sciences Photography, University of Washington, Seattle.

FIG. 43. A cardinalfish, *Apogon imberbis,* described from the Mediterranean Sea by Cuvier. *Mém. Mus. Hist. Nat.* (Paris), 1815c, vol. 1, pl. 11 (lower). Photograph by Stanley A. Shockey, Classroom Support Services, University of Washington, Seattle.

FIG. 44.
François Péron (1775–1810). Engraving by Lambert from a drawing by Charles Alexander Lesueur. After Péron 1807–16, text vol. 1, frontispiece. Photograph by Stanley A. Shockey, Classroom Support Services, University of Washington, Seattle.

under the direction of Tilesius, with numerous additions by the editor and long excerpts from Steller's observations, contains not only the fishes of the Black Sea, the Baltic, the Arctic Ocean, and the North Pacific Ocean, but also fishes of the lakes and rivers of this whole vast empire. In particular, one reads interesting facts on the fishes of the Black Sea, which Pallas himself must have observed when he lived in the Crimea. However, by no means all the species of this immense territory are covered in this work. Only 240 species are included, distributed among thirty-eight genera, all Linnaean except for three: *Comephorus* Lacepède *(elaeorhous)*, *Agonus* Schneider *(phalangistes)*, and *Labrax* Pallas *(labrax)*. They are divided into only two orders, the Spiraculata or Chondropterygii, and the Branchiata, which comprises all the others; and these two orders, with the reptiles represented under the name Pulmonata, form only one class, called the Monocardia or cold-blooded animals [see table 13].

As prosperity increases and the love of science grows in the United States, a better study of its natural products is being made, and although formerly it was the Europeans who went to collect samples, now it is the natives or the Europeans living in the country who are collecting samples, and doing it more extensively and with greater accuracy than could the voyaging naturalists.

Thus, in the eighteenth century we had little more on the fishes of North America than the work by Catesby and what Pennant included in his *Arctic Zoology.* But in 1815 Dr. Mitchill, a scholarly naturalist from New York, published a natural history of the fishes caught in the environs of that city, in which he described 149 fishes, classified according to the Linnaean system, with small but well-done illustrations of 60 of the more interesting species.[30] Because he adopted only two of the genera established since Linnaeus, *Bodianus (bodians)* and *Centronotus (centronotes),* he placed his species somewhat arbitrarily; among the pikes *(ésox),* for example, there are some rather heterogeneous species. Neither did he always manage to sort out the correct nomenclature in the often very confused writings of the European naturalists; but in his descriptions he himself provided the means of rectifying errors he overlooked, and his memoir is certainly the best that has appeared in this century on the fishes of the New World. He has since written about some new species in the periodical literature.[31]

The example of Dr. Mitchill has encouraged other naturalists—especially Lesueur,[32] a French painter famous for having been the faithful companion of Péron on his voyage to southern lands,[33] who has been living in the United States and has published descriptions of several beautiful species, with very detailed drawings, in the *Journal of the Academy of Natural Sciences of Philadelphia*[34] and in other periodicals [fig. 45].

Rafinesque, the writer on Sicilian ichthyology, was brought to the United States to occupy a chair at the Academy of Lexington in Kentucky and immediately began to study the fishes of that region. He described three new genera in the *Journal of the Academy of Natural Sciences of Philadelphia*[35] and proposed seventeen additional new genera in a paper printed in the *Journal de Physique* of Paris,[36] to which he added several more in a publication titled *Annals of Nature.*[37] Finally, in a natural history of fishes of the Ohio River and its tributaries,[38] which he published at Lexington in 1820, he added more genera and described 111 species, a large number of which are new and had escaped the attention of Mitchill and Lesueur.

Doubtless this vast continent of America, with its long and irregular coastlines and its large lakes and immense rivers, has still more rich contributions to make to ichthyology, and one can only hope that the naturalists who live there will pursue their research with the enthusiasm

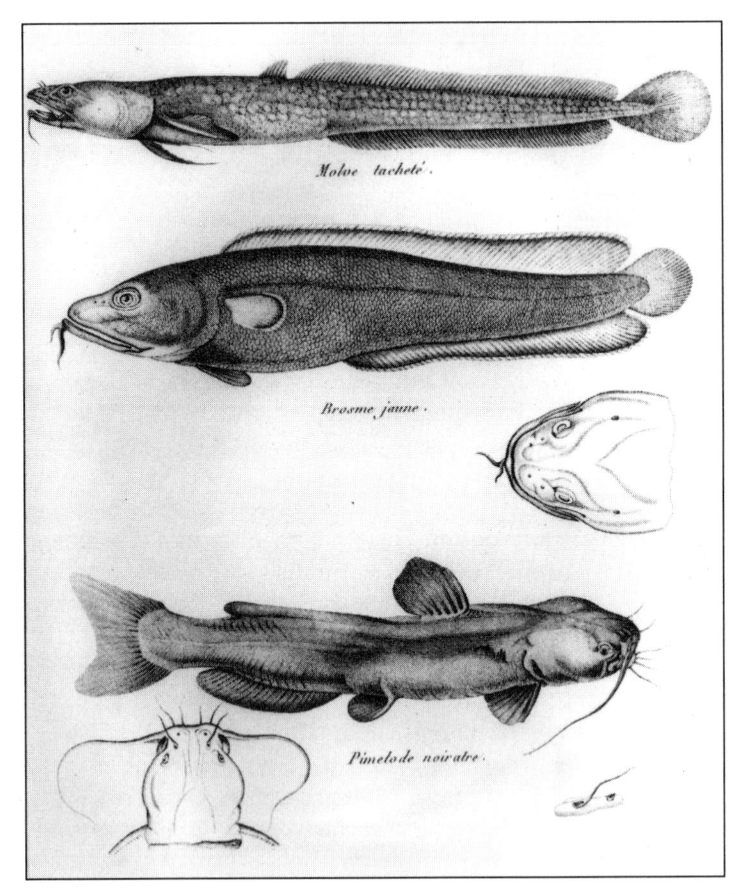

FIG. 45. Fishes from the lakes of Canada described by Charles Alexander Lesueur. *Mém. Mus. Nat. Hist.* (Paris), 1819, vol. 5, pl. 16. Photograph by Stanley A. Shockey, Classroom Support Services, University of Washington, Seattle.

that has characterized them for some years now; they will thereby provide a good return to the Old World for what they have received from it in instruction and enlightenment.

Today a similar zeal motivates Englishmen living in the Indies and in Australia (Nouvelle-Hollande) that has already produced excellent effects. In addition to the great work by Russell mentioned above, a natural history of the fishes of the Ganges was published in 1822 by Mr. Hamilton [formerly] Buchanan,[39] containing 267 species, excellent illustrations [fig. 46], careful descriptions, and some very interesting details of their habits. It is the finest contribution to ichthyology ever received from a distant land. The author simply followed the orders of Linnaeus, or rather of Pennant;[40] but he adopted the genera of Lacepède and added

FIG. 46. Fishes from Francis Hamilton, *An Account of the Fishes Found in the River Ganges and Its Branches,* 1822, vol. 2, pl. 14. Courtesy of Eveline Nave Overmiller and the Library of Congress, Washington, D.C.

several new ones. Like all recent authors, and especially those writing at a distance from literary sources, he sometimes failed to understand Lacepède's nomenclature, or even Bloch's. Therefore some of the genera and species he proposed as new are not new; but his work loses none of its value from an accident that has befallen so many others.

Discoveries in natural history are today considered an essential part of the discoveries to be made by the large nautical expeditions, and the latest voyages of the Russians and French have fulfilled this objective in an exemplary manner.

The account of von Krusenstern's expedition[41] published by Tilesius presents descriptions and drawings of twenty species of fishes.[42] Captain Baudin's expedition has also procured a great quantity of new fishes,[43] thanks to the zeal of Péron and Lesueur, but their descriptions and drawings have not been published and it is not known what has happened to them since the death of Péron. Fortunately the fishes themselves are preserved at the Muséum National d'Histoire Naturelle, Paris, and we shall avail ourselves of them for our work.

The French government has taken steps to ensure that in the future the works of our naturalists are not thus lost to the public. The zoological part of the voyage of de Freycinet has already appeared, with magnificent plates of colored drawings of sixty-two species of fishes, among other animals.[44] Quoy and Gaimard, the naturalists on this expedition, reported a much larger number, and we shall avail ourselves of them also. The engraving of the plates for Duperrey[45] and d'Urville's[46] voyage has already commenced. Among them are several beautiful fish species, collected by Lesson and Garnot, and drawn with great accuracy by Lesson.

NOTES

1. [Constantine Samuel Rafinesque-Schmaltz, born in Constantinople in 1783, died in 1840, was highly industrious, accomplishing during his lifetime a great deal of work relating to almost every subject.] His *Caratteri di alcuni nuovi generi e nuove specie di animali e piante della Sicilia,* was published at Palermo, 1810a, in octavo [see Holthuis and Boeseman 1977]; the dedication is dated 1 April. [For a sketch of Rafinesque's life and a bibliography of his works, see Fitzpatrick 1982; for an English translation of some of his early works, see Cain 1990; see also Call 1895, 1899; Wheeler 1988; Adler 1989, 25–26.]

2. [Rafinesque, see n. 1 above], *Indice d'ittiologia Siciliana, ossia catalogo metodico dei nomi Latini, Italiani e Siciliani dei pesci che si rivengono in Sicilia, disposti secondo un metodo naturale,* Messina, 1810b in octavo; the dedication is dated 15 May [1810].

3. Franciscus Cupani, born in Sicily in 1657, died in 1711, joined the order of Minims in 1681, student of Boccone [see chap. 5, n. 20], had prepared for his *Panphyton Siculum* as many as 700 plates, which are said to be kept in the library of the prince de la Catolica. [Filippo] Buonanni [1638–1725] had begun to publish the work in 1713. There are proofs of 168 of these plates in the Banks library [British Library, Bloomsbury]. [But nearly all are devoted solely to plants. Exceptions include a few minerals and some animals, all shown on the margins of the plates, incidental to the plants; the animals depicted include a caterpillar and a moth (pl. 39); three gastropods (pl. 42); two gastropods (pl. 48); a bee (pl. 70); a crab (pl. 106); a fish, "Gobio uarius ex cruentato atrate maculatus" (pl. 113); three gastropods (pl. 118); another fish, "Asellus mas adolescens Smiridotu" (pl. 121); a praying mantis (pl. 143); a lobsterlike crustacean (pl. 147); and three gastropods and two bivalves (pl. 152) (see Cupani 1713).]

4. Domenico Viviani, professor [of natural history] at Genoa [born at Log-

nano Levanto in 1772, died at Genoa in 1840, provided collections for Barthélemy Faujas de Saint-Fond (1741–1819) to produce a] catalog of the fishes of the river Genoa and the Gulf of Spezzia, published in *Ann. Mus. Hist. Nat.* (Paris), vol. 8 [see Viviani 1806].

5. [Marchese Massimiliano] Spinola [count di Tassarolo], a naturalist [who also donated fishes from Genoa, born at Toulouse in 1780, died at Tassarolo, Alessandria, in 1857], described a seabass *(serran)*, a cardinalfish *(apogon)*, a flounder *(pleuronecte)*, a member of the centracanthid genus *Spicara (mendole)*, an anglerfish *(lophie)*, and so on; he also published a catalog of the Ligurian names of several fishes, in *Ann. Mus. Hist. Nat.* (Paris), vol. 10 [see Spinola 1807].

6. [Michel Esprit] Giorna [1741–1809], professor at Turin, described the crestfish *(lophote)* of Lacepède in *Mem. R. Accad. Sci. Torino*, vol. 16, 1809; I [Cuvier] described this fish from an individual that was more nearly whole, in *Ann. Mus. Hist. Nat.* (Paris), vol. 20 [see Cuvier 1813].

7. [Franco Andrea] Bonelli, also a professor of natural history at the University of Turin [born at Cuneo in 1784, died at Turin in 1830; see Mearns and Mearns 1988, 83–85] reported on a ribbonfish *(Trachipterus* or *gymnètre)* in *Mem. R. Accad. Sci. Torino*, vol. 24 [1820].

8. [Adolph Wilhelm] Otto [1786–1845], professor at Breslau, described several fishes of the Mediterranean in his *Conspectus animalium quorundam maritimorum nondum editorum*, Breslau, 1821.

9. [Camillo] Ranzani [1775–1841], professor at Bologna and *primicier* canon of the cathedral of that city, described in the *Opuscoli Scientifici* of Bologna a ribbonfish *(gymnètre)* that he calls *Epidesmus maculatus* [see Ranzani 1818].

10. [Valenciennes], a monograph on the hammerhead shark *(marteaux)*, *Mém. Mus. Hist. Nat.* (Paris), vol. 9 [1822]; description of the wreckfish *(cernié* or *Polyprion)*, *Mém. Mus. Hist. Nat.* (Paris), vol. 11 [1824]. [Achille Valenciennes, whose father had been an aide to Daubenton since 1784, was born in the Muséum National d'Histoire Naturelle, Paris, in 1794 and spent his entire life associated with that institution. His formal education cut short by the early death of his father, he became a *préparateur* at the museum in 1812 and aided Geoffroy Saint-Hilaire, Lamarck, Cuvier, and others with their zoological collections. He eventually became an *aide-naturaliste* associated with the chair of fishes and reptiles. Early in his career he was given the task of classifying the animals described by Alexander Humboldt on his journey to South America (1799–1803), and it was this relationship with Humboldt that later led to Valenciennes's admission into the Royal Academy of Sciences in 1844. Although best known as an ichthyologist—his major scientific achievement was his collaboration with Cuvier on the great twenty-two-volume *Histoire naturelle des poissons* (see Pietsch 1985)—he made contributions in many other areas of zoology, including monographs on mollusks and zoophytes. He died in Paris in 1865. For more on Valenciennes, see Monod 1963; Appel 1976; Daget 1994.]

11. [Cuvier, in the memoirs of the Muséum d'Histoire Naturelle, Paris]: vol. 1, on the meagre *(maigre* or *fegaro, Sciaena umbra)* [1815a]; the argentine *(argentine)* and cardinalfish *(apogon)* [1815c]; the cuskeel *(ophidium imberbe)* and razorfish *(razon)* [1815d]; the damselfish *(castagnau)*, wrasses of the genus *Crenilabrus (cré-*

nilabres), etc. [1815e]; and various porgies *(spares),* the sprat *(melet),* etc. [1815f]. Vol. 3, on frogfishes *(chironectes)* [1817]. Vol. 4, on porcupinefishes *(diodons)* [1818a], and the pacus *(mylètes)* [1818b]. Vol. 5, on various other characiforms *(salmones)* and the bonefish *(glossodonte)* [1819b].

12. Fortunato Luigi Naccari [1793–1860], vice-consul of the two Sicilies at Chioggia, and librarian of the seminary in that city, published a memoir in 1822 titled "Ittiologia Adriatica, ossia catalogo de' pesci del golfo e lagune di Venezia," *Gior. Fisic. Chim. Stor. Nat.* (Pavia), decad. 2, vol. 5, to which a supplement titled "Aggiunta all'ittiologia Adriatica" was added in 1825, published in the *Giornale dell'Italiana Letterature* [compiled by the Società di Letterati Italiani, Padua].

13. [Giovanni] Domenico Nardo [1802–77], also from Chioggia, published an article called "Osservazioni ed aggiunte all'Adriatica ittiologia," *Gior. Fisic. Chim. Stor. Nat.* (Pavia), decad. 2, vol. 7, 1824.

14. Nardo's prospectus, published in *Isis,* vol. 20 [1827], is titled "Prodromus observationum et disquisitionum ichthyologiae Adriaticae." Its contents are arranged according to the order presented in my *Règne animal* [see Cuvier 1816].

15. *Fauna Orcadensis, or The Natural History of the Quadrupeds, Birds, Reptiles, and Fishes of Orkney and Shetland,* by the Rev. [George] Low [1746–95], from a manuscript in the possession of [William] Elford Leach [see chap. 16, n. 58], Edinburgh, 1813, in quarto.

16. [George Montagu, 1753–1815], in *Mem. Wernerian Nat. Hist. Soc.,* vol. 1, 1811 [describes a trichiurid (pls. 2–3), a pipefish (pl. 4), a cuskeel, a cyclopterid, and a blenny (pl. 5)]. [On Montagu, see Cleevely 1978; Mearns and Mearns 1988, 263–70.]

17. [Louis Jurine (1751–1819), "Histoire abrégée des poissons du Lac Léman" (1825), published in] *Mém. Soc. Phys. Hist. Nat. Genève,* vol. 3.

18. [Etienne Geoffroy Saint-Hilaire, on the fishes of the Nile]: the bichir *(polyptère), Ann. Mus. Hist. Nat.* (Paris), vol. 1 [1802a]; the flatfish *(achire barbu), Ann. Mus. Hist. Nat.* (Paris), vol. 1 [1802b]; and the characiforms *(salmones)* of the Nile, *Ann. Mus. Hist. Nat.* (Paris), vol. 14 [1809a]. [Etienne Geoffroy Saint-Hilaire, born at Etampes in 1772, was employed at the Jardin des Plantes as assistant attendant and demonstrator. In 1793 he was appointed professor of zoology at the Muséum and after the nomination of Lacepède to the chair of reptiles and fishes, he was made attendant of mammals and birds. From 1798 to 1799 he took part in the scientific expedition to Egypt with Bonaparte's army, bringing large collections back to France, including fishes from the Red Sea and the Nile, and in particular the bichir *(polyptère),* the discovery of which alone would have justified the expedition. In 1808 he was put in charge of a mission to Portugal, again bringing collections back to France, mostly from the Museum of Ajuda in Lisbon. He died at Paris in 1844.]

19. [The great monograph series titled] *Description de l'Egypte* is being published by order of the French government, and since 1809 it has grown to several volumes in folio of text and as many volumes of plates in atlas format; the part on natural history makes up three of these atlases: [*Description de l'Egypte, ou Recueil des observations et des recherches qui ont été faites en Egypte pendant l'expédition de l'armée française,* edited by E. F. Jomard, with a historical preface by M. Fourier,

Paris, 1809–30, the text in nine volumes, the atlas in eleven volumes (see Description of Egypt, 1809–30)]. There is an octavo edition of the text [in twenty-four volumes], Paris, Panckoucke, 1821 to the present [1829]. [There is also a two-volume facsimile of the plates from the *Description de l'Egypte,* edited with introduction and notes by Gillispie and Dewachter 1987.]

20. [Pierre Joseph Redouté the Younger (1759–1840), *Description de l'Egypte,* in the plates of "Antiquities," vol. 2, 1812.]

21. [Isidore Geoffroy Saint-Hilaire, son of Etienne Geoffroy Saint-Hilaire, 1805–61, on the fishes of the Nile, the Red Sea, and the Mediterranean, in *Description de l'Egypte,* vol. 1, pt. 1, of the section on natural history, 1827a, b.]

22. [Lacepède, see chap. 12, n. 1] on a ray *(raie),* three frogfishes *(lophies),* a boxfish *(ostracion),* a pufferfish *(tétrodon),* a pipefish *(syngnathe),* a wrasse *(labre),* a surgeonfish *(prionure),* and so on, *Ann. Mus. Hist. Nat.* (Paris), vol. 4 [1804a]. [On Péron, see n. 33 below.]

23. *Nouveau Dictionnaire d'histoire naturelle, appliquée aux arts, principalement à l'agriculture et à l'économie rurale et domestique: par une société de naturalistes et d'agriculteurs: avec des figures tirées des trois règnes de la nature,* Paris, Déterville [1803–4, twenty-four volumes in octavo. A second edition in thirty-six octavo volumes was published at Paris from 1816 to 1819].

Dictionnaire des sciences naturelles, dans lequel on traite méthodiquement des différens êtres de la nature, considérés soit en eux-mêmes, d'après l'état actuel de nos connoissances, soit relativement à l'utilité qu'en peuvent retirer la médecine, l'agriculture, le commerce et les arts. suivi d'une biographie des plus célèbres naturalistes. Par plusieurs professeurs du Jardin du Roi, et des principales écoles de Paris, Strasbourg and Paris, F. G. Levrault. [This dictionary was edited by Frédéric Cuvier (1773–1838), with a prospectus by Georges Cuvier and an introduction by the comte de Fourcroy. Vols. 1–3 appeared in 1804; vols. 4 and 5 and a few copies of vol. 6 were issued in 1805–6, but publication was then suspended until 1816, when these volumes were brought up to date by means of supplements and the work was completed in sixty volumes, plus twelve volumes of plates and one of portraits, 1816–30.]

Dictionnaire classique d'histoire naturelle, par Messieurs Audouin, Isid. Bourdon, Ad. Brongniart, de Candolle, Daudebard de Férussac, A. Desmoulins, Drapiez, Edwards, Flourens, Geoffroy de Saint-Hilaire, A. de Jussieu, Kunth, G. de Lafosse, Lamouroux, Latreille, Lucas fils, Presle-Duplessis, C. Prévost, A. Richard, Thiébaut de Berneaud et Bory de Saint-Vincent, Paris, Baudouin Frères [1822–31, seventeen volumes in octavo].

24. [Bosc (see chap. 12, n. 5) made important contributions to both the first and second editions of the *Nouveau Dictionnaire d'histoire naturelle* (see n. 23 above).]

25. [Jean Baptiste Georges Marie Bory de Saint-Vincent, born at Agen in 1780 and died at Paris in 1846, a nephew of Lacepède and a student of Faujas de Saint-Fond, was a major contributor to the *Dictionnaire classique d'histoire naturelle* (see n. 23 above).]

26. Anselme Gaëtan Desmarest [protégé of Lacepède and student of Cuvier, was born at Paris in 1784 and died at Alfort in 1838. Appointed professor of zoology at the Veterinary School of Alfort (1815), member of the Royal Academy

of Medicine, correspondent of the Royal Academy of Sciences, Paris (1825), and member of several other scholarly societies in France and abroad, he was the author of numerous works, including] a series of articles published separately [1823] under the title "Décades ichthyologiques" [and contributions to vols. 22–24 of the first edition, and all volumes of the second edition, of the *Nouveau Dictionnaire d'histoire naturelle* (see n. 23 above)].

27. [Hippolyte Cloquet (1787–1840) contributed numerous articles to the *Dictionnaire des sciences naturelles* (see n. 23 above).]

28. In 1810 Pallas [see chap. 8, n. 41] published a memoir on the genus *Labrax* [a junior synonym of *Hexagrammos* Steller, in Tilesius von Tilenau 1809; see Eschmeyer 1990, 184, 206], in which he described six new species (*Mém. Acad. Impér. Sci. St. Pétersb.*, vol. 2). In that same year, [Wilhelm Gottlieb] Tilesius von Tilenau [see n. 42 below], also described a member of this genus, as well as a new species of codfish that he calls *Gadus vachnia* [a junior synonym of *Gadus macrocephalus*] (*Mém. Acad. Impér. Sci. St. Pétersb.*, vol. 2). In 1811 Tilesius described a stickleback *(gastéroste)*, a gunnel *(blennie)*, a lamprey *(lamproie)*, a flounder *(pleuronecte)*, and two sculpins *(Cottus hemilepidotus* and *Synanceia cervus)* (*Mém. Acad. Impér. Sci. St. Pétersb.*, vol. 3); in 1813a, four poachers *(agonus)*, a minnow *(cyprin)*, a grouper *(épinéphélus)*, and a sandfish *(trachinus)* [= *Trichodon trichodon*] (*Mém. Acad. Impér. Sci. St. Pétersb.*, vol. 4); and in 1820, a triggerfish *(baliste)* without pelvic fins, which he calls *Balistapus* (*Mém. Acad. Impér. Sci. St. Pétersb.*, vol. 7).

29. [Pallas, see chap. 8, n. 41], *Zoographia Rosso-Asiatica*, published at St. Petersburg, 1811–14, three volumes in quarto. The printing and distribution of this work has been much delayed, apparently owing to the loss of the copperplates, but I have received a copy through the kindness of the president of the Academy of Sciences of St. Petersburg. [For dating of Pallas's *Zoographia Rosso-Asiatica*, see Sherborn 1934, 1947; Sclatter 1947; Stresemann 1951; Svetovidov 1976; see also chap. 8, n. 42.]

30. [Pennant, see chap. 8, n. 38. Samuel Latham Mitchill, born at North Hempstead on Long Island in 1764, died in New York in 1831, received his medical diploma from the University of Edinburgh in 1786. Returning to the United States, he lived in New York, where he taught the natural sciences. In 1801 he was elected to the United States Senate and moved to Washington, D.C., where he remained until 1813, devoting his free time to the study of fishes. In 1815 he published], in *Trans. Lit. Phil. Soc. N.Y.*, vol. 1, a natural history of the fishes of New York; the year before, he produced an essay, New York, twenty-eight pages in duodecimo [on this same subject].

31. [Mitchill, see n. 30 above] on a common eel *(anguille)*, a codfish *(gade)*, and a salmonid *(salmone)*, published in *J. Acad. Nat. Sci. Philad.*, vol. 1 [1818]; and in *Ann. Lyceum Nat. Hist. N.Y.*, vol. 1, 1824, on a new genus, *Saccopharynx*, the same that was described by [John] Harwood in *Phil. Trans. Roy. Soc. Lond.*, vol. 118, 1827, under the name *Ophiognathus*.

32. [Charles Alexander Lesueur was born at Le Havre in 1778. Embarking as a simple gunner's aide on the corvette *Géographe*, on its voyage around the world (1800–1804) under the command of Captain Nicolas Thomas Baudin (see n. 43 below), he showed such a remarkable talent for drawing fishes and other marine

animals that Baudin relieved him of his military duties and gave him the title of artist of the expedition. In 1816 he left for the United States, where he collected fishes as he traveled through the valley of the Great Lakes and the Saint Lawrence River. He took up residence in Philadelphia, where he became one of the most assiduous members of the Philosophical Society and the Academy of Natural Sciences. He returned to Le Havre in 1837 and directed the museum there until his death in 1846.]

33. [François Péron, born at Cérilly in 1775, died there in 1810, embarked as naturalist on Nicolas Baudin's (see n. 43 below) voyage of discovery to New Holland aboard the *Géographe* (1800–1804), during which time he, with the help of Lesueur (see n. 32 above), gathered large collections of natural objects. Upon his return he was charged with writing a narrative of the voyage, the first volume of which appeared in 1807 (see Péron 1807–16).]

34. [Lesueur] on three species of rays *(raies)* [1817a]; five of eels *(anguilles)* [1817b]; two of codfishes *(gades)* [1817c]; one of a minnow *(cyprin)* [1817d]; four killifishes of the genus *Hydrargira (hydrargires)* [1817f]; and the whole genus of *Catostomus (catostomus)* [1817e], which he separated from the minnows *(cyprins)*, and of which he describes seventeen species; on several sharks *(squales)*—including the angelshark *(Squatina)*, a group he calls *Platirostra*—two herrings *(clupés)*, and two whitefishes *(corégones)* [1818a]; three tarpons *(mégalopes)*, which he groups under the name *Hiodon (hiodon)* [1818b]; four pikes *(ésoces)* [1818c]; some fishes from upper Canada, including six long-whiskered catfishes *(pimélodes)*, a sturgeon *(esturgeon)*, a toadfish *(batrachoïde)*, a codfish of the genus *Brosme (brosme)*, and two species of the gadid genus *Molva (lingues)* [1819]; three needlefishes *(orphies)* [1821c]; three drums *(sciènes)* [1822b]; five species attributed to the genus *Cichla (cichla)* [1822a]; two flying fishes *(exocets)* [1821b]; several small fishes akin to the live-bearers *(poecilie)* [1821a]; and a large specimen of a shark he calls *Squalus elephas* [1822c]; six rays *(raies)* or fishes like them [1824]; two new blennies *(blennies)* [1825a]; a new subgenus of *Salmo* that he calls *Harpadon* [1825b]; four moray eels *(muraenophis)* [1825c]; and a lizardfish *(saurus)* [1825d].

35. [Rafinesque, see n. 1 above] on three new genera, *Pomochis, Sarchirus,* and *Exoglossum,* in *J. Acad. Nat. Sci. Philad.*, vol. 1 [1818].

36. Rafinesque, a survey of seventy new genera of animals, discovered in the interior of the United States of America in 1818, *J. Physique* (Paris), vol. 88 [1819].

37. [Rafinesque, a synopsis of new genera and species of animals and plants discovered in North America, 1820a.]

38. [Rafinesque], *Ichthyologia Ohiensis,* a natural history of the fishes inhabiting the Ohio River and its tributaries, published at Lexington, Kentucky, 1820b, in octavo [see Call 1899].

39. *An Account of the Fishes Found in the River Ganges and Its Branches* by Francis Hamilton (formerly Buchanan), M.D., Edinburgh, 1822, in quarto, with an atlas of thirty-nine plates. [Hamilton (formerly Buchanan), 1762–1829, was a medical officer in the service of the East India Company, who traveled by order of the governor of India through Mysore, Canara, and Malabar in 1800 to study, among other things, the natural history of these countries (for more on Hamilton, see Mearns and Mearns 1988, 97–99).]

40. [On Pennant, see chap. 8, n. 38.]

41. Captain [Adam Ivan von] Krusenstern (now Admiral), left Kronstadt on 7 August 1803, made port in England, the Canaries, and Brazil, rounded Cape Horn, visited the Marquesas, Washington Island, sailed up to Kamchatka, left there for Japan, returned to Kamchatka, crossed the China Sea, and returned by way of the Sunda Strait, the Cape, St. Helena, and the north of Scotland. He returned to Kronstadt on 9 August 1806. [Krusenstern, Russian navigator, hydrographer, and admiral, was born at Haggud, Estonia, in 1770, and died at Reval in 1846; the 1803–6 voyage was the first Russian expedition to circumnavigate the world *(Encyclopaedia Britannica,* 13:508).]

42. [Wilhelm Gottlieb Tilesius von Tilenau, born at Milhausen in 1769 and died there in 1857, was a physician who sailed as naturalist and artist aboard the *Nadjedjeda* on the Russian expedition of von Krusenstern (see n. 41 above). His account of the expedition was published in 1813b.]

43. Captain [Nicolas Thomas] Baudin [born at Saint-Martin on the island of Ré in 1750, died at Port Louis, Mauritius, in 1803] left Le Havre on 19 October 1800 with the corvettes *Géographe* and *Naturaliste;* he called at the Canaries in November and Mauritius in March–April 1801, visited the southwest coasts of New Holland, stayed at Timor, and returned to Van Diemen's Land, stayed at Port Jackson, revisited various parts of New Holland, and returned home by way of Mauritius and the Cape, reaching Europe on 16 April 1804. The account of this voyage [by François Péron, see n. 33 above] appeared in two volumes of text, in quarto, and two atlas volumes, Paris, 1807 and 1816; the navigational and geographical part, by [Louis Claude de Saulces de] Freycinet [see n. 44 below], in one volume in quarto, with atlas volume, was published in 1815. The king's cabinet received from this expedition more than two hundred species of fishes, but often they were small individuals. [For more on Baudin, see Horner 1987.]

44. Captain [Louis Claude de Saulces de] Freycinet [born at Montélimar in 1779, died near Loriol, Drôme, in 1842, served as naval ensign aboard the *Naturaliste* on Baudin's voyage of 1800–1804 (see n. 43 above). Later] commanding the corvette *Uranie,* [he] left Toulon on 17 September 1817, sailed by way of the Canaries to Rio de Janeiro, from there to the Cape, Mauritius, Timor, Rauwac near New Guinea, the Mariannas, and the Sandwich Islands, and returned by way of Port Jackson and Tierra del Fuego. The *Uranie* ran aground in the Falkland Islands, and he returned on an American vessel by way of Montevideo and Rio de Janeiro, reaching Le Havre on 13 November 1820. The king's cabinet received about 150 species of fishes from the expedition, and [Jean René Constant] Quoy and [Joseph Paul] Gaimard [see chap. 16, nn. 87, 88] presented illustrations of 63 of them, published in quarto, with an atlas in folio, Paris, 1824.

45. Captain [Louis Isidore] Duperrey [born at Paris in 1786, joined the navy at the age of sixteen and served as sublieutenant on the *Uranie,* under the command of Freycinet on a voyage around the world, from 1817 to 1820, during which he distinguished himself by his observations on magnetism]. Commanding the corvette *Coquille,* [he] left Toulon on 11 August 1822, went to Brazil and the Falkland Islands, doubled Cape Horn, visited the coast of Chile and Peru, the Friendly Islands [Tonga], New Ireland, Waigiu, and the Moluccas, doubled the southern

tip of Van Diemen's Land, sailed to Port Jackson and from there to New Zealand, the Carolines, and New Guinea, and returned by way of Java, Mauritius, the Cape, St. Helena, and Ascension. He reached Marseilles on 24 April 1825. The king's cabinet is obliged to this voyage for 288 species of fishes, collected by [René Primevère] Lesson [see chap. 16, n. 90] and [Prosper] Garnot [see chap. 16, n. 65]. [The zoological results of the voyage were published by Lesson and Garnot (assisted by Félix Edouard Guérin-Méneville, 1799–1874) in two volumes, 1826 and 1831; but the section on fishes was produced solely by Lesson. Duperrey, elected to the Royal Academy of Sciences of Paris in 1842 and becoming its president in 1850, died at Paris in 1865.]

46. [Captain Jules Sébastian César Dumont d'Urville, born at Condé-sur-Noireau in 1790, was second in command to Louis Isidore Duperrey on the *Coquille* (1822–25) and led the expedition of the *Astrolabe* (1826–29). In 1837–40 he led the *Astrolabe* and the *Zélée* on a voyage to the Southern Ocean and reached the Antarctic continent. Appointed rear admiral in 1841, he died in 1842, along with his wife and son, in a train accident between Paris and Saint-Germain-en-Laye. For more on Dumont d'Urville, see Rosenman 1992.]

TABLE 12. Here is the classification of Rafinesque-Schmaltz [as presented in his *Indice d'ittiologia Siciliana* of 1810 (9–49)].

Prima sotto-classe. POMNIODI			
Prima divisione. GIUGULARI [Jugulares]			
Prima sezione. CORISOSTALMI			
I. Ordine. BLENNIDI			
Blennius	*Phycis*	*Gaidropsarus*	
II. Ordine. GADINI			
Gadus	*Onus*	*Strinsia*	
III. Ordine. TRACHINIDI			
Callionymus	*Trachinus*	*Oxycephas*	
Uranoscopus	*Corystion*		
IV. Ordine. CURTISI			
Chrysostroma			
Seconda sezione. PLEUROSTAMI			
V. Ordine. AGHIRINI			
Symphurus			
VI. Ordine. PLERONETTI			
Solea	*Scophthalmus*	*Bothus*	
Seconde divisione. TORACICI [Thoracici]			
Prima sezione. Emisferonoti			
VII. Ordine. SELENIDI			
VIII. Ordine. ZEUSIDI			
Zeus	*Capros*		
IX. Ordine. EQUEDINI			
X. Ordine. CHETODONIDI			
XI. Ordine. ACANTURINI			
XII. Ordine. OLACANTINI			
Seconde sezione. TOSSONOTI			
XIII. Ordine. PERCIDI			
Lepipterus	*Sciena*	*Centropomus*	*Aylopon*
Perca	*Lopharis*	*Holocentrus*	*Lutianus*
XIV. Ordine. SCARIDI			
Scarus			
XV. Ordine. ACANTI			
Centronotus	*Naucrates*	*Notognidion*	
Hypacanthus	*Centracanthus*	*Gasterosteus*	
XVI. Ordine. SCOMBERINI			
Scomber			
XVII. Ordine. SPARIDI			
Trachurus	*Symphodus*	*Diplodus*	*Apogon*
Tracorus	*Labrus*	*Dipterodon*	*Scorpena*
Lepodus	*Spicara*	*Gonenion*	
Cheilinus	*Sparus*	*Mullus*	

Continued on next page

Table 12—*continued*

Terza sezione. ORTONOTI			
XVIII. Ordine. DACTIPLI			
Dactylopterus	*Peristedion*	*Lepadogaster*	
Trigla	*Octonus*		
XIX. Ordine. ECHENEIDI			
Echeneis			
XX. Ordine. CORIFENIDI			
Coryphena	*Lepimphis*	*Cottus*	*Gobius*
XXI. Ordine. ISTIOFORIDI			
Tetrapturus			
XXII. Ordine. CEPOLIDI			
Cepola	*Lepidopus*		
XXIII. Ordine. GINNETRIDI			
Argyctius	*Cephalepis*		
XXIV. Ordine. GINNURINI			
Terza divisione. Addominali [Abdominales]			
Prima sezione. TOSSOGASTRI			
XXV. Ordine. POLLINEMIDI			
XXVI. Ordine. SALMONIDI			
Salmo	*Osmerus*		
XXVII. Ordine. CLUPIDI			
Clupea			
XXVIII. Ordine. CIPRINIDI			
Mugil	*Cyprinus*		
Seconde sezione. ORTOGASTRI			
XXIX. Ordine. POLITTERINI			
*Polypterus**			
XXX. Ordine. SAIRIDINI			
Sayris			
XXXI. Ordine. ESOCIDI			
Sphyrena	*Esox*	*Sudis*	
XXXII. Ordine. NOTACANTINI			
*Notacantus**			
XXXIII. Ordine. CENTRISCHINI			
Centriscus			
XXXIV. Ordine. LORICARINI			
*Loricaria**			
XXXV. Ordine. SILURIDI			
Macroramphosus			
XXXVI. Ordine. ESOCETINI			
Exocetus	*Myctophum*	*Atherina*	
Tirus	*Argentina*		
XXXVII. Ordine. AMIDI			
*Amia**			

Continued on next page

Table 12—*continued*

XXXVIII. Ordine. BUTIRINIDI
*Butirinus**
XXXIX. Ordine. COLUMBRINIDI
XL. Ordine. OLOSTOMIDI
Quarta divisione. APODI [Apodes]
Prima sezione. MACROSOMI
XLI. Ordine. SIGNATIDI
Typhle *Hippocampus* *Nerophis*
Siphostoma *Syngnathus*
XLII. Ordine. TRIURIDI
XLIII. Ordine. TRICHIURINI
*Trichiurus**
XLIV. Ordine. GINNOTINI
Carapus *Ophisurus* *Oxyrus*
XLV. Ordine. ANGUILLIDI
Anguilla
XLVI. Ordine. OFIDINI
Ophidium *Ammodytes* *Scarcina*
XLVII. Ordine. ZIFIDI
Xiphias
XLVIII. Ordine. COMEFORINI
Seconde sezione. BRACHISOMI
XLIX. Ordine. STROMATINI
Stromateus *Luvarus*
L. Ordine. OSTRACIDI
Ostracion
LI. Ordine. ODONTINI
Tetrodon *Diodon* *Orthragus* *Diplanchias*
LII. Ordine. ORBIDI
Seconde sotto-classe. ATELINI
Prima divisione. POMANCHIDI
LIII. Ordine. STERNOTTIDI
*Sternoptyx**
LIV. Ordine. STURIONIDI
Sturio
LV. Ordine. COGRIDI
Cogrus
Seconde divisione. OMNANCHIDI
LVI. Ordine. MORMIRIMI
LVII. Ordine. CHIMERINI
Piescephalus
LVIII. Ordine. BALISTINI
Balistes *Capriscus*
LIX. Ordine. LOFIDI
Lophius

Continued on next page

Table 12—*continued*

LX. Ordine. ECHELINI			
Echelus			
LXI. Ordine. CLOPSIDINI			
Chlopsis	*Nettastoma*		
LXII. Ordine. ZITTERINI			
Xypterus			
Terza divisione. GINNANCHIDI			
Prima sezione. DIPLANCHIDI			
LXIII. Ordine. MONOTTERIDI			
Pterurus			
LXIV. Ordine. DALOFIDINI			
Dalophis			
LXV. Ordine. MURENIDI			
Murena			
Seconde sezione. POLIANCHIDI			
LXVI. Ordine. CHONDROTTERI			
Dalatias	*Isurus*	*Rhina*	*Tetroras*
Carcharias	*Cerictius*	*Pristis*	*Galeus*
Heptranchias	*Squalus*	*Aodon*	*Sphyrna*
Alopias	*Oxynotus*	*Etmopterus*	*Hexanchus*
LXVII. Ordine. PLATOSOMI			
Raja	*Dipturus*	*Uroxis*	
Leiobatus	*Mobula*	*Apterurus*	
Torpedo	*Cephaleutherus*	*Dasyatis*	
LXVIII. Ordine. LAMPREDINI			
Petromyzon			
Terza sezione. ETTERITTI			
LXIX. Ordine. ATTERIDI			
Oxystomus	*Helmictis*		
LXX. Ordine. ANOFTALMINI			
Cecilia			
LXXI. Ordine. MISSINIDI			
*Myxine**			

[*Genera not cited originally by Rafinesque-Schmaltz (1810b) but added by Cuvier.]

TABLE 13. [Here is the classification of Pallas as it appears in the third volume of his *Zoographia Rosso-Asiatica:* "Animalia monocardia seu frigidi sanguinis imperii Rosso-Asiatici" (1811–14, 3:5–428).] The genera are not very numerous and are distributed without further subdivision according to certain analogies that are external but rather unimportant.

ANIMALIUM MONOCARDIA

Ordo I. PULMONATA

- Genus I. Ranae *(Rana)*
- Genus II. Testudines *(Testudo)*
- Genus III. Lacertae *(Lacerta)*
- Genus IV. Colubri *(Coluber et Vipera)*
- Genus V. Angues *(Anguis)*

Ordo II. SPIRACULATA

- Genus VI. Rajae *(Raja)*
- Genus VII. Chimaera *(Chimaera)*
- Genus VIII. Squali *(Squalus)*
- Genus IX. Petromyzontes *(Petromyzon)*

Ordo III. BRANCHIATA

- Genus X. Muraenae *(Muraena)*
- Genus XI. Cycloptae *(Cyclopterus)*
- Genus XII. Anarhichae *(Anarhichas)*
- Genus XIII. Siluri *(Silurus)*
- Genus XIV. Acipenseres *(Acipenser)*
- Genus XV. Phalangistae *(Phalangistes)*
- Genus XVI. Syngnathi *(Syngnathus)*
- Genus XVII. Elaeorhous *(Elaeorhous)*
- Genus XVIII. Cotti *(Cottus)*
- Genus XIX. Callionymus *(Callionymus)*
- Genus XX. Gobii *(Gobius)*
- Genus XXI. Cobitides *(Cobitis)*
- Genus XXII. Blennii *(Blennius)*
- Genus XXIII. Ophidion *(Ophidion)*
- Genus XXIV. Gadi *(Gadus)*
- Genus XXV. Clupeae *(Clupea)*
- Genus XXVI. Scombri *(Scomber)*
- Genus XXVII. Mugiles *(Mugil)*
- Genus XXVIII. Mullus *(Mullus)*
- Genus XXIX. Ammodytes *(Ammodytes)*
- Genus XXX. Gasteracanthi *(Gasteracanthus)*
- Genus XXXI. Trigla *(Trigla)*
- Genus XXXII. Trachini *(Trachinus)*
- Genus XXXIII. Scorpaenae *(Scorpaena)*
- Genus XXXIV. Percae *(Perca)*
- Genus XXXV. Sciaenae *(Sciaena)*
- Genus XXXVI. Coracini *(Coracinus)*
- Genus XXXVII. Labri *(Labrus)*
- Genus XXXVIII. Spari *(Sparus)*
- Genus XXXIX. Labraces *(Labrax)*
- Genus XL. Cyprini *(Cyprinus)*
- Genus XLI. Esoces *(Esox)*
- Genus XLII. Salmones *(Salmo)*
- Genus XLIII. Pleuronectae *(Pleuronectes)*

14 Improvements in Classification

In addition to these authors who have served ichthyology by bringing to light considerable numbers of fishes, there are those who have sought to place the classification of the class more at the level of students, or to improve it either by setting it up according to new relationships or by introducing more numerous and more precise subdivisions. Unfortunately, most of these authors frequently have not taken natural relationships into consideration.

After Rafinesque published his ichthyological classification in his catalog of Sicilian fishes, mentioned above, he published in 1815 a slightly different approach in a general work titled *Analyse de la nature.*[1]

He began as before with the supposition that whenever Lacepède alleged the absence of opercula or [branchiostegal] rays in fishes, it must be true. He continued to place together the fishes to which he attributed these negative characters, whether they have a bony or a cartilaginous skeleton, the chondropterygians as well as the others. He then divided the fishes that are supposed to have these structures, as did Linnaeus, according to the position of the pelvic fins. He thus obtained eight orders, which he subdivided into thirty families, each containing two or three subfamilies, into which he introduced 377 genera [see table 14]. All the genera of Lacepède are included in this number without further examination, which reproduces all of Lacepède's errors and double usages. The added genera themselves are often a double usage; and because their characters and the species they are meant to embrace are not defined, it is difficult to get a clear idea of which genera the author had not listed in his preceding works. Moreover, it suffices to see the close juxtaposition of well-known genera, such as *Polypterus (polyptère)* with *Acipenser (esturgeon)* and *Mormyrus (mormyre)* with *Chimaera (chimère),* to judge the extent to which he continued to remain a stranger to natural classification. It is nonetheless true that several of the genera he recognized seem worthy of being retained.

Monsieur de Blainville published his classification of fishes in 1816, along with a general classification of the animal kingdom.[2] He reprinted it in 1822 in his *Principes d'anatomie comparée,* but presenting it in inverse order and adding Greek names to his subdivisions [fig. 47].[3] His classification differs from Gmelin's[4] only in that the Chondropterygii, which he called Dermodontes, are distinguished from other fishes called Gnathodontes because the teeth adhere only to the skin; and the Branchiostegi, which he called Heterodermes, are distinguished from ordinary fishes called Squammodermes because the skin (according to the author) is of "variable structure." Furthermore, subsequent subdivision is based, as in Linnaeus, on the presence or absence of pelvic fins and whether they are jugular, thoracic, or abdominal in position, which destroys any natural order; for example, the swordfish *(xiphias)* is separated from the mackerels *(scombres),* while the toadfish *(batrachus)* is grouped with codfishes *(gades)* and the flounders *(pleuronectes),* and the cutlassfishes *(trichiures)* with the sand lances *(ammodites)* and knifefishes *(gymnotes),* and so on. Families are based only on characters of body shape: ordinary, catfishlike *(siluroïde),* long and narrow, or long and slightly narrow. But the author cited only a few genera under each division, as if to provide examples, and did not give a complete list; therefore for several genera there is doubt about the place the author assigned them. He at least had the advantage of not using the erroneous characters taken from the opercula and rays, which, after Lacepède, were introduced into several classifications [see table 15].

It was in 1816 that my table of the animal kingdom appeared, but I had indicated the basis of my classification as early as 1815.[5] In this proposal I deleted the order Branchiostegi, and from some of those genera I created the Plectognathi, based on a particular mode of articulation of the jaws. The others were distributed among the orders of bony fishes, for which I reestablished the divisions first established by Artedi, based on the nature of the dorsal spines. The Acanthopterygii became one order; but for the Malacopterygii I saw no difficulty in classifying its families according to the position of the pelvic fins. Demonstrating the error of characters taken from opercula and rays, I was able to bring together many fishes that the proponents of these characters had left widely separated, notably all those in the families of the Anguilliformes. Generally, I strive to compose natural families. I have established twenty-one of them [see table 16]. All the genera have been submitted to a new examination, disencumbered of the species that seemed to me not to belong to them, and subdivided into subgenera, so as to aid the distribution and recognition of the species. I have no doubt that this part of my

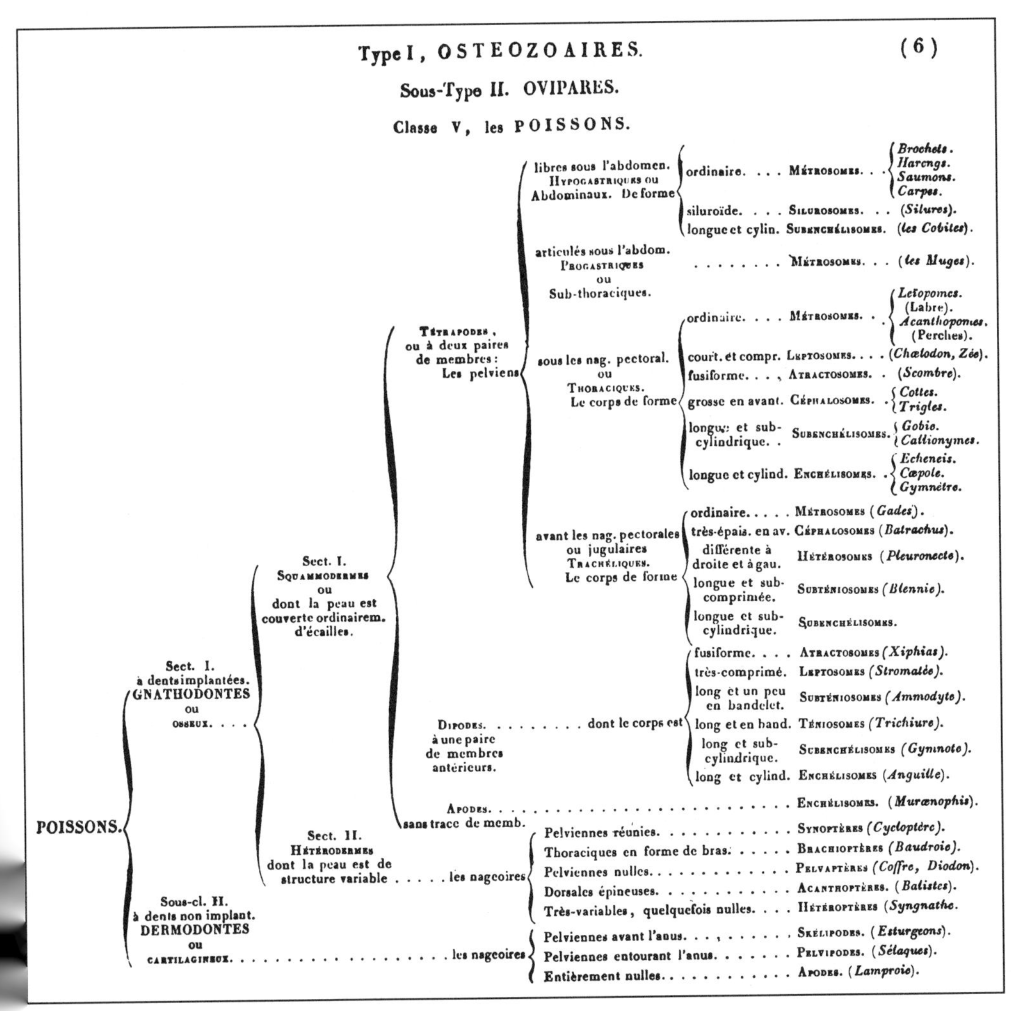

FIG. 47. Henri Marie Ducrotay de Blainville's classification of fishes as presented in his *De l'organisation des animaux, ou Principes d'anatomie comparée*, 1822, table 6.

work will be recognized as original, having been based on direct observation of the objects, and not simply an extract from other ichthyologists.

Goldfuss also believed he should publish a classification of fishes, with Greek names for the divisions.[6] To do so, he simply took Gmelin's divisions and combined the Jugulares and the Thoracici under the name Sternopterygii, also expanding the Chondropterygii to include the Branchiostegi [see table 17]. Instead of Apodes, he used the name Peropterygii, and instead of Abdominales, he proposed the Gasteropterygii. Each order is subdivided into families according to general shape, that of the head, that of the mouth, or some other exterior character, but in such a way that the silversides *(athérine)*, for example, are placed between the live-bearers *(poecilie)* and minnows *(cyprins)*, and the longfin herring *(gnathobolus)*, which is almost a true herring *(hareng)*, next to the slender sunfish *(pomatias)*, which is in fact a mola *(lune, Orthagoriscus)*.

Quite recently (1827), Risso judged it necessary to arrange fishes in a classification of his own.[7] To that end he used the orders of Linnaeus, adding to their number my Plectognathi and Lophobranchii. He subdivided the orders of common fishes, as did Forster, according to their soft or spiny dorsal rays, and divided some genera, taken mostly from my *Règne animal*, into a number of families that he calls natural, several of which are taken from the same source; but his first distribution, based on Linnaeus, forced him to disperse some of them in a way that has little to do with their title [see table 18].

One sees that the classifications of most of these ichthyologists, as varied as they appear in their arrangement, are no more than repetitions under different names of the classification of Linnaeus, changed in only a few of them by the introduction of these allegedly imperfect classes, following Lacepède in being based on the supposed absence of some part of the branchial coverings, and in others by characters taken from the nature of the spines, as had been done by Artedi. It was therefore quite likely that these classifications would separate organisms that should be close together, and that they would list characters that could not be found in the objects themselves.

Oken[8] tried another way: he is known to have undertaken to resolve a great philosophical problem of the idealists, that of deducing a priori from the general idea of being the whole diversity of particular beings, which he believed he could do by combinations of ideas of different degrees. Having come to fishes [fig. 48], he must have also sought to deduce by this process, from the general idea of fish, the idea of all particular fishes, and the combinations to which he had recourse, descending from degree to degree, form a sort of classification. He has already published three or four different attempts, but none of them seem to us to group the genera according to relationships that natural classification might acknowledge. We do not see how one could even assign precise characters to his subdivisions.

In the third volume of his *Lehrbuch der Naturphilosophie*, published in 1811,[9] he limited himself to dividing the fishes, or what he calls his *Fleischthiere* (animals in which flesh dominates), according to the predominance he attributed in them to each part of the body, into *Bauchfische* (abdominal types), *Brustfische* (thoracic types), *Gliederfische* (limbed types), and *Kopffische* (head types); he compared them, respectively, to infusoria or mollusks, to univalves, and to cuttlefishes or jellyfishes [see table 19].

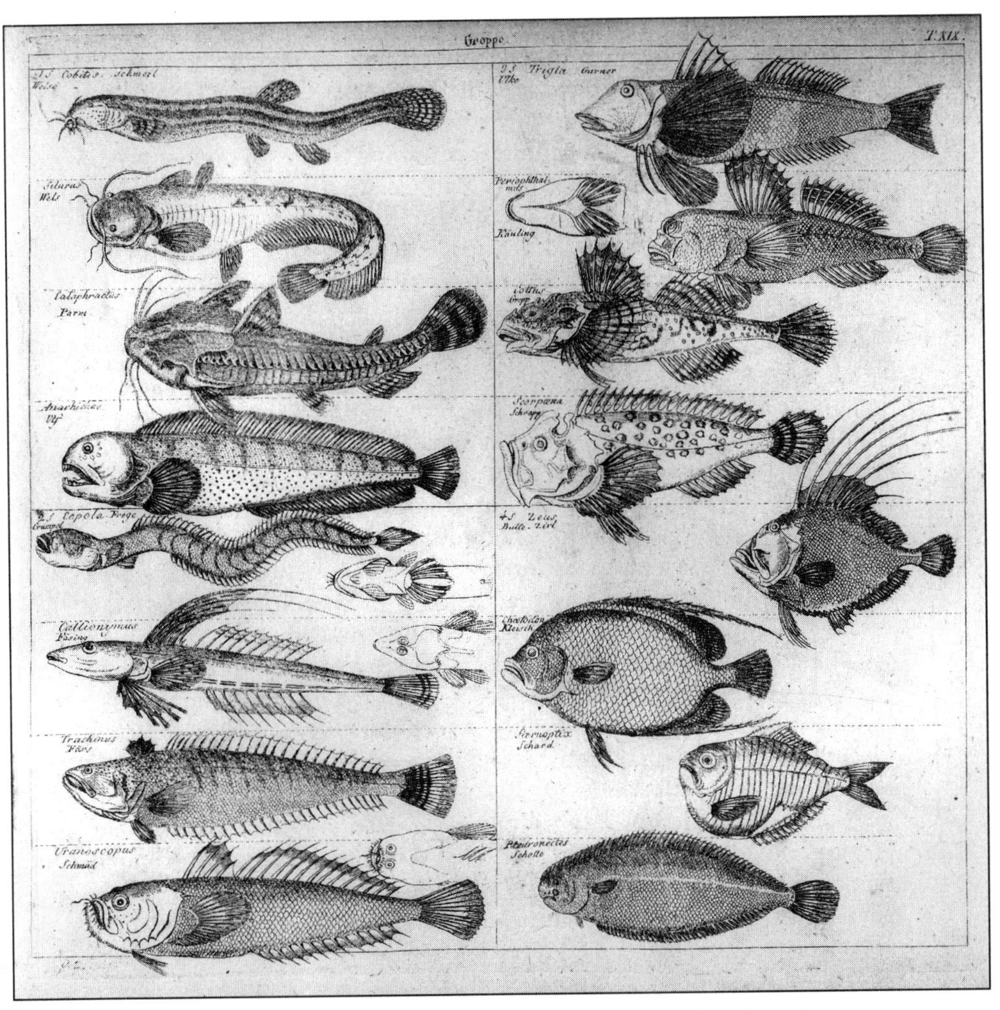

FIG. 48. Fishes from Lorenz Oken's *Lehrbuch der Naturgeschichte*, 1815–16, pt. 3, Zoologie, Atlas, pl. 19. Courtesy of Leslie Overstreet and the Special Collections Department, Smithsonian Institution Libraries, Washington, D.C.

In 1816, in the main body of his *Lehrbuch der Naturgeschichte*,[10] Oken arranged this class into seven orders so as to represent, according to him, the seven classes into which he divided the animal kingdom. Each of the seven orders is then divided into four suborders or families, and each family into four genera, which makes 112 genera [see tables 20 and 21]. But in the table at the beginning of the same work,[11] desiring to push his idealistic method further, he reduced the number of his orders to four, corresponding to the four classes of vertebrates; each order has four suborders, corresponding to the four orders; each suborder has four genera, corresponding to the four suborders, which makes a triple tetratomy (if one may use the word), and reduced the genera to sixty-four [see table 22].

Then, in a treatise on natural history for schools, published in 1821,[12] he divided the class into five orders, according to the predominance he believed he saw in them of seed, sex, entrails, flesh, or sense organs. The first four orders are each divided into three suborders, and in the first three orders each suborder is divided into nine genera; in the fourth order each suborder is divided into four tribes, each of which has three genera, with the same relationships as the four orders. Finally, the fifth classification, which Oken published in French in 1821,[13] is divided into five orders based on the five senses [see table 23].

We need not judge these attempts from a metaphysical standpoint or evaluate the solidity of their foundations; that is for metaphysicians to do, not naturalists. But as for results, anyone can see that they do not agree with the true relationships of organisms, and although the last classification diverges less from nature than the preceding ones, it will never be possible in a natural method to place the swordfish *(xiphias)* next to the sturgeon *(esturgeon)* and the grenadier or rattail *(lepidoleprus)* next to the suckermouth armored catfish *(loricaire)*, and so on.

It can be seen, moreover, that a great stroke of good fortune would have been necessary for the genera, as established by preceding authors, to lend themselves to arrangements of such punctilious symmetry. Thus Oken was obliged sometimes to combine a number of genera into only one, sometimes to subdivide a genus into several, and this was especially true in his third attempt, in which he recognized only sixty-four genera. His groupings are not always felicitous. When he placed, for example, the squirrelfish *(holocentrum, sogho)* with the porgies *(canthères)* under the genus *Cichla*, it is evident that he consulted neither apparent relationships nor true analogies.

NOTES

1. [Rafinesque (see chap. 13, n. 1), *Analyse de la nature, ou Tableau de l'univers et des corps organisés*], Palermo, 1815, in octavo.

2. [Henri Marie Ducrotay de Blainville's (1777–1850) classification of the animal kingdom, titled "Prodrome d'une nouvelle distribution systématique du règne animal," was originally published in *J. Physique* (Paris), vol. 83, 1816. The part on fishes, found on pp. 254–55, is reproduced here in table 15].

3. [See de Blainville 1822, especially his table 6.]

4. [Johann Friedrich Gmelin (see chap. 10, n. 15), in the thirteenth edition of Linnaeus's *Systema naturae*, vol. 1, pt. 3, pp. 1126–1516, 1789 (for a full citation of Gmelin's edition of the *Systema naturae*, see Linnaeus 1788–93).]

5. [Cuvier first published his classification of the animal kingdom in 1816,

Paris, four volumes in quarto; the part on fishes is found in 2:104–351. The bases for his conclusions, however, were outlined earlier in a paper on the composition of the upper jaw of fishes and its use in classification, published] in *Mém. Mus. Nat. Hist. Nat.* (Paris), vol. 1 [1815b.]

6. [Georg August Goldfuss (1782–1848) included a classification of fishes in the second volume of his *Handbuch der Zoologie,* published at Nuremburg in 1820.]

7. [Risso's (see chap. 12, n. 19) classification appeared in his "Histoire naturelle des poissons de la Méditerranée qui fréquentent les côtes des Alpes Maritimes, et qui vivent dans la golfe de Nice," 1827b, 99–112.]

8. [Lorenz Oken, German naturalist and philosopher, was born at Bohlsbach bei Offenburg, Baden, in 1779. After graduating from the University of Fribourg in 1804, he held various teaching posts at Göttingen, Jena, Munich, and Erlangen. In 1832 he secured a post at the recently founded University of Zurich, where he taught until his death in 1851. For more on Oken, see Klein 1974.]

9. Oken, *Lehrbuch der Naturphilosophie,* in German, pt. 3, pp. 301–4 [1809–11].

10. Oken, *Lehrbuch der Naturgeschichte,* in German, 2:12–13 [1815–16].

11. [Oken, *Lehrbuch der Naturgeschichte,* 1815–16, 2:i–iv.]

12. [Oken, *Naturgeschichte für Schulen,* 1821a, xxviii–xxix.]

13. [Oken, *Esquisse du système d'anatomie, de physiologie et d'histoire naturelle,* 1821b, 56–58.]

TABLE 14. This is the classification of Rafinesque-Schmaltz [as it appears in his *Analyse de la nature*], Palermo, 1815, in octavo [79–94].

Classe ICHTHYOSIA

1. Sous-classe. HOLOBRANCHIA

I. Ordre. DERIPIA. Les Jugulaires
- 1. Sous-ordre. Chorizopsia
  - 1. Famille. BLENNIDIA
    - 1. Sous-famille. Monactylia
      - 1. *Dactyleptus* (*Murenoide*, Lacepède)
      - 2. *Pteraclidus* (*Oligopodus*, Lacepède)
    - 2. Sous-famille. Polactylia
      - 3. *Blennius*
      - 4. *Phycis*
      - 5. *Pholidus*
      - 6. *Enchelyopus*
      - 7. *Pacamus*
      - 8. *Ictias*
      - 9. *Dropsarus*
  - 2. Famille. GADINIA
    - 1. Sous-famille. Merluccia
      - 1. *Gadus*
      - 2. *Merluccius*
      - 3. *Trisopterus*
      - 4. *Strinsia*
      - 5. *Brosme*
    - 2. Sous-famille. Trachinia
      - 6. *Batrictius* (*Batrachoides*, Lacepède)
      - 7. *Platicephalus*
      - 8. *Ceracantha*
      - 9. *Taunis*
      - 10. *Calliomoris*
      - 11. *Callionymus*
      - 12. *Oxycephas*
      - 13. *Uranoscopus*
      - 14. *Trachinus*
      - 15. *Corystion*
  - 3. Famille. BRACHOMIA
    - 1. *Chrysotroma*
    - 2. *Kurtus*
- 2. Sous-ordre. Pleuropsia
  - 4. Famille. PLEURONECTIA
    - 1. Sous-famille. Achiria
      - 1. *Achirus*
      - 2. *Symphurus*
      - 3. *Monochirus*
    - 2. Sous-famille. Diplochiria
      - 4. *Pleuronectes*
      - 5. *Scophthalmus*
      - 6. *Bothus*
      - 7. *Plagiusa*

II. Ordre. THORAXIPIA. Les Thoraciques
- 1. Sous-ordre. Leptosomia
  - 5. Famille. CHETODONIA
    - 1. Sous-famille. Leiobranchia
      - 1. *Chetodon*
      - 2. *Chetodipterus*
      - 3. *Teuthis*
      - 4. *Acanthinion*
    - 2. Sous-famille. Odobranchia
      - 5. *Pomadasys*
      - 6. *Enoplosus*
      - 7. *Holacantha*
      - 8. *Pomacantha*
      - 9. *Pomacentrus*
  - 6. Famille. ZEDIA
    - 1. Sous-famille. Glyphisodia
      - 1. *Glyphisodon*
      - 2. *Acanthopodus*
      - 3. *Acanthurus*
      - 4. *Aspisurus*
      - 5. *Nasonus* (*Naso*, Lacepède)

Continued on next page

Table 14—*continued*

2. Sous-famille. Aplodia		
6. *Zeus*	8. *Selene*	
7. *Argyreiosus*	9. *Alectis* (*Gallus*, Lacepède)	
3. Sous-famille. Leiodia		
10. *Chrysostosus*	11. *Capros*	
7. Famille. PETALOMIA		
1. Sous-famille. Cepolidia		
1. *Cepola*	4. *Pterops*	5. *Tasica*
2. *Trachypterus*	(*Bostrychoides*,	6. *Lepidopus*
3. *Bostrictis*	Lacepède)	
2. Sous-famille. Gymnetria		
7. *Gymnetrus*	10. *Hiatula*	13. *Gymnurus*
8. *Nemipus*	11. *Argyctius*	14. *Tenioides*
9. *Regalecus*	12. *Cephalepis*	
2. Sous-ordre. Toxonotia		
8. Famille. ATRACTOMIA		
1. Sous-famille. Scomberia		
1. *Scomber*	3. *Orcynus*	
2. *Polipturus*	(*Scomberoides*,	
(*Scomberomorus*,	Lacepède)	
Lacepède)	4. *Trachinotus*	
2. Sous-famille. Caranxia		
5. *Caranx*	10. *Centropodus*	15. *Baillonus*
6. *Tricropterus*	11. *Notognidion*	16. *Cesiomorus*
7. *Centracantha*	12. *Centrolophus*	17. *Centronotus*
8. *Hypacantha*	13. *Gasterosteus*	
9. *Hypodis*	14. *Naucrates*	
9. Famille. POMODIA		
1. Sous-famille. Notacandia		
1. *Lepicantha*	3. *Cephimnus*	
2. *Gastrogonus*	(*Gymnocephalus*, Bloch)	
2. Sous-famille. Percidia		
4. *Lepipterus*	12. *Panotus*	16. *Lopharis*
5. *Holocentrus*	(*Tenianotus*,	17. *Centropomus*
6. *Perca*	Lacepède)	18. *Cephacandia*
7. *Micropterus*	13. *Lutianus*	(*Cephalacanthus*,
8. *Epinephelus*	14. *Johonius*	Lacepède)
9. *Sciena*	15. *Aylopon*	19. *Trachychthis*
10. *Bodianus*	(*Anthias*,	20. *Epigonus*
11. *Pomatomus*	Bloch)	
10. Famille. LEIOPOMIA		
1. Sous-famille. Osteostomia		
1. *Leiognathus*	2. *Scarus*	3. *Ostorincus*
2. Sous-famille. Trichopodia		
4. *Monodactylus*	5. *Trichopodus*	6. *Osphronemus*

Continued on next page

Table 14—*continued*

3. Sous-famille. Monotia
- 7. *Trachurus*
- 8. *Cesio*
- 9. *Gonurus*
- 10. *Lepodus*
- 11. *Harpe*
- 12. *Cheilinus*
- 13. *Pimelepterus*
- 14. *Hologymnosus*
- 15. *Lepomus*
- 16. *Pomagonus*
- 17. *Macropodus*
- 18. *Kyphosus*
- 19. *Ophicephalus*
- 20. *Sparus*
- 21. *Diplodus*
- 22. *Mesopodus*
- 23. *Spicara*
- 24. *Labrus*
- 25. *Cynedus*
- 26. *Pogonias*
- 27. *Acaramus*
- 28. *Symphodus*
- 29. *Lonchurus*
- 30. *Gomphosus*
- 31. *Centrogaster*
- 32. *Plectorincus*

4. Sous-famille. Mullidia
- 33. *Dipterodon*
- 34. *Gonenion*
- 35. *Leiostomus*
- 36. *Clodipterus* (*Cheilodipterus,* Lacepède)
- 37. *Mullus et Apogon*
- 38. *Macrolepis*

3. Sous-ordre. Orthonotia

11. Famille. LOPHIONOTA

1. Sous-famille. Istiophoria
- 1. *Istiophorus*
- 2. *Tetrapturus*
- 3. *Guebucus*
- 4. *Makaira*

2. Sous-famille. Coryphenia
- 5. *Macrurus*
- 6. *Coryphena*
- 7. *Hemipteronotus*
- 8. *Micropodus* (*Cheilio,* Lacepède)
- 9. *Megaphalus* (*Gobiesox,* Lacepède)
- 10. *Pomacanthis*
- 11. *Leptopus*
- 12. *Oxima*
- 13. *Equetus* (*Eques,* Bloch)
- 14. *Branchiostegus* (*Coryphenoides,* Lacepède)
- 15. *Eleotris* (*Gobiomorus,* Lacepède)
- 16. *Epiphthalmus* (*Gobiomoroides,* Lacepède)
- 17. *Lepimphis*

12. Famille. PLECOPODIA
- 1. *Gobius*
- 2. *Plecopodus* (*Gobioides,* Lacepède)
- 3. *Umbra*
- 4. *Lepadogaster*
- 5. *Piescephalus*
- 6. *Cyclopterus*
- 7. *Lumpus*
- 8. *Liparius*

13. Famille. CEPHOPLIA

1. Sous-famille. Echenidia
- 1. *Echeneis*

2. Sous-famille. Cephalotia
- 2. *Cottus*
- 3. *Aspidophorus*
- 4. *Percis* (*Aspidophoroides,* Lacepède)
- 5. *Aygula* (*Coris,* Lacepède)
- 6. *Scorpena*

III. Ordre. GASTRIPIA

1. Sous-ordre. Brachistomia

Continued on next page

Table 14—*continued*

14. Famille. DACTYLINIA		
1. Sous-famille. Triglidia		
1. *Prionotus*	3. *Peristedion*	
2. *Trigla*	4. *Octonus*	
2. Sous-famille. Dimeredia		
5. *Dactylopterus*	6. *Cirrhitus*	
3. Sous-famille. Polinemia		
7. *Polydactylus*	9. *Clodactylus*	
8. *Polynemus*	(*Cheilodactylus,*	
	Lacepède)	
15. Famille. DERMOPTERIA		
1. *Salmo*	4. *Characinus*	6. *Gasterodon*
2. *Osmerus*	5. *Anostoma*	(*Serrasalmus,*
3. *Coregonus*		Lacepède)
16. Famille. CYPRINIA		
1. Sous-famille. Gasterogonia		
1. *Xysterus*	5. *Gasteroplecus*	7. *Thrissa*
2. *Dorsuarius*	(*Serpe,* Lacepède)	(*Clupanodon,*
3. *Meneus*	6. *Clupea*	Lacepède)
4. *Buronus*		8. *Mystus*
(*Buro,* Lacepède)		
2. Sous-famille. Gymnopomia		
9. *Cyprinus*	14. *Prinodon*	18. *Argentina*
10. *Gonorincus*	(*Cyprinodon,*	19. *Hydrargyra*
11. *Megalops*	Lacepède)	20. *Stolephorus*
12. *Myctophum*	15. *Maturacus*	21. *Gonipus*
13. *Gonostoma*	16. *Edonius*	22. *Tirus*
	17. *Atherina*	
3. Sous-famille. Lepomia		
23. *Exocetus*	26. *Myxonum*	27. *Trichonotus*
24. *Chanos*	(*Mugiloides,*	(*Mugilomorus,*
25. *Mugil*	Lacepède)	Lacepède)
		28. *Soranus*
17. Famille. OPLOPHORIA		
1. Sous-famille. Loricaria		
1. *Plecostomus*	3. *Cordorinus*	5. *Cataphractus*
(*Loricaria,*	(*Corydoras,*	6. *Pogonathus*
Linnaeus)	Lacepède)	
2. *Hypostomus*	4. *Doras*	
2. Sous-famille. Siluridia		
7. *Silurus*	12. *Pimelodus*	16. *Centranodon*
8. *Platistus*	13. *Malapterurus*	17. *Macroramphosus*
9. *Bagrus*	14. *Plotosus*	18. *Clarias*
10. *Macropteronotus*	15. *Ageneiosus*	19. *Aspredo*
11. *Trachysurus*		

Continued on next page

Table 14—*continued*

18. Famille. CYLINDROSOMIA
    - 1. *Anableps*; 2. *Amiatus* (*Amia,* Linnaeus); 3. *Misgurnus*; 4. *Cobitis*; 5. *Fundulus*

2. Sous-ordre. Macrostomia
- 19. Famille. SIAGONIA
  - 1. Sous-famille. Sphyrenidia
    - 1. *Sphyrena*; 2. *Sudis*; 3. *Sayris* (*Scombresox,* Lacepède); 4. *Tripteronotus*
  - 2. Sous-famille. Esoxidia
    - 5. *Esox*; 6. *Raphistoma* (*Belone,* Gronovius; 7. *Lepisosteus*; 8. *Synodus*; 9. *Megalops*; 10. *Elops*; 11. *Stomias*
  - 3. Sous-famille. Notacanthia
    - 12. *Notacanthum*; 13. *Odamphus*; 14. *Onopionus*
- 20. Famille. SIPHOSTOMIA
  - 1. Sous-famille. Colubrinia
    - 1. *Butyrinus*; 2. *Colubrinus*; 3. *Guaris*
  - 2. Sous-famille. Aulostomia
    - 4. *Aulostomus*; 5. *Fistularia*; 6. *Solenostoma*; 7. *Macrorincus*; 8. *Centriscus*

IV. Ordre. APODIA

1. Sous-ordre. Osteodermia
- 21. Famille. APHYOSTOMIA
  - 1. *Syngnathus*; 2. *Typhlinus*; 3. *Siphostoma*; 4. *Hippocampus*; 5. *Phyllophorus*; 6. *Homolenus*; 7. *Nerophis*
- 22. Famille. OSTEODIA
  - 1. Sous-famille. Ostracidia
    - 1. *Ostracion*; 2. *Gonodermus*
  - 2. Sous-famille. Odopsia
    - 3. *Tetrodon*; 4. *Orthragus* (*Cephalus,* Shaw); 5. *Diplanchias*; 6. *Diodon*; 7. *Cephalopsis*
  - 3. Sous-famille. Orbidia
    - 8. *Orbidus* (*Spheroide,* Lacepède); 9. *Oonidus* (*Ovoide,* Lacepède)

II. Sous-ordre. Malacodermia
- 23. Famille. PANTOPTERIA
  - 1. Sous-famille. Stromatia
    - 1. *Rhombus*; 2. *Stromateus*; 3. *Luvarus*; 4. *Tangus*; 5. *Heptaca*; 6. *Piratia*

Continued on next page

Table 14—*continued*

2. Sous-famille. Xyphidia			
	7. *Anarhicas*	9. *Opictus*	11. *Macrognathus*
	8. *Comephorus*	10. *Xyphias*	12. *Odontognathus*
3. Sous-famille. Anguillinia			
	13. *Eleuthurus*	18. *Anguilla* (*Murena*, Lacepède)	21. *Pterops* (*Bostrychoide*, Lacepède)*
	14. *Mastacembalus*	19. *Triurus*	
	15. *Scarcina*	20. *Ictiopogon* (*Botrychus*, Lacepède)*	
	16. *Ammodytes*		
	17. *Ophidium*		
24. Famille. PEROPTERIA			
1. Sous-famille. Gymnotia			
	1. *Gymnotus*	3. *Apteronotus*	5. *Neleus*
	2. *Carapus*	4. *Dameus*	
2. Sous-famille. Trichiuria			
	6. *Trichiurus*	8. *Diepinotus*	
	7. *Nemochirus*	9. *Symphocles*	
3. Sous-famille. Ophisuria			
	10. *Notopterus*	12. *Leptocephalus*	
	11. *Ophisurus*	13. *Oxyurus*	
V. Ordre. ELTROPOMIA			
25. Famille. POMANCHIA			
1. Sous-famille. Sternoptigia			
	1. *Sternoptix*	2. *Melanictis*	
2. Sous-famille. Sturionia			
	3. *Polypterus*	5. *Polyodon*	
	4. *Accipenser*	6. *Pegasus*	
VI. Ordre. CHISMOPNEA			
26. Famille. BRANCHISMEA			
1. Sous-famille. Chimeria			
	1. *Chimera*	2. *Mormyrus*	
2. Sous-famille. Balistia			
	3. *Balistes*	5. *Vetula*	
	4. *Capriscus*	6. *Epimonus*	
3. Sous-famille. Lophidia			
	7. *Lophidius* (*Lophius*, Linnaeus)	8. *Chironectes*	
		9. *Conomus*	
27. Famille. MEIOPTERIA			
1. Sous-famille. Echelia			
	1. *Echelus*	2. *Stylephorus*	
2. Sous-famille. Chlopsidia			
	3. *Chlopsis*	5. *Xypterus*	
	4. *Nettastoma*	6. *Monopterus*	
VII. Ordre. Tremapnea			
28. Famille. OPHICTIA			

Continued on next page

Table 14—*continued*

1. Sous-famille. Apteridia		
1. *Branderius*	3. *Gymnopsis*	
(*Cecilia,* Lacepède)	(*Gymnomurena,*	
2. *Anopsus*	Lacepède)	
(*Murenoblenna,*	4. *Helmictis*	
Lacepède)	5. *Oxystomus*	
2. Sous-famille. Murenidia		
6. *Rincoxis*	10. *Murena*	
7. *Zebriscium*	(*Gymnothorax,*	
8. *Pterurus*	Bloch; *Murenophis,*	
9. *Dalophis*	Lacepède)	
3. Sous-famille. Catremia		
11. *Synbranchus*	12. *Sphagebranchus*	
(*Unibranchapertura,*		
Lacepède)		
29. Famille. PLAGIOSTOMIA		
1. Sous-famille. Antacea		
1. *Carcharias*	7. *Galeus*	12. *Oxynotus*
2. *Heptranchias*	8. *Sphyrnias*	13. *Squatina*
3. *Alopias*	9. *Hexanchus*	14. *Pristis*
4. *Isurus*	10. *Dalatias*	15. *Aodon*
5. *Cerictius*	11. *Squalus*	16. *Etmopterus*
6. *Tetroras*		
2. Sous-famille. Platosomia		
17. *Rhinobatus*	22. *Torpedo*	27. *Sephenia*
18. *Platopterus*	23. *Dipturus*	28. *Megabatus*
(*Raja,* Linnaeus)	24. *Mobula*	29. *Dasyatis*
19. *Leiobatus*	25. *Ictaetus*	30. *Uroxys*
20. *Epinotus*	26. *Cephaleutherus*	31. *Apturus*
21. *Lymnea*		
30. Famille. CYCLOSTOMIA		
1. Sous-famille. Lampredia		
1. *Petromyzon*	2. *Lampreda*	3. *Pricus*
2. Sous-famille. Myxinia		
4. *Gastrobranchus*	5. *Myxine*	

*These two supposed genera are already included in his seventh family.

TABLE 15. Here is the classification of de Blainville as it was printed in the *Journal de Physique* 83:254–55, 1816, and, in a somewhat different format, in his *Organisation des animaux, ou Principes d'anatomie comparée* [1822; his table 6].

CLASSE V. POISSONS
Sous-classe I. GNATHODONTES ou Osseux
Section I. SQUAMMODERMES
Ordre I. TÉTRAPODES
Sous-ordre I. HYPOGASTRIQUES ou Abdominaux
MÉTROSOMES [ordinary]
Brochets *Saumons* *Carpes*
Harengs
SILUROSOMES [siluroid]
Silures
SUBENCHÉLISOMES [long and cylindrical]
Cobites
Sous-ordre II. PROGASTRIQUES ou Sub-thoraciques
MÉTROSOMES [ordinary]
Muges
Sous-ordre III. THORACIQUES
MÉTROSOMES [ordinary]
Léïopomes *Acanthopomes*
(Labre) (Perches)
LEPTOSOMES [short and compressed]
Chaetodon *Zée*
ATRACTOSOMES [fusiform]
Scombre
CÉPHALOSOMES [head large]
Cottes *Trigles*
SUBENCHÉLISOMES [long and subcylindrical]
Gobie *Callionymes*
ENCHÉLISOMES [long and cylindrical]
Echeneis *Caepole* *Gymnètre*
Sous-ordre IV. TRACHÉLIQUES ou Jugulaires
MÉTROSOMES [ordinary]
Gades
CÉPHALOSOMES [head large]
Batrachus
HÉTÉROSOMES [asymmetrical]
Pleuronecte
SUBTÉNIOSOMES [long and subcompressed]
Blennie
SUBENCHÉLISOMES [long and subcylindrical]
Ordre II. DIPODES
ATRACTOSOMES [fusiform]
Xiphias

Continued on next page

Table 15—*continued*

LEPTOSOMES [highly compressed]
Stromatée
SUBTÉNIOSOMES [long and slightly compressed]
Ammodyte
TÉNIOSOMES [long and highly compressed]
Trichiure
SUBENCHÉLISOMES [long and subcylindrical]
Gymnote
ENCHÉLISOMES [long and cylindrical]
Anguille
Ordre III. APODES
ENCHÉLISOMES [long and cylindrical]
Muraenophis
Section II. HÉTÉRODERMES
SYNOPTÈRES [pelvic fins united]
Cycloptère
BRACHIOPTÈRES [pectoral fins armlike]
Baudroie
PELVAPTÈRES [pelvic fins absent]
Coffre *Diodon*
ACANTHOPTÈRES [dorsal fin spiny]
Balistes
HÉTÉROPTÈRES [fins variable, sometimes absent]
Syngnathe
Sous-classe II. DERMONDONTES ou Cartilagineux
SKÉLIPODES [pelvic fins anterior to anus]
Esturgeons
PELVIPODES [pelvic fins surrounding anus]
Sélaques
APODES [pelvic fins absent]
Lamproie

TABLE 16. This is my classification as it appears in my *Règne animal* of 1816 [see Cuvier 1816, vol. 2]. I must say, however, that when it was published I had no knowledge of Rafinesque's works. I have tried to improve on it in the present work [see table 24].

POISSONS

CHONDROPTÉRYGIENS

CHONDROPTÉRYGIENS à branchies fixes [cartilaginous fishes with fixed gills]

- Suceurs
  - *Lamproies*
    - *Lamproies* proper
    - *Ammocètes*
  - *Gastrobranches*
- Sélaciens
  - *Squales*
    - *Roussettes*
    - *Squales* proper
      - *Requins*
      - *Lamies*
      - *Marteaux*
      - *Milandres*
      - *Emissoles*
      - *Grisets*
      - *Pélerins*
      - *Cestracions*
      - *Aiguillats*
      - *Humantins*
      - *Leiches*
    - *Anges*
    - *Scies*
  - *Raies*
    - *Rhinobates*
    - *Rhinas*
    - *Torpilles*
    - *Raies* proper
    - *Pastenagues*
    - *Mourines*
    - *Céphaloptères*
  - Chimerès
    - *Chimères* proper
    - *Callorinques*

CHONDROPTÉRYGIENS à branchies libres [cartilaginous fishes with free gills]

- Sturioniens
  - *Esturgeons*
  - *Polyodons*

OSSEUX

PLECTOGNATHES

- Gymnodontes
  - *Diodons*
  - *Tétrodons*
  - *Mules*
- Sclérodermes
  - *Balistes*
    - *Balistes* proper
    - *Monacanthes*
    - *Alutères*
    - *Tetracanthes*
  - *Coffres*

LOPHOBRANCHES

- *Syngnathes*
  - *Syngnathes* proper
  - *Hippocampes*
  - *Solénostomes*
- *Pégases*

MALACOPTÉRYGIENS ABDOMINAUX

- Salmones
  - *Saumons*
    - *Saumons* proper
      - *Truites*
    - *Eperlans*
    - *Ombres*
    - *Argentines*
    - *Characins*
      - *Curimates*
      - *Anostomes*
      - *Serrasalmes*
      - *Piabuques*
      - *Tétragonoptères*
      - *Raüs*
      - *Hydrocyns*
      - *Cytharines*
    - *Saurus*
    - *Scopèles*
    - *Aulopes*
    - *Serpes*
  - *Sternoptyx*
- Clupes
  - *Harengs*

Continued on next page

Table 16—*continued*

    - *Harengs* proper
    - *Mégalopes*
    - *Anchois*
    - *Thrisses*
    - *Odontognathes*
    - *Pristigastres*
    - *Notoptères*
  - *Elopes*
  - *Chirocentres*
  - *Erythrins*
  - *Amies*
  - *Vastrés*
  - *Lépisostées*
  - *Polyptères*
- Esoces
  - *Brochets*
    - *Brochets* proper
    - *Galaxies*
    - *Microstomes*
    - *Stomias*
    - *Chauliodes*
    - *Salanx*
    - *Orphies*
    - *Scombrésoces*
    - *Demibecs*
  - *Exocets*
  - *Mormyres*
- Cyprins
  - *Carpes*
    - *Carpes* proper
    - *Barbeaux*
    - *Goujons*
    - *Tanches*
    - *Cirrhines*
    - *Brèmes*
    - *Labéons*
    - *Ables*
    - *Gonorhynques*
  - Loches
    - *Anableps*
    - *Poecilies*
    - *Lebias*
    - *Cyprinodons*
- Siluroïdes
  - *Silures*
    - *Silures* proper
      - *Silures* sensu stricto
      - *Schilbés*
    - *Machoirans*
      - *Pimelodes*
        - *Shals*
        - *Pimelodes* proper
        - *Bagres*
      - *Agénéioses*
      - *Doras*
    - *Hétérobranches*
      - *Macroptéronotes*
      - *Hétérobranches* proper
    - *Plotoses*
    - *Callichtes*
  - *Malaptérures*
  - *Asprèdes*
  - *Loricaires*
    - *Hypostomes*
    - *Loricaires* proper

MALACOPTÉRYGIENS SUBBRACHIENS
- Gadoïdes
  - *Gades*
    - *Morues*
    - *Merlans*
    - *Merluches*
      - *Lotes*
      - *Mustèles*
    - *Brosmes*
    - *Phycis*
    - *Raniceps*
  - *Lépidolèpres*
  - *Macroures*
- Poisson plats
  - *Pleuronectes*
    - *Plies*
    - *Flétans*
    - *Turbots*
    - *Soles*
    - *Monochires*
    - *Achires*

Continued on next page

Table 16—*continued*

- Discoboles
  - *Lépadogastres*
    - *Lépadogastres* proper
    - *Gobiésoces*
  - *Cycloptères*
    - *Lumps*
    - *Liparis*
  - *Echenéis*
  - *Ophicéphales*
- MALACOPTÉRYGIENS APODES
  - Anguilliformes
    - *Anguilles*
      - *Anguilles* proper
        - *Anguilles* sensu stricto
        - *Congres*
      - *Ophisures*
      - *Murènes*
        - *Gymnomurènes*
    - *Sphagébranches*
      - *Aptérichtes*
    - *Synbranches*
    - *Alabes*
    - *Gymnotes*
      - *Gymnotes* proper
      - *Carapes*
      - *Aptéronotes*
    - *Leptocéphales*
    - *Donzelles*
      - *Donzelles* proper
      - *Fierasfers*
    - *Equilles*
- ACANTHOPTÉRYGIENS
  - Taenioïdes
    - *Rubans*
    - *Lophotes*
    - *Régalecs*
    - *Gymnètres*
    - *Trachyptères*
    - *Gymnogastres*
    - *Ceintures*
    - *Jarretières*
    - *Styléphores*
  - Gobioïdes
    - *Blennies*
      - *Blennies* proper
      - *Pholis*
      - *Salarias*
      - *Clinus*
      - *Gonnelles*
      - *Opistognathes*
    - *Anarhiques*
    - *Gobies*
      - *Gobies* proper
      - *Gobioïdes*
      - *Taenioïdes*
      - *Périophtalmes*
      - *Eléotris*
    - *Sillago*
    - *Callionymes*
      - *Trichonotes*
      - *Coméphores*
  - Labroïdes
    - *Labres*
      - *Labres* proper
      - *Girelles*
      - *Crénilabres*
      - *Sublets*
      - *Chéilines*
      - *Filous*
      - *Gomphoses*
    - *Rasons*
    - *Chromis*
    - *Scares*
    - *Labrax*
  - Percoïdes
    - A dorsale unique [dorsal fin single]
      - A mâchoires protractiles [jaws protractile]
        - *Picarels*
    - A dents tranchantes [teeth sharp]
        - *Bogues*
    - A dents en pavé [teeth like paving stones]
        - *Spares*

Continued on next page

Table 16—*continued*

Sargues
Daurades
Pagres
A dents en crochets [teeth hooked]
Dentés
Lutjans
Diacopes
Cirrhites
Bodians
Serrans
Plectropomes
A dents en velours [teeth covered in velvet]
Canthères
Cicles
Pristipomes
Scolopsis
Diagrammes
Chéilodactyles
Microptères
Grammistes
Priacanthes
Polyprions
Soghos
Gremilles
Stellifères
Rascasses
Rascasses proper
Synancées
Ptérois
Taenianotes
A dorsale double [dorsal fin double]
A dorsales très-séparées [dorsal fins widely separated]
Ventrales abdominales [pelvic fins abdominal]
Athérines
Sphyrènes
Paralépis
Ventrales subbrachiennes [pelvic fins beneath the gills]
Mulles
Pomatomes
Muges
A dorsales rapprochées [dorsal fins close together]
A tête armée [head armed]
Perches
Perches proper
Centropomes
Enoploses
Sandres
Esclaves
Apogons
Sciènes
Cingles
Ombrines
Lonchures
Sciènes proper
Pogonias
Otolithes
Ancylodons
Percis
Vives
A tête cuirassée [head hardened]
Uranoscopes
Trigles
Trigles proper
Malarmats
Pirabèbes
Céphalacanthes
Lépisacanthes
Chabots
Chabots proper
Aspidophores
Platycéphales
Batracoïdes
A pectorales en forme de bras [pectoral fins armlike]
Baudroies
Baudroies proper
Chironectes
Malthées
Scombéroïdes
A deux dorsales [dorsal fin double]
Scombres
Maquereaux

Continued on next page

Table 16—*continued*

- - - *Thons*
    - *Germons*
    - *Caranx*
    - *Citules*
    - *Sérioles*
    - *Pasteurs*
  - Vomers
    - *Sélènes*
    - *Gals*
    - *Argyreyoses*
  - Tétragonures
- A première dorsale divisée en épines [first dorsal fin spinulose]
  - Rhinchobdelles
    - *Macrognathes*
    - *Mastacembles*
  - Epinoches
    - *Epinoches* proper
    - *Gastrés*
    - *Centronotes*
    - *Liches*
    - *Ciliaires*
- A dorsale unique [dorsal fin single]
  - Dents en velours [teeth covered in velvet]
    - *Dorées*
      - *Dorées* proper
      - *Capros*
      - *Equula*
      - *Ménés*
    - *Atropus*
    - *Trachichtes*
    - *Chrysotoses*
    - *Espadons*
      - *Espadons* proper
      - *Voiliers*
    - *Coryphènes*
      - *Centrolophes*
      - *Leptopodes*
      - *Coryphènes* proper
      - *Oligopodes*
  - Dents tranchantes [teeth sharp]
    - *Sidjans*
    - *Acanthures*
      - *Aspisures*
      - *Prionures*
    - *Nasons*
- Squammipennes
  - A dents en soie ou en velours [teeth covered with silk or velvet]
    - *Chaetodons*
      - *Chaetodons* proper
        - *Chaetodons* sensu stricto
        - *Chelmons*
        - *Platax*
        - *Héniochus*
        - *Ephippus*
      - *Holacanthes*
    - *Acanthopodes*
    - *Osphromènes*
      - *Osphromènes* proper
      - *Trichopodes*
    - *Archers*
    - *Kurtes*
    - *Anabas*
    - *Caesio*
    - *Castagnoles*
  - A dents sur une seule rangée [teeth in a single row]
    - *Stromatées*
      - *Fiatoles*
      - *Sésérinus*
    - *Piméleptères*
    - *Kyphoses*
    - *Plectorhynques*
    - *Glyphisodons*
    - *Pomacentres*
    - *Amphiprions*
    - *Premnas*
  - A deux dorsales [dorsal fin double]
    - *Temnodons*
    - *Chevaliers*
    - *Polynèmes*
- Bouches en flute [mouth flutelike]
  - *Fistulaires*
    - *Fistulaires* proper
    - *Aulostomes*
  - *Centrisques*
    - *Centrisques* proper
    - *Amphisiles*

TABLE 17. Here is the classification of Goldfuss [as it appears in volume 2 of his *Handbuch der Zoologie,* published in 1820 (v–x)].

Classe. PISCES (Fische)

Erste Ordnung. GASTEROPTERYGII. Bauchflosser [pelvic fins abdominal]

I. Familie. LEPTOCEPHALA. Schmalköpfe [narrow-headed forms]

Clupea	*Amia*	*Salmo*	*Esox*
Elops	*Poecilia*	*Coregonus*	*Sudis*
Chirocentrus	*Atherina*	*Characinus*	*Polypterus*
Synodus	*Cyprinus*	*Scopelus*	*Lepisosteus*

II. Familie. RHYNCHOCEPHALA. Schnabelköpfe [beaked forms]

Centriscus	*Mormyrus*	*Acanthonotus*	*Fistularia*

III. Familie. CYRTOCEPHALA. Stussköpfe [curved-headed forms]

Mugil	*Sphyraena*	*Exocetus*	*Polynemus*

IV. Familie. PLATYCEPHALA. Breitköpfe [wide-headed forms]

Loricaria	*Cobitis*	*Malapterurus*	*Silurus*
Cataphractus	*Anableps*	*Platystacus*	

Zweite Ordnung. PEROPTERYGII. Kahlbäuche [pelvic fins absent]

I. Familie. OPHIOIDES. Schlangenfische [snakelike fishes]

Leptocephalus	*Ammodytes*	*Rhynchobdella*	*Ophidium*

II. Familie. ENCHELYOIDES. Aale [eels]

Gymnothorax	*Anguilla*	*Gymnotus*
Apterichthys	*Trichiurus*	

III. Familie. XIPHONOTI. Schwerdtrücken [highly compressed forms]

Gnathobolus	*Pomatias*	*Stromateus*
Gymnogaster	*Rhombus*	*Sternoptyx*

IV. Familie. MACRORHYNCHI. Grossmäuler [large-mouthed forms]

Anarrhichas

Dritte Ordnung. STERNOPTERYGII. Brustflosser [pelvic fins thoracic]

I. Familie. ORTHOSOMATA. Barsche [perches]

Gadus	*Labrus*	*Xyrichthys*	*Holocentrus*
Mullus	*Ophicephalus*	*Sparus*	*Coryphaena*
Sciaena	*Amphacanthus*	*Lutjanus*	
Perca	*Scarus*	*Bodianus*	

II. Familie. TAENIOSOMATA. Bandfische [ribbonfishes]

Regalecus	*Tachypterus*	*Cepola*	*Lepidoleprus*
Gymnetrus	*Lepidopus*	*Macrourus*	*Lophotes*

III. Familie. LEPTOSOMATA. Schmalfische [narrow fishes]

Pleuronectes	*Monocentris*	*Zeus*	*Toxotes*
Pimelopterus	*Gasterosteus*	*Atropus*	*Kurtus*
Glyphisodon	*Scomber*	*Acantharus*	*Brama*
Plectorhynchus	*Tetragonurus*	*Monoceros*	*Anabas*
Premnas	*Xiphias*	*Chaetodon*	

Continued on next page

Table 17—*continued*

IV. Familie. CEPHALOTES. Dickköpfe [thick-headed forms]			
Batrachus	*Blennius*	*Percis*	*Scorpaena*
Uranoscopus	*Gobius*	*Callionymus*	*Cottus*
Echeneis	*Trachinus*	*Trigla*	
Vierte Ordnung. CHONDROPTERYGII. Knorpelfische [cartilaginous fishes]			
I. Familie. MICROSTOMATA. Kleinmäuler [small-mouthed forms]			
Gnathodon	*Balistes*	*Solenostomus*	*Polyodon*
Ostracion	*Syngnathus*	*Pegasus*	*Accipenser*
II. Familie. CYCLOSTOMATA. Saugmäuler [sucker-mouthed forms]			
Gastrobranchus	*Petromyzon*		
III. Familie. MACROSTOMATA. Grossmäuler [large-mouthed forms]			
Cyclopterus	*Lepadogaster*	*Chironectes*	*Lophius*
IV. Familie. PLAGIOSTOMATA. Queermäuler [oblique-mouthed forms]			
Chimaera	*Rhinobates*	*Raja*	*Squalus*

TABLE 18. Risso's classification as outlined in the second edition [of his *Ichthyologie de Nice,* volume 3 of *Histoire naturelle des principales productions de l'Europe méridionale et particulièrement de celles des environs de Nice et des Alpes Maritimes* (see Risso 1827b, 99–112)].

POISSONS

I. Série. POISSONS CHONDROPTÉRYGIENS

I. Ordre. CHONDROPTÉRYGIENS À BRANCHIES FIXES

I. Famille. PÉTROMYZIDES

Pétromyzon, Lamproie

II. Famille. SQUALIDES

Scyllium, Roussettes	*Acanthias, Aiguillat*
Carcharias, Requin	*Centrina, Humantin*
Lamia, Lamie	*Scymnus, Liche*
Zygaena, Marteau	*Squatina, Squatine*
Mustellus, Emissolle	*Pristis, Scie*
Notidanus, Griset	

III. Famille. RAIEDES

Torpedo, Torpille	*Myliobatis, Mourine*
Raia, Raie	*Cephaloptera, Céphaloptère*
Trygon, Pastenague	

II. Ordre. CHONDROPTÉRYGIENS À BRANCHIES LIBRES

IV. Famille. ESTRUGEONIDES

Acipenser, Esturgeon

V. Famille. BAUDROIDES

Lophius, Baudroie

II. Série. POISSONS OSSEUX

III. Ordre. PLECTOGNATHES

I. Famille. GYMNODONTES

Cephalus, Lune

II. Famille. BALISTIDES

Balistes, Baliste	*Ostracion, Coffre*

IV. Ordre. LOPHOBRANCHES

Syngnathus, Syngnathe	*Scyphius, Scyphius*
Hippocampus, Hippocampe	

V. Ordre. APODES

I. Division. APODES MALACOPTÉRYGIENS

I. Famille. MURÉNIDES

Murena, Murène	*Anguilla, Anguille*
Murenophis, Murénophis	*Conger, Congre*
Sphagebranchus, Sphagebranche	*Leptocephalus, Leptocéphale*

II. Famille. OPHISURIDES

Ophisurus, Ophisure

II. Division. APODES ACANTHOPTÉRYGIENS

III. Famille. XIPHOIDES

Xiphias, Espadon	*Ophidium, Ophidie*
Ammodytes, Ammodyte	

VI. Ordre. JUGULAIRES

I. Division. JUGULAIRES MALACOPTÉRYGIENS

Continued on next page

Table 18—*continued*

I. Famille. GADOIDES
- *Onos, Onos*
- *Lotta, Lotte*
- *Mora, Mora*
- *Merlucius, Merluche*
- *Phycis, Phycis*
- *Morua, Morue*
- *Merlangus, Merlan*

II. Famille. BLENNIOIDES
- *Blennius, Blennie*
- *Salarias, Salarias*
- *Clinus, Cline*
- *Tripterygion, Triptérygion*

III. Famille. LÉPIDOLÉPRIDES
- *Lepidoleprus, Lépidolèpre*

IV. Famille. PLEURONECTIDES
- *Hippoglossus, Fletan*
- *Solea, Sole*
- *Rhombus, Turbot*
- *Monochirus, Monochire*

II. Division. JUGULAIRES ACANTHOPTÉRYGIENS

V. Famille. TRACHINIDES
- *Trachinus, Vive*
- *Uranoscopus, Uranoscope*
- *Callionymus, Callionyme*

VI. Famille. DIANIDES
- *Diana, Diane*

VII. Ordre. THORACIQUES

I. Division. THORACIQUES MALACOPTÉRYGIENS

I. Famille. ECHÉNÉIDES
- *Echeneis, Echénéis*

II. Famille. GOBIOIDES
- *Lepadogaster, Lépadogastère*
- *Gobius, Gobie*

III. Famille. FIATOLOIDES
- *Aphia, Aphie*
- *Fiatola, Fiatole*

IV. Famille. TÉNIOIDES
- *Lepidopus, Lépidope*
- *Lophotes, Lophote*
- *Cepola, Ruban*
- *Gymnetrus, Gymnètre*
- *Bogmarus, Vogmare*

II. Division. THORACIQUES ACANTHOPTÉRYGIENS

V. Famille. LABROIDES
- *Labrus, Labre*
- *Julis, Girelle*
- *Crenilabrus, Crénilabre*
- *Coricus, Sublet*
- *Novacula, Razon*

VI. Famille. CORYPHÉNOIDES
- *Centrolophus, Centrolophe*
- *Oligopus, Oligope*
- *Coryphena, Coryphène*
- *Lampris, Chrysostose*
- *Ausonia, Ausonie*

VII. Famille. SPAROIDES
- *Chromis, Chromis*
- *Smaris, Picarel*
- *Boops, Bogue*
- *Sargus, Sargue*
- *Charax, Charax*
- *Aurata, Dorade*
- *Pagrus, Pagre*
- *Dentex, Denté*
- *Cantharus, Canthare*

Continued on next page

Table 18—*continued*

VIII. Famille. SCORPÉNIDES	
Holocentrus, Soldado	*Ailopon, Ailopon*
Scorpena, Scorpène	*Zeus, Dorée*
Serranus, Serran	*Capros, Capros*
IX. Famille. TÉTRAGONURIDES	
Tetragonurus, Tétragonure	
X. Famille. MUGILIDES	
Apogon, Apogon	*Pomatomus, Pomatome*
Mullus, Mulle	*Mugil, Muge*
XI. Famille. TRIGLIDES	
Trigla, Trigle	*Dactylopterus, Pirapède*
Peristedion, Malarmat	
XII. Famille. PERCHIDES	
Cottus, Chabot	*Umbrina, Ombrine*
Perca, Perche	*Sciaena, Sciène*
XIII. Famille. SCOMBÉROIDES	
Scomber, Maquereau	*Caranx, Caranx*
Thynnus, Thon	*Citula, Citule*
Orcynus, Germon	*Seriola, Sériole*
XIV. Famille. CENTRONOTIDES	
Gasterosteus, Epinoche	*Lichia, Lichie*
Centronotus, Centronote	
XV. Famille. SQUAMIPENNES	
Chaetodon, Chétodon	*Lepterus, Leptère*
Brama, Castagnolle	
VIII. Ordre. Abdominaux	
I. Division. ABDOMINAUX MALACOPTÉRYGIENS	
I. Famille. CYPRINIDES	
Cyprinus, Carpe	*Leuciscus, Able*
Barbus, Barbeau	
II. Famille. EXOCÉIDES	
Stomias, Stomias	*Scomberosox, Scombresoce*
Chauliodes, Chauliode	*Exocetus, Exocet*
Belone, Orphie	
III. Famille. CLUPÉOIDES	
Macrostoma, Macrostome	*Engraulis, Anchois*
Alepocephalus, Alépocéphale	*Alpismaris, Alpesmer*
Clupanodon, Clupanodon	
IV. Famille. SALMONIDES	
Salmo, Truite	*Saurus, Saurus*
Argentina, Argentine	*Scopelus, Scopèle*
II. Division. ABDOMINAUX ACANTHOPTÉRYGIENS	
V. Famille. ATHÉRINIDES	
Atherina, Athérine	*Paralepis, Paralepis*
Sphyrena, Sphyrène	*Microstoma, Microstome*
VI. Famille. CENTRISCIDES	
Centriscus, Centrisque	

TABLE 19. [The classification of] Oken [as it appears in his] *Lehrbuch der Naturphilosophie* [1809–1811], 3:301–4.

Fleischthiere [flesh animals, i.e., fishes].

1. Ordnung. Bauchfische [abdominal fishes]
   Fishes without scales
2. Ordnung. Brustfische [thoracic fishes]
   Fishes with scales: perch, pike, salmon, etc.
3. Ordnung. Gliederfische [limbed fishes]

Fistularia	*Pegasus*	*Ostracion*
Centriscus	*Syngnathus*	*Diodon*, etc.

4. Ordnung. Kopffische [head fishes]
   Cartilaginous fishes: lampreys, rays, sharks

TABLE 20. [A brief summary of the classification of] Oken [as presented in the main body of the text of his] *Lehrbuch der Naturgeschichte* [1815–16], 2:12–13.

1. Grätenfische [bony fishes]

- A. Regelmässige [regular]
  - I. Bauchflossen verrückt [pelvic fins displaced]
    - I. Ordnung. Quallenfische, Aale [jellyfish fishes, eels]
      - *Anguilla*, etc.
    - II. Ordnung. Lechfische, Dorsche [worm fishes, codfishes]
      - *Blennius*, *Gadus*, *Scomber*
    - III. Ordnung. Kerffische, Brassen [insect fishes, breams]
      - *Labrus*, *Sciaena*
  - II. Bauchflossen hinten—Bauchflosser [pelvic fins abdominal]
    - IV. Ordnung. Fischfische, Rapfen [fish fishes, minnows]
      - *Mugil*, *Cyprinus*
    - V. Ordnung. Lurchfische, Heuche [amphibian fishes, pikes]
      - *Cobitis*, *Silurus*, *Salmo*, *Esox*
- B. Unregelmässige [irregular]
    - VI. Ordnung. Vogelfische, Groppe [bird fishes, sculpins]
      - *Callionymus*, *Gobius*, *Chaetodon*, *Pleuronectes*

2. Knorpelfische [cartilaginous fishes]

- VII. Ordnung. Suckfische, Knurfe [mammal fishes, rays]
  - *Acipenser*, *Diodon*, *Raja*
  - *Lophius*, *Petromyzon*, *Squalus*

TABLE 21. Here is the table of Oken's second ichthyological method [a more detailed outline of his classification as it appears in the main body of the text of his *Lehrbuch der Naturgeschichte* (1815–16, 2:13–178), in which he recognizes seven orders].

Order / Family	1.	2.	3.	4.
I. Ordnung. Quallenfische—Aale [jellyfish fishes, eels]				
1. Muränen [morays]	1. *Apterichthys*	2. *Synbranchus*	3. *Sphagebranchus*	4. *Muraena*
2. Aale [eels]	1. *Anguilla*	2. *Gymnotus*	3. *Ophidium*	4. *Ammodytes*
3. Schmälte	1. *Trichiurus*	2. *Leptocephalus*	3. *Regalecus*	4. *Anarhichas*
4. Näle	1. *Caepola*	2. *Gymneter*	3. *Lepidopus*	4. *Centronotus*
II. Ordnung. Lechfische—Dösche [worm fishes, codfishes]				
1. Truschen [burbots]	1. *Blennius*	2. *Phycis*	3. *Pteraclis*	4. *Gadus*
2. Klege	1. *Echeneis*	2. *Eleotris*	3. *Gobiomoroides*	4. —
3. Thunne [tunas]	1. *Scomber*	2. *Trachinotus*	3. *Caranx*	4. *Pomatomus*
4. Stichlinge [sticklebacks]	1. *Gasterosteus*	2. *Centronotus*	3. *Lepisacanthus*	4. *Centrogaster*
III. Ordnung. Kerffische—Warsche [insect fishes, breams]				
1. Persinge [perches]	1. *Sciaena*	2. *Bodianus*	3. *Perca*	4. *Holocentrus*
2. Schratzen	1. *Gymnocephalus*	2. *Anthias*	3. *Lutjanus*	4. *Grammistes*
3. Frache [wrasses]	1. *Labrus*	2. *Calliodon*	3. *Ophicephalus*	4. *Sparus*
4. Horke [sea breams]	1. *Mullus*	2. *Scarus*	3. *Coryphaena*	4. *Macrourus*
IV. Ordnung. Fischfische—Rapfen [fish fishes, minnows]				
1. Ossane [mullets]	1. *Mugil*	2. *Mugilomorus*	3. *Acanthonotus*	4. *Exocoetus*
2. Guren [threadfins]	1. *Polynemus*	2. *Polydactylus*	3. *Cirrites*	4. *Cheilodactylus*
3. Renge [herrings]	1. *Clupea*	2. *Mene*	3. *Siche* (*Cyprinus cultratus*)	4. *Gasteropelecus*
4. Karsche [carps]	1. *Atherina*	2. *Argentina*	3. *Synodus*	4. *Cyprinus*
V. Ordnung. Lurchfische—Heuche [amphibian fishes, pikes]				
1. Grundeln [loaches]	1. *Cobitis*	2. *Anableps*	3. *Poecilia*	4. *Amia*
2. Waller [catfishes]	1. *Silurus*	2. *Platystomus*	3. *Doras*	4. *Loricaria*

Continued on next page

Table 21—*continued*

3. Salmen [salmon]			
1. *Serrosalmo*	2. *Characinus*	3. *Corregonus*	4. *Salmo*
4. Schurte [pikes]			
1. *Elops*	2. *Sphyraena*	3. *Chauliodes*	4. *Esox*
VI. Ordnung. Vogelfische—Groppe [bird fishes, sculpins]			
1. Grümpel			
1. *Callionymus*	2. *Percis*	3. *Uranoscopus*	4. *Trachinus*
2. Ulke			
1. *Gobius*	2. *Cottus*	3. *Scorpaena*	4. *Trigla*
3. Butte [flounders]			
1. *Pleuronectes*	2. *Zeus*	3. *Chaetodon*	4. *Stromateus*
4. Schniffe [snipefishes]			
1. *Centriscus*	2. *Mormyrus*	3. *Fistularia*	4. *Stylephorus*
VII. Ordnung. Suckfische—Knurfe [mammal fishes, rays]			
1. Querder [ammocoetes]			
1. *Myxine*	2. *Lampetra*	3. *Syngnathus*	4. *Pegasus*
2. Morke			
1. *Cyclopterus*	2. *Balistes*	3. *Ostracion*	4. *Gnathodon (Diodon, Tetrodon)*
3. Schirke [sturgeons]			
1. *Sturio*	2. *Polyodon*	3. *Xiphias*	4. *Istiophorus*
4. Krospel			
1. *Lophius*	2. *Raja*	3. *Squalus*	4. *Chimaera*

TABLE 22. This third method of Oken [which appears in the introduction to his *Lehrbuch der Naturgeschichte* (1815–16, 2:i–iv) and which recognizes only four orders] lists fishes as follows.

V. Klasse. Fleischthiere, Fische [flesh animals, fishes]

1. Ordnung. Fischfische, Aale [fish fishes, eels]

1. Aalaale, Muränen [eel eels]			
1. *Apterichthys*	2. *Synbranchus*	3. *Sphagebranchus*	4. *Muraena*
2. Heuchaale, Näle [pike eels]			
1. *Anguilla*	2. *Gymnotus*	3. *Ophidium*	4. *Ammodytes*
3. Warschaale, Schmälte [sparid eels]			
1. *Trichiurus*	2. *Leptocephalus*	3. *Cepola*	4. *Gymneter*
4. Knurfaale, Alfe [ray eels]			
1. —	2. *Anarrhichas*	3. *Xiphias*	4. *Zisius*

II. Ordnung. Lurchfische, Heuche [amphibian fishes, pikes]

1. Aalheuche, Dösche [eel pikes, cods]			
1. *Gadus*	2. *Echeneis*	3. *Gasterosteus*	4. *Scomber*
2. Heuchheuche, Groppe [pike pikes, sculpins]			
1. *Callionymus*	2. *Cottus*	3. *Gobius*	4. *Cyclopterus*
3. Warschheuche, Butte [sparid pikes, flounders]			
1. *Pleuronectes*	2. *Zeus*	3. *Chaetodon*	4. *Stromateus*
4. Knurfheuche, Schurte [ray pikes]			
1. *Cobitis*	2. *Silurus*	3. *Salmo*	4. *Esox*

III. Ordnung. Vogelfische, Warsche [bird fishes, sparids]

1. Aalwarsche, Guren [eel sparids]			
1. *Scorpaena*	2. *Trigla*	3. *Polynemus*	4. *Exocoetus*
2. Heuchwarsche, Persinge [pike sparids, perches]			
1. *Sciaena*	2. *Perca*	3. *Schrassen*	4. *Mullus*
3. Warschwarsche, Frache [sparid sparids]			
1. *Labrus*	2. *Sparus*	3. *Scarus*	4. *Coryphaena*
4. Knurfwarsche, Rapfen [ray sparids, minnows]			
1. *Mugil*	2. *Clupea*	3. *Atherina*	4. *Cyprinus*

IV. Ordnung. Suckfische, Knurfe [mammal fishes, rays]

1. Aalknurfe, Schniffe [eel rays]			
1. *Centriscus*	2. *Fistularia*	3. *Stylephorus*	4. *Syngnathus*
2. Heuchknurfe, Morke [pike rays]			
1. *Mormyrus*	2. *Balistes*	3. *Ostracion*	4. *Gnathodon*
3. Warschknurfe, Schirke [sparid rays]			
1. *Pegasus*	2. *Sturio*	3. *Chimaera*	4. *Lophius*
4. Knurfknurfe, Kröfpel [ray rays]			
1. *Myxine*	2. *Lampetra*	3. *Raja*	4. *Squalus*

TABLE 23. Here is the table of Oken's fourth ichthyological method [as published in his *Esquisse du système d'anatomie, de physiologie et d'histoire naturelle*, 1821b, 56–58].

X. Classe. OSSIERS [bony fishes]

I. Ordre. POISSONS GERMIERS [germinator fishes]

I. Tribu. Poissons spermiers [seed germinators]

Anguilles [eels]

1. *Apterichthys*	4. *Muraena*	7. *Ophidium*
2. *Sphagebranchus*	5. *Anguilla*	8. *Leptocephalus*
3. *Synbranchus*	6. *Gymnotus*	9. *Ammodytes*

II. Tribu. Poissons oviers [egg germinators]

Cépoles [bandfishes]

1. *Lophotus*	4. *Cepola*	7. *Stylephorus*
2. *Gymnetrus*	5. *Trachypterus*	8. *Lepidopus*
3. *Regalecus*	6. *Gymnogaster*	9. *Trichiurus*

III. Tribu. Poissons fétiers [envelope germinators]

Morues [codfishes]

1. *Pleuronectes*	4. *Macrourus*	7. *Centronotus*
2. *Echeneis*	5. *Phycis*	8. *Blennius*
3. *Platycephalus*	6. *Gadus*	9. *Anarrhichas*

II. Ordre. POISSONS SEXIERS [sex-bearing fishes]

IV. Tribu. Poissons reiniers [kidney sex-bearers]

Boulereaux [gobies]

1. *Gobius*	4. *Comephorus*	7. *Trachitus*
2. *Periophthalmus*	5. *Trichionotus*	8. *Trigla*
3. *Eleotris*	6. *Callionismus*	9. *Lepisacanthus*

V. Tribu. Poissons femelliers [female sex-bearers]

Maquereaux [mackerels]

1. *Chaetodon*	4. *Vomer*	7. *Rhynchobdella*
2. *Stromateus*	5. *Zeus*	8. *Gasterosteus*
3. *Eques*	6. *Coryphaena*	9. *Scomber*

VI. Tribu. Poissons masculiers [male sex-bearers]

Perches [perches]

1. *Otolithes*	4. *Cichla*	7. *Labrus*
2. *Sciaena*	5. *Serranus*	8. *Scarus*
3. *Perca*	6. *Dentex*	9. *Sparus*

III. Ordre. POISSONS ENTRAILLIERS [entrail-bearing fishes]

VII. Tribu. Poissons intestiers [intestinal entrail-bearers]

Silures [catfishes]

1. *Cobitis*	4. *Pimelodes*	7. *Doras*
2. *Anableps*	5. *Malapterurus*	8. *Heterobranchus*
3. *Poecilia*	6. *Silurus*	9. *Cataphractus*

VIII. Tribu. Poissons veiniers [blood vessel entrail-bearers]

Brochets [pikes]

1. *Antherina*	4. *Erythrinus*	7. *Sternoptyx*
2. *Spiraena*	5. *Lepisosteus*	8. *Gasteropelecus*
3. *Polypterus*	6. *Esox*	9. *Salmo*

IX. Tribu. Poissons pulmoniers [pulmonary entrail-bearers]

Carpes [carps]

1. *Mullus*	4. *Clupea*	7. *Gonorhynchus*
2. *Mugil*	5. *Elops*	8. —
3. —	6. *Exocoetus*	9. *Cyprinus*

Continued on next page

Table 23—*continued*

IV. Ordre. POISSONS CARNIERS [flesh-bearing fishes]		
X. Tribu. Poissons ossiers [bony flesh-bearers]		
Baudroies [anglerfishes]		
I. Famille. Germières [seed flesh-bearers]		
1. *Lepadogaster*	2. —	3. *Cyclopterus*
II. Famille. Sexières [sex flesh-bearers]		
4. *Uranoscopus*	5. *Cottus*	6. *Batrachus*
III. Famille. Entraillières [entrail flesh-bearers]		
7. *Taenionotus*	8. *Synanceia*	9. *Scorpaena*
IV. Famille. Carnières [flesh flesh-bearers]		
10. *Malthe*	11. *Antennarius*	12. *Lophius*
XI. Tribu. Poissons masculiers [male flesh-bearers]		
Fistulaires [cornetfishes]		
I. Famille. Germières [seed-bearers]		
1. *Syngnathus*	2. —	3. *Solenostoma*
II. Famille. Sexières [sex-bearers]		
4. *Pegasus*	5. —	6. —
III. Famille. Entraillières [entrail-bearers]		
7. *Fistularia*	8. *Aulostoma*	9. —
IV. Famille. Carnières [flesh-bearers]		
10. *Centriscus*	11. *Amphisile*	12. *Mormyrus*
XII. Tribu. Poissons nerviers [nerve-bearing flesh-bearers]		
Esturgeons [sturgeons]		
I. Famille. Germières [seed-bearers]		
1. *Balistes*	2. *Triacanthus*	3. *Ostracion*
II. Famille. Sexiers [sex-bearers]		
4. *Tetrodon*	5. *Diodon*	6. *Orthragoriscus*
III. Famille. Entrailliers [entrail-bearers]		
7. *Platystacus*	8. *Loricaria*	9. *Lepidoleprus*
IV. Famille. Carnières [flesh-bearers]		
10. *Polyodon*	11. *Acipenser*	12. *Xiphias*
V. Ordre. POISSONS SENSIERS [sense-bearing fishes]		
XIII. Tribu. Poissons sensiers [sense-bearing fishes]		
Squales [sharks]		
1. Peaussier [skin sensitive]		
Myxine		
2. Languier [taste sensitive]		
Petromyzon		
3. Nasier [nasal sensitive]		
Chimaera		
4. Oreillier [ear sensitive]		
Raja		
5. Oculier [eye sensitive]		
Squalus		

15 Anatomical and Physiological Contributions of the Early Nineteenth Century

If the philosophy of nature did not contribute much to improving our knowledge of the natural history of fishes with regard to classification, it did encourage anatomical research that brought useful results.

The osteology of fishes had scarcely been touched by the beginning of the nineteenth century. It was in 1800 that Autenrieth, in his anatomy of the plaice *(plie)*, began to examine the analogy between the skeletal parts of fishes and those of the higher classes;[1] on the hyoid apparatus in particular he offered ideas that were later presented by others as new. In 1807 Geoffroy Saint-Hilaire compared the bones that support the pectoral fin with those of the shoulder, the upper arm, the lower arm, and the wrist bones of the higher animals[2] and brought to light the varieties and different uses of the thin bone behind the shoulder.[3] He also studied the apparatus that carries the branchial membrane and thought it was formed by the union of certain parts of the hyoid apparatus, the sternum, and the rib cartilages.[4] He regarded the opercula as parietals separated from the cranium.[5]

In 1811 Rosenthal began his works on the osteology of fishes with a memoir in which he described the bones in the head with great accuracy, but he was not so fortunate in understanding their analogy.[6] Since then (1812–22) he has published four collections of ichthyotomical plates in which he has carefully represented the skeletons of a fair number of fishes for which no osteology had until then been published.[7]

At that time I had also done much work on this subject and had gathered more than three hundred skeletons of fishes. I published in 1812,[8] 1815,[9] and 1816[10] my ideas on the osteology of this class, as well as various examples taken from individual species [fig. 49].

Some years before, Burtin and Duméril explained the relation between the cranium and the vertebrae.[11] In 1807 Oken tried to apply this

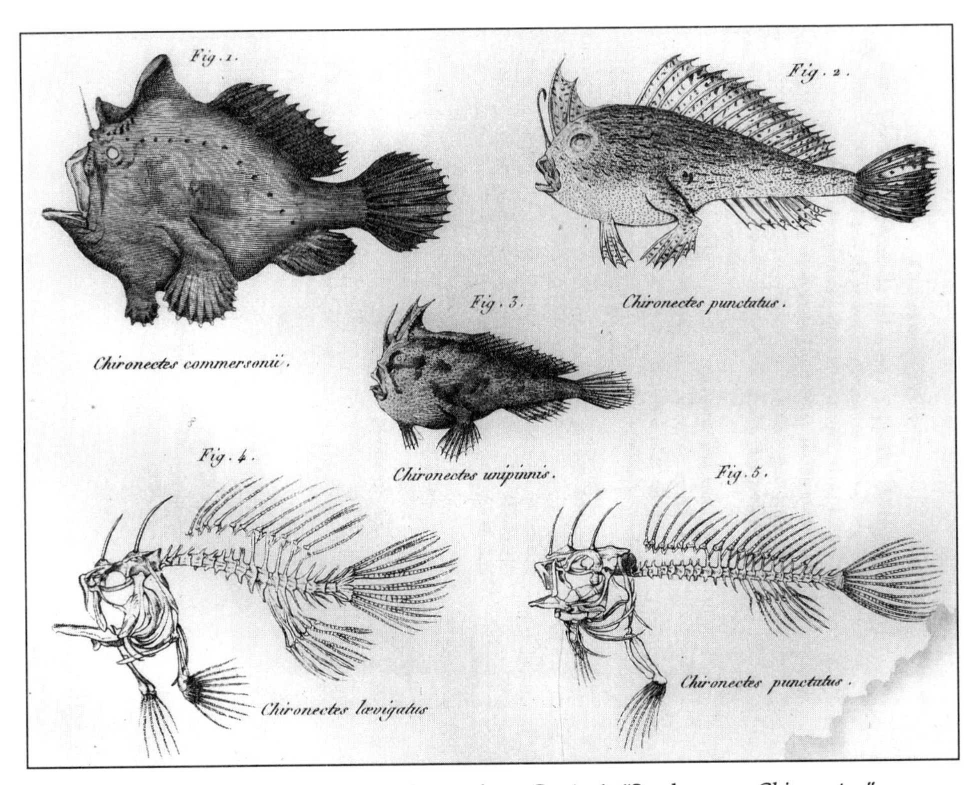

FIG. 49. Frogfishes and their skeletons from Cuvier's "Sur le genre *Chironectes*." *Mém. Mus. Nat. Hist. Nat.* (Paris), 1817, vol. 3, pl. 18. Photograph by Stanley A. Shockey, Classroom Support Services, University of Washington, Seattle.

idea to the structure of the head of animals according to the principles of his philosophy of nature; he thought it was formed of three vertebrae, but he examined it only in the quadrupeds.[12] Spix developed these views and modified the details in his great work *Cephalogenesis*, printed in 1815.[13] He represented the crania of several fishes and provided separate drawings of the bones composing them. He was the first to advance the idea that the opercular elements correspond to the ossicles of the middle ear.

Geoffroy Saint-Hilaire, who arrived at essentially the same theory on the opercular bones, published his hypothesis in 1818 in his *Philosophie anatomique* [fig. 50]. In the same publication, he developed his earlier ideas on the derivation of the branchial apparatus, which he regarded as analogous to the sternum, the hyoid bone, the larynx, and the trachea and its bronchi. In particular, he gave an exact description and enumeration of the parts composing this apparatus.[14]

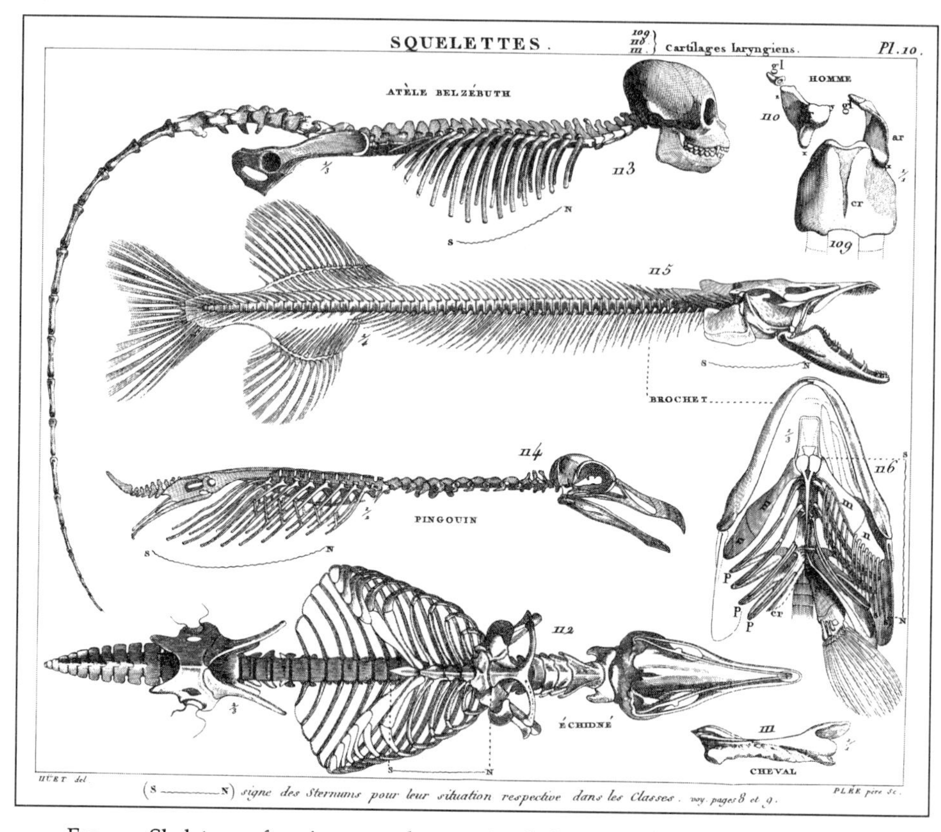

FIG. 50. Skeletons of various vertebrate animals from Etienne Geoffroy Saint-Hilaire's *Philosophie anatomique*, 1818, pl. 10.

In the same year (1818), Bojanus published some determinations on the bones of the head of fishes that were rather different from mine and Geoffroy Saint-Hilaire's.[15] Others appeared in 1820 by Fenner,[16] and in 1822 by Arendt; the latter proposed these theories in a special treatise on the cranium of the pike *(brochet)*.[17]

In 1818 Carus published his *Lehrbuch der Zootomy*, including a general description of the fish skeleton and some of his own ideas on the branchial apparatus, which he considered to be the only analogue of the thorax.[18] He did not discuss the nature of the opercular elements.

Also in that year, Schultze included many interesting facts on the osteology of fishes, particularly their vertebrae, in a memoir on the osteogenesis and subsequent development of the vertebral column in general.[19]

In 1820 Weber, in his treatise on the ears of various animals,[20] proposed the idea that the bony ossicles of the middle ear are those that in the common carp *(carpe)*, the catfish *(silure)*, and others, are placed between the cranium and the top of the swim bladder and that in fact communicate with the cavity that contains the labyrinth. The following year, Bojanus published a memoir in favor of this new theory.[21]

But in 1822 and 1824 Geoffroy Saint-Hilaire again took up the whole subject of the composition of the head and reiterated his opinion on the opercula. He preceded his work with a general theory on the composition of the vertebra, which he regarded as composed of nine pieces, or rather twelve; the head itself is a series of seven vertebrae and consequently contains eighty-four bones. The author made a special application of this theory to the head of the seabass *(Serranus gigas)* [fig. 51].[22]

The fish skeleton as a whole was the subject of two works published in Holland in 1822: Van der Hoeven's dissertation[23] and Bakker's osteography.[24] The latter work is accompanied by handsome lithographed illustrations representing various bony parts of several fishes. These two authors considered the opercular apparatus to be peculiar to fishes, but as regards the hyoid and branchial apparatus, they agree with Geoffroy Saint-Hilaire.

The second volume of *System der vergleichenden Anatomie* by Meckel, printed in 1824, also contains a well-written summary of fish osteology, and one hopes to find summaries as instructive on the other parts of their organization in the volumes to come. The author does not adopt the idea of the fusion of the sternum with the hyoid bone, or the dividing of the lower jaw to form the opercula, or other hypotheses of this kind, and in general does not feel obliged to find, bone for bone, the same elements in all animals; he even gives evidence that this concordance does not exist.[25]

Thus it is that fish osteology, in some ways born during the present time, has been brought to a high level of perfection.

Fish myology has not been nearly so well researched and is virtually limited to what I have said in my classes on comparative anatomy and what Carus has published more recently in his *Lehrbuch der Zootomy;* but I have conducted many studies on this subject for my comparative anatomy, from which I shall give an excerpt in the present work.

There has been more research on fish neurology. Weber published a comparative anatomy of the sympathetic nerve and showed an illustration of the encephalon of the common carp *(carpe)*.[26] A dissertation on the fish brain by Apostolos Arsakis appeared at Halle in 1813, in which the encephala of several species are described and illustrated and new ideas are proposed on the analogy of their tubercles.[27] The late Kuhl also described and illustrated several fish encephala in his materials for a comparative anatomy, printed in 1820.[28] That same year, the fish brain

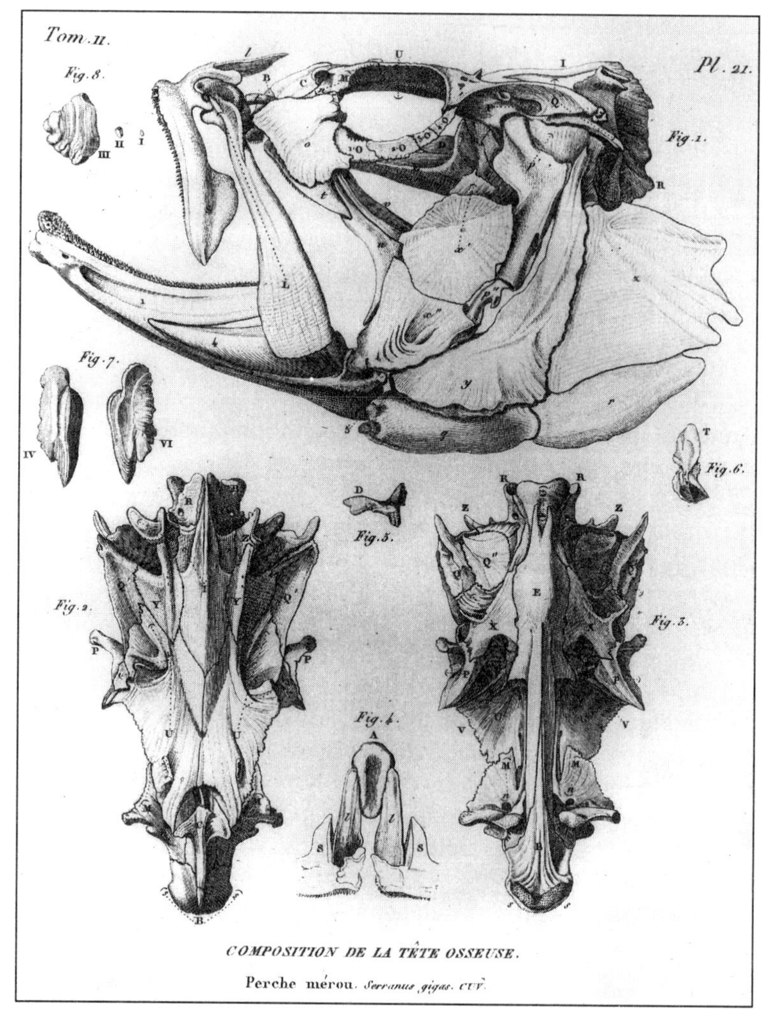

FIG. 51. Etienne Geoffroy Saint-Hilaire's osteology of the head of a seabass identified as *Serranus gigas*. *Mém. Mus. Hist. Nat.* (Paris), 1824a, vol. 11, pl. 21. Photograph by Stanley A. Shockey, Classroom Support Services, University of Washington, Seattle.

was also discussed in Fenner's thesis cited above. The work by Serres on the brain published from 1824 to 1827[29] and that by Desmoulins and Magendie on the nervous system in 1825[30] also discussed the encephala of many fishes, the latter containing research on the distribution of the nerves.

Fish sense organs were also the object of interesting observations. Soemmerring the Younger, in his work on the eye of animals, provided instructive cross sections of the fish eye.[31] Massalien[32] and Jurine[33] described the eye of the tuna *(thon);* Sir Everhard Home, that of *Squalus maximus.*[34]

Weber in the work cited above brought forth new and precise details on the fish inner ear and its relation to the exterior ear.[35] Detailed arrangements of this organ have been observed in a grenadier *(lépidoléprus)* by Otto,[36] and in an elephantfish *(mormyre),* a long-whiskered catfish *(pimélode),* and a freshwater hatchetfish *(serpe)* by Heusinger.[37]

Observations and drawings of the ear of some species are to be found in the dissertation on hearing by Pohl.[38] Geoffroy Saint-Hilaire has given us his own ideas on the otoliths of the auricular sac.[39] Duméril has advanced his views on the source of their sense of smell.[40] Geoffroy Saint-Hilaire later proposed rather different ones,[41] and also wrote on the analogy of the bones surrounding the nares.

Bailly examined the nature and mechanism of the thin dorsal-fin spines that the anglerfish *(baudroie)* carries on its head.[42] Geoffroy Saint-Hilaire discussed in particular the branchial sac in the anglerfish *(baudroie).*[43]

As for studies of fish thoracic and abdominal splanchnology, one is more likely to find them reported in anatomical monographs than in special treatises. There are many of these particular descriptions. Duméril wrote about the splanchnology of lampreys *(lamproies)* in general,[44] and Rathke wrote about that of the river lamprey *(lamproie de rivière)* in such a way as to leave nothing more to be desired on the subject of this curious genus.[45] Sir Everhard Home and de Blainville wrote about the splanchnology of the great sharks *(squales)* of northern seas.[46] Rathke wrote on that of the lumpfish *(lump).*[47]

Sir Everhard Home, in his magnificent work titled *Lectures on Comparative Anatomy,*[48] described and illustrated the stomachs and intestines of about thirty species, both European and foreign [fig. 52]. He also discussed the heart, the gills, and the reproductive organs of some of them; his observations on the lamprey *(lamproie),* the hagfish *(myxine),* and various sharks *(squales)* are particularly worthy of attention.

But the most important treatise on the abdominal viscera of fishes that has appeared recently is by Rathke, of Danzig, on the intestinal canal and reproductive organs. He described the visceral parts of fifty-six species, all from the Baltic Sea.[49] The same author published interesting memoirs on the liver, the portal vein system, and the auricle of the fish heart[50] and a series of fine observations on their genital organs and the way they develop.[51]

Tiedemann and Doellinger have discussed the heart. The former showed this organ in thirty-one species;[52] the latter wrote about the heart from a more general point of view and believed he had found a cavity

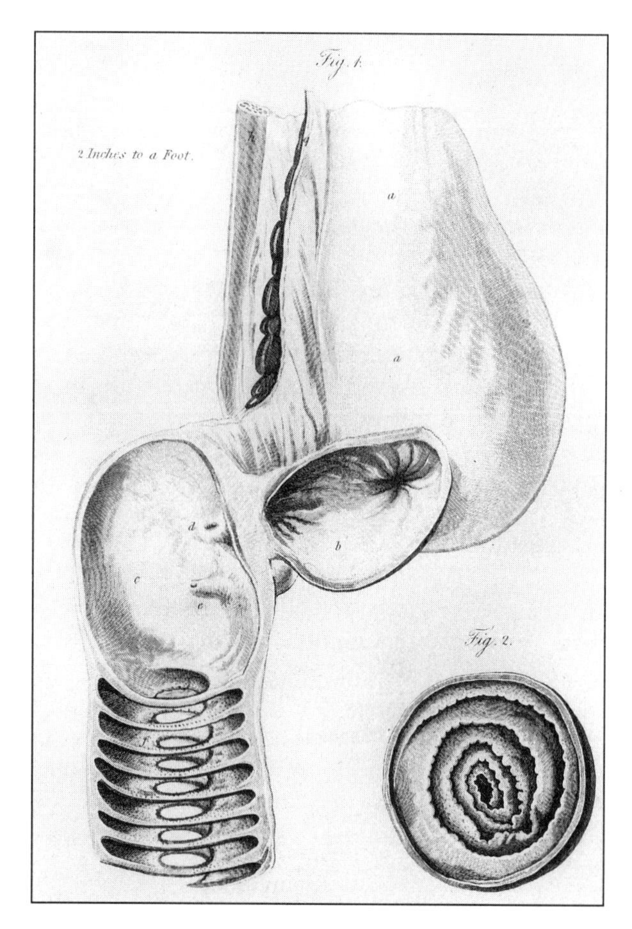

FIG. 52.
The stomach of a shark, showing the structure of the spiral valve, from Everhard Home's "An Anatomical Account of the *Squalus maximus.*" *Phil. Trans. Roy. Soc. Lond.*, 1809, vol. 99, pl. 8. Photograph by Stanley A. Shockey, Classroom Support Services, University of Washington, Seattle.

similar to the right ventricle in birds but that plays no part in the circulation.[53]

Quite recently (1827), Fohmann[54] discussed in great detail the lymphatic vessels in fishes and their relation to the veins.

The secretions and secretory organs of fishes have been studied with great care. To the information already given by Hunter[55] and others on the electric organs of the electric ray *(torpille)* and the knifefish *(gymnote)*, Geoffroy Saint-Hilaire added information on the same organs in the catfish *(silure)*,[56] and Rudolphi soon after gave a more detailed description.[57] Von Humboldt has performed extensive and valuable experiments on the knifefish *(gymnote)*.[58]

On the swim bladder, there are the observations of Delaroche[59] and my own [fig. 53];[60] on the air in the swim bladder, the experiments of Biot[61] and Configliachi;[62] and on its functions, a special memoir by Tre-

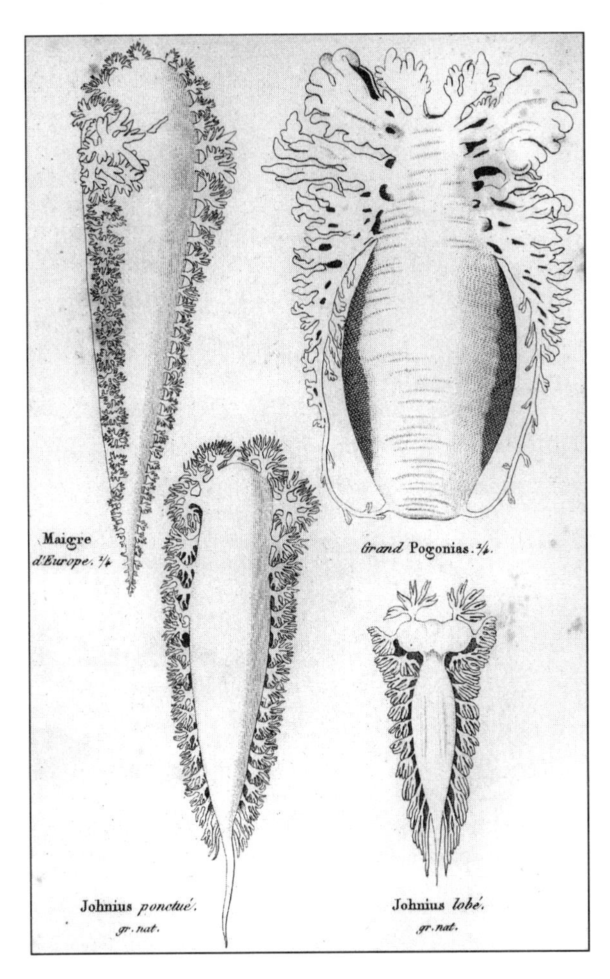

FIG. 53.
Various swim bladders of drums, family Sciaenidae. After Cuvier and Valenciennes, *Histoire naturelle des poissons*, 1828–49, vol. 5, pl. 139. Courtesy of Leslie Overstreet and the Special Collections Department, Smithsonian Institution Libraries, Washington, D.C.

viranus, who attributed to it the faculty of warning of changes in the weather.[63] Provençal and von Humboldt have also examined the air in the swim bladder and combined their observations with a very close investigation of the effect of fishes on the air they breathe.[64] Erman revealed very interesting experiments on the decomposition of atmospheric air in the intestines of the loach *(misgurn)* and the kind of respiration that results from it.[65]

Also, tests have been conducted on the chemical composition of various organs in these animals. Fourcroy and Vauquelin performed a chemical analysis of carp milt.[66] Chevreul analyzed their bones, their cartilage (1811a), and even the liquid contained in their intervertebral cavities (1811b).[67]

NOTES

1. Johann Heinrich Ferdinand von Autenrieth [1772–1835] was professor [of anatomy] and chancellor of the University of Tübingen. [His anatomy of the flatfish, *Pleuronectes platessa,* was] published in *Arch. Zool. Zoot., Wiedemann,* vol. 1 [1800].

2. [Etienne Geoffroy Saint-Hilaire (see chap. 13, n. 18), on the bones of the pectoral fins of fishes compared with those of the forelimbs of other vertebrates], *Ann. Mus. Hist. Nat.* (Paris), vol. 9 [1807a].

3. [Geoffroy Saint-Hilaire, on the *"os furculaire,"* one of the parts of the pectoral fin] *Ann. Mus. Hist. Nat.* (Paris), vol. 9 [1807b].

4. [Geoffroy Saint-Hilaire, on the general form of the sternum] *Ann. Mus. Hist. Nat.* (Paris), vol. 10 [1807c].

5. [Geoffroy Saint-Hilaire, reflections on the parts of the bony head of vertebrates, and in particular on those of the cranium of birds] *Ann. Mus. Hist. Nat.* (Paris), vol. 10 [1807d].

6. [Friedrich Christian Rosenthal (d. 1829), on the skeleton of fishes], *Arch. Physiol.* (Reil), vol. 10 [1811].

7. Rosenthal, *Ichthyotomical Tafeln,* in German, Berlin, 1812–22, [four parts] in quarto.

8. [Cuvier, on the composition of the cranium of vertebrates], *Ann. Mus. Hist. Nat.* (Paris), vol. 19 [1812].

9. [Cuvier, on the elements of the upper jaw of fishes], *Mém. Mus. Nat. Hist. Nat.* (Paris), vol. 1 [1815b].

10. In the plates of my animal kingdom [Cuvier's osteological illustrations of fishes in *Règne animal,* vol. 2, 1816].

11. [Burtin is probably François Xavier de Burtin, 1743–1818. On the relationship between the cranium and the vertebrae, I have been unable to locate a publication either by Burtin and Duméril or by Burtin alone. On Duméril, see chap. 12, n. 16.]

12. [Oken, see chap. 14, n. 8] on the importance of the bones of the head, Jena, 1807.

13. [Johann Baptist von] Spix of the Academy of Sciences of Munich [born at Höchstädt in 1781, participated in an expedition to South America and later, from 1817 to 1820, explored the various provinces of Brazil, from which he brought back representatives of 116 species of fishes; his drawings of the fishes were later donated to the Muséum National d'Histoire Naturelle, Paris. His great work, titled] *Cephalogenesis* [on the bones of the cranium, was published in] 1815, Munich, large in folio. [He died at Munich in 1826 of an illness contracted in Brazil. For more on Spix, see Adler 1989, 23.]

14. *Philosophie anatomique:* the determination and identification of the bony parts of the respiratory organs, by Geoffroy Saint-Hilaire, Paris, 1818.

15. Ludwig Heinrich Bojanus [b. 1776], author of an excellent monograph on the European turtle [1819–21], was a member of the Academy of Sciences of St. Petersburg; formerly professor at Vilna, he died in 1827. [His work on the elements of the cranium of fishes appeared in] *Isis,* vol. 2, 1818; later, in vol. 9 of the

same journal, he published a second note on this subject, 1821b. [For more on Bojanus, see Adler 1989, 20–21.]

16. C. W. H. Fenner, on the cranium of fishes, Jena, 1820.

17. [Eduard Arendt, b. 1790] on the structure of the cranium of the pike *(esocis lucci)*, Regiomonti, 1822.

18. Carl Gustav Carus [1789–1869], professor at the surgical academy of Dresden, [published his] anatomical treatise in German [1818].

19. [Carl August Sigmund Schultze (1795–1877) published his studies on the development of the vertebrae of fishes in] *Deut. Arch. Physiol.* (Meckel), vol. 4, 1818.

20. [Ernst Heinrich] Weber [1795–1878], professor of comparative anatomy, Leipzig; his *De aure animalium aquatilium* [was published at Leipzig in 1820].

21. [Bojanus, see n. 15 above], *Isis*, vol. 8, 1821a.

22. See the memoirs of Geoffroy Saint-Hilaire on the vertebra, *Mém. Mus. Hist. Nat.* (Paris), vol. 9 [1822]; on the opercular elements of fishes, *Mém. Mus. Hist. Nat.* (Paris), vol. 11 [1824a]; and on the composition of the cranium of man and fishes, *Ann. Sci. Nat.* (Paris), vol. 3, 1824b.

23. Jan van der Hoeven [1801–68], professor of philosophy at Leiden University, in a dissertation on the fish skeleton, Leiden, 1822, in octavo.

24. Gerbrand Bakker [1771–1828], professor at Gröningen, in his *Osteographia piscium*, Gröningen, 1822, in octavo, with a collection of eleven plates in quarto.

25. [Johann Friedrich] Meckel [1781–1833], professor at Halle, *System der vergleichenden Anatomie;* so far, only two volumes have appeared, Halle, 1821 and 1824 [but four more volumes were published from 1828 to 1833, also at Halle]. Riester and Alphonse Sanson have just published a French translation [in ten volumes], Paris, 1828–38].

26. [Weber, see n. 20 above], *Anatomia comparata nervi sympathici*, Leipzig, 1817, in octavo.

27. [Apostolos Arsakis, a Greek physician, who wrote a dissertation on the brain of fishes], *Commentatio de piscium cerebro et medulla spinali*, Halle, 1813 [a second edition was published at Leipzig, 1836].

28. [Heinrich Kuhl, 1797–1821], *Beiträge zur Zoologie und vergleichenden Anatomie*, Frankfurt am Main, 1820, [two parts] in quarto [pt. 2 was coauthored by Jan Coenraad van Hasselt (see chap. 16, n. 45)].

29. [Antoine Etienne Renaud Augustin Serres, 1786–1868] comparative anatomy of the brain in the four classes of vertebrates, Paris, two volumes of text in octavo, 1827, with an atlas [of sixteen plates printed in 1824].

30. [Louis Antoine Desmoulins (1794–1828) and François Magendie (1783–1855)] anatomy of the nervous system of vertebrate animals, Paris, 1825, two volumes in octavo, with an atlas [of thirteen plates].

31. [Detmar Wilhelm Soemmerring, 1793–1871], descriptions of cross sections through the eyes of animals, Göttingen, 1818, in folio.

32. [Ferdinandus Christophilus Massalien (b. 1789), in a] dissertation on the eyes of the mackerel, the tuna, and the cuttlefish, Berlin, 1815, in quarto.

33. [Louis Jurine, 1751–1819] on some details of the eye of the tuna, *Mém. Soc. Phys. Hist. Nat. Genève*, vol. 1, 1821.

34. [Everhard Home (1756–1832), in his] *Lectures on Comparative Anatomy,* 3:246–48, 1823.

35. [Weber, see n. 20 above], *De aure animalium aquatilium,* Leipzig, 1820, in quarto.

36. [Adolph Wilhelm Otto (1786–1845), on the organ of hearing in fishes, 1826.]

37. [Carl Friedrich von Heusinger (1792–1883), on the organ of hearing in fishes], *Arch. Anat. Physiol.* (Meckel), vol. 9, 1826.

38. [Christian Eduard] Pohl, in a dissertation on the organs of hearing in animals, Vienna, 1818, in quarto.

39. [Geoffroy Saint-Hilaire (see chap. 13, n. 18), on the composition of the cranium of man and animals, 1824b.]

40. [André Marie Constant Duméril, see chap. 12, n. 16], his memoir [on olfaction in fishes], read at the Institut de France in 1807, was published [that same year] along with the author's collected works on [zoology and] comparative anatomy [Paris, 1807b, in octavo].

41. [Geoffroy Saint-Hilaire (see chap. 13, n. 18), on the structure and function of the olfactory apparatus of fishes], *Ann. Sci. Nat.* (Paris), vol. 6, 1825.

42. [Etienne Marin Bailly (b. 1796) on the fishing apparatus of the anglerfish], *Ann. Sci. Nat.* (Paris), vol. 2 [1824].

43. [Geoffroy Saint-Hilaire, see chap. 13, n. 18], *Ann. Mus. Hist. Nat.* (Paris), vol. 10 [1807e].

44. [Duméril (see chap. 12, n. 16), in a] dissertation on fishes most closely related to invertebrate animals, Paris, 1812, in quarto, and in his collected memoirs on [zoology and] comparative anatomy [1807b].

45. [Martin Heinrich Rathke, 1793–1860] on the internal anatomy of the "pricka" or river lamprey, in German, Danzig, 1825b, in quarto.

46. [Home (see n. 34 above), an] anatomical description of *Squalus maximus* Linnaeus, in the *Phil. Trans. Roy. Soc. Lond.,* 1809. [Originally published in English, this paper was later translated into French and augmented] with notes by de Blainville in *J. Physique* (Paris), 1810. [See also de Blainville's] memoir on the basking shark *(squale pélerin), Ann. Mus. Hist. Nat.* (Paris), vol. 18 [1811].

47. [Rathke (see n. 45 above) on the anatomy of *Cyclopterus lumpus*] *Deut. Arch. Physiol.* (Meckel), vol. 7 [1822].

48. [Home, see n. 34 above] *Lectures on Comparative Anatomy,* London, six volumes, large in quarto, 1814–28.

49. [Rathke (see n. 45 above) on the abdominal viscera of fishes], *Neue. Schr. Naturf. Gesell. Danzig* (Halle), vol. 1, 1824.

50. [Rathke, on the chambers of the heart of fishes], *Deut. Arch. Anat. Physiol.* (Meckel), vol. 9, 1826.

51. [Rathke, on the reproductive organs of fishes], *Neue. Schr. Naturf. Gesell. Danzig* (Halle), vol. 1, 1825a.

52. [Friedrich Tiedemann (1781–1861) on the] anatomy of the fish heart, in German, Landshut, 1809, in quarto.

53. [Ignaz Doellinger (1770–1841) on the anatomy of the fish heart, in German], *Ann. Wetterau. Gesell. Natur.* (Frankfurt), vol. 2, 1811.

54. [Vincent] Fohmann [1794–1837], the first part of a larger work on the lymphatic system of vertebrates, Leipzig and Heidelberg, 1827, in folio, with sixteen lithographed plates.

55. [Hunter (see chap. 11, nn. 20 and 44), anatomical observations on the torpedo and on *Gymnotus electricus*, 1774, 1775.]

56. [Geoffroy Saint-Hilaire (see chap. 13, n. 18) on the electric organs of the electric ray, the knifefish, and a catfish], *Ann. Mus. Hist. Nat.* (Paris), vol. 1 [1802c].

57. [Karl Asmund Rudolphi (1771–1832) on electric fishes, in pt. 2, pp. 137–44, of his collected anatomical works, published by the] Royal Academy of Sciences of Berlin, 1826.

58. [Alexander von Humboldt on the electric eel, in pt. 2 (pp. 49–92) of *Voyage aux régions equinoxiales du nouveau continent, fait en 1799–1804*, by von Humboldt and Bonpland (see Humboldt 1811c). Baron Friedrich Heinrich Alexander von Humboldt was born at Berlin in 1769. After studying at the University of Frankfurt, then at Göttingen, where he learned anatomy, anthropology, and archaeology, he traveled throughout Europe. Inheriting a large fortune at the death of his mother in 1796, he went to Paris, where at the Jardin des Plantes he met Bougainville, Cuvier, and Geoffroy Saint-Hilaire, and he was named correspondent of the Muséum in 1798. In that same year, with Aimé Jacques Alexandre Goujaud Bonpland (1773–1858), a naval surgeon who became his traveling companion, he sailed from Marseilles to Spain, and through the good offices of the Saxon ambassador in Madrid, Baron Forell, they had an audience with King Charles IV of Spain, who took an interest in their plan to explore the Spanish possessions in South America. Under the sponsorship of the king, they embarked on the *Pizarro* in May 1799 and for five years explored Venezuela, Brazil, Colombia, Peru, and Mexico, making observations and extensive collections of the flora and fauna. Earning the respect and gratitude of all for his encyclopedic knowledge and his diplomacy in the service of science, he died in Berlin in 1859.]

59. [Delaroche (see chap. 12, n. 17) on the swim bladder of fishes], *Ann. Mus. Hist. Nat.* (Paris), vol. 14 [1809b].

60. [Cuvier, comments on Delaroche's studies of the swim bladder of fishes], *Ann. Mus. Hist. Nat.* (Paris), vol. 14 [1809].

61. [Jean Baptiste Biot (1774–1862) on the nature of the gases found in the swim bladder of fishes], *Mém. Phys. Chim. Soc. Arcueil* (Paris), vols. 1 and 2 [1807]; he found nitrogen and oxygen in all proportions, from nitrogen alone to 87/100 oxygen.

62. [Pietro Configliachi, 1779–1844] on the gases found in the swim bladder of fishes, Pavia, in quarto, 1809.

63. [Gottfried Reinhold Treviranus (1776–1837) on the function of the swim bladder of fishes] in vol. 2 [of a much larger collection] of miscellaneous works by Treviranus on anatomy and physiology, Göttingen, [four volumes bound in two], in quarto, 1816–21.

64. [Provençal (probably Jean Michel Provençal, 1781–1845) and von Humboldt (see n. 58 above), studies on the respiration of fishes, *J. Physique* (Paris), vol. 69 (1809).]

65. [Paul Erman (1764–1851) on respiration in the loach, *Ann. Physik.* (Gilbert, Leipzig), vol. 30 (1808).]

66. [Antoine François de Fourcroy (1755–1809) and Louis Nicolas Vauquelin (1763–1829) on the chemical composition of carp milt], *Ann. Mus. Hist. Nat.* (Paris), vol. 10 [1807].

67. [Michel Eugène Chevreul (1786–1889) on the chemical composition of the cartilage (1811a) and intervertebral fluid (1811b) of a shark, *Squalus peregrinus*], *Ann. Mus. Hist. Nat.* (Paris), vol. 18.

16 Materials for the *Histoire Naturelle des Poissons*

This is as faithful a summary as we can make of the works that have brought ichthyology to its present state. These works will serve as source material and as a point of departure, and for our work we shall endeavor to take from them all that is accurate and useful, while being careful to acknowledge each author to whom acknowledgment is due. But we shall also add many other materials that have not heretofore been made public. We consider it our duty to recognize these, either by citing the sources of the augmentations that this new natural history of fishes is bringing to the science or by expressing our appreciation to the people who have rendered assistance.

I myself for many years now have been gathering a portion of this material.[1] As early as 1788–89, on the Normandy coast, I described, dissected, and sketched almost all the fishes of the English Channel, and some of the observations I made at that time have been used in my *Tableau élémentaire de l'histoire naturelle des animaux*[2] and for my *Leçons d'anatomie comparée*.[3]

In 1803, during a stay of several months at Marseilles, I continued this kind of research on the fishes of the Mediterranean. I did the same in 1809 and 1810 at Genoa and in 1813 at various places in Italy, and I published examples of these observations in the first volumes of the memoirs of the Muséum National d'Histoire Naturelle, Paris.[4]

It was then especially that I began to perceive the degree to which all the existing ichthyologies were yet imperfect—in the number of fishes, in their relations, in the criticism of synonyms, and even in the characters assigned to the species.

Therefore I sought an opportunity to make a general and comparative study of the whole class of fishes, and I found it in preparing the large collection that the late Péron brought back from the Indian Ocean.[5]

Lacepède and Duméril were willing to permit me to take charge of this task, and I included in my arrangement the fishes from the king's cabinet, those from the stadholder's cabinet, those of Commerson's, which Duméril fortunately recovered and set in order, those the late Delaroche had brought back from Ibize,[6] and those the late Delalande brought back from Toulon.[7]

It was after this first inspection that I wrote, during the tumultuous years 1814–15, the fish part of my *Règne animal*, published in 1816.[8] It must have been evident to all my readers that in this book the classification, the characters of the genera, the division of genera into subgenera, and the criticism of species resulted from the study of nature itself, and one could readily see that preceding works deserved many corrections.

Since then, in concert with my colleagues, professors of ichthyology, I have ceaselessly used every means at our disposal to augment this part of the king's cabinet. Thus the ministers of the marine, the officers under their orders, and the heads of colonies have consistently supported my efforts and those of the administration of the Muséum National d'Histoire Naturelle, Paris, and the collection has grown in a few years to a surprising number, at least four times larger than collections shown in the latest works.

These large augmentations are due principally to the explorers who, since 1816, thanks to a convention proposed by the minister of the interior and approved by the late king, have traveled at government expense to various parts of the world.

Our first contribution, for which we have to thank both Péron and Lesueur,[9] included the Atlantic Ocean, the Cape, Ile de France and Bourbon, a part of the Moluccas, and the coast of New Holland.

All the other seas have successively given their quotas. The late Delalande went to Brazil in 1817 and to the Cape of Good Hope in 1820, where that indefatigable preparator made collections as astonishing for their number as for their preservation.[10] Auguste de Saint-Hilaire, a knowledgeable botanist, during a long visit to Brazil, neglected no part of natural history, and for fishes in particular he provided some fine additions to Delalande's collection.[11] His Highness Prince Maximilien de Neuwied kindly lent us several fishes gathered in the same country,[12] and we have seen many interesting ones drawn by the late Spix, which his heirs saw fit to submit to us before the forthcoming publication they intend to make of them.[13]

Cayenne is a place where we have always had collectors stationed in some way or another. In addition to the fishes formerly collected by Richard[14] and Le Blond,[15] we have recently received some through the good offices of Poiteau,[16] while he was head of agriculture in that colony, and of Leschenault[17] and Doumerc,[18] who traveled there in 1824.

Thus we have had ample means of shedding light on Marcgrave's fishes and those that Bloch published from the drawings made under the direction of Prince Maurits of Nassau.

The Antilles and the whole Gulf of Mexico have furnished us information no less abundantly. Plée, that courageous voyager who died a victim of the sufferings brought on by his sojourn of six or seven years in those terrible climates, gathered as many as five collections there, some from Martinique and Guadeloupe, others from Puerto Rico and the whole coastline of Colombia.[19] Remarkable for both the size and the preservation of specimens, these collections are accompanied by precise notes on the habits of the species, their characteristics, and the common names given them in different places.

Lefort, chief physician at Martinique,[20] and Achard, pharmacist,[21] sent to us from Martinique and Guadeloupe samples with even their colors as fresh as if they had just been caught. Ricord brought us from Santo Domingo a rather large number, also well preserved.[22] Poey y Aloy, educated in natural science and living in Havana, brought us samples from the island of Cuba.[23] We have also been sent a collection of fine drawings of fishes from the coasts of Mexico, made for the late king of Spain, by Mocigno.[24]

It has thus been easy for us to recognize all of Plumier's fishes and to rectify many of Bloch's errors as regards these fishes. All the fishes that Parra described in Cuba are also to be found in our collections, and we have been able to verify and complete his work.[25]

Even the fishes in the high valleys of the Cordilleras are not unknown to us. The illustrious and learned explorer von Humboldt has been willing to send us some of those that he described in his zoological observations.[26]

Our sources for the North American coasts have also been multiplied. The famous naturalist Bosc, who was the French consul to Carolina,[27] has transmitted to us the fishes he has collected and the drawings he has made of them, some of which have already been published by Lacepède, but in a manner that needed clarification from nature.

We are also indebted to Milbert, a capable artist who has lived a long time in New York, for a considerable number of fishes.[28] He has sent us nearly all the species described by Dr. Mitchill and many other species collected from the coasts, rivers, and lakes of that part of the world.

Lesueur added a number of interesting species, especially from the fresh waters of the interior, some of which he described in the scientific journals of that country.[29] We have also received specimens from DeKay, a young naturalist from New York who studied at the Muséum National d'Histoire Naturelle, Paris and has maintained an affection for this fine establishment.[30] Dr. Mitchill himself sent some samples;[31] in particular, he sent the museum administration some handwritten memoirs that we have used.

The fishes of Newfoundland have been carefully observed and described by de La Pilaye, who has freely transmitted to us his notes and drawings, from which we have taken some useful information.[32] Recently Mr. Richardson was willing to show us fishes that were taken during the last voyage of Captain Franklin to North America.[33]

Africa is the most difficult part of the world in which to travel with the apparatus necessary for making large collections; however, Roger, governor of the French settlements in Senegal, assembled for us a series of the fishes of this river,[34] all the more interesting to us because we could compare it with the fishes Geoffroy Saint-Hilaire had collected on the Nile.[35] This collection, along with the species of the rivers at the Cape brought back by Delalande[36] and fishes that Marceschaux, French consul at Tunis,[37] has just caught for us in the lake of Bizerte, has given us some idea of the freshwater populations of this vast country.

As for the oriental seas, we have a small collection of dried fishes made by the late Sonnerat,[38] which he gave us in 1814, but also a very large collection amassed during several years at Pondicherry and Ile de France and Bourbon by Leschenault.[39] This has enabled us to verify most of the fishes of Commerson and Russell. Mathieu, a very learned artillery officer, has sent several rare and well-preserved species from Ile de France.[40]

Diard[41] and Duvaucel,[42] during a rather long visit to Sumatra and Java, collected a good number of fishes; and the generous loans and gifts that the famous Temminck[43] has made us of fishes collected in those islands by Kuhl[44] and van Hasselt [fig. 54],[45] and their drawings of the fishes, have fulfilled all we could desire in this regard. These two young and unfortunate observers had also been to the Moluccas, and their collections along with Péron's have begun to clarify Valentijn's and Renard's drawings for us and to convince us that these drawings, although they may be outlandish, nevertheless all represent real objects.

Reinwardt, learned professor of natural history at Leiden,[46] has been no less generous than Temminck and has communicated to us all he collected on his difficult voyage to the [East] Indian archipelago.

We include, among the most valuable consignments we have received, the fishes of the Ganges and its tributaries, which Duvaucel, my stepson, has collected with the greatest zeal, including even some fishes from the rivers of Nepal.[47] These shipments, added to the immense collections of quadrupeds, birds, reptiles, insects, and skeletons and anatomic preparations that he has sent to the king's cabinet, will cause him always to be remembered here with gratitude. Were it not for our misfortune in losing this interesting young man, no less witty and educated

FIG. 54. Heinrich Kuhl (1797–1821), *left*, and Jan Coenraad van Hasselt (1797–1823), *right*. Courtesy of M. Boeseman, L. B. Holthuis, I. Henneke, and the Rijksmuseum van Natuurlijke Historie, Leiden, Holland.

than he was enthusiastic in his research—a misfortune due partly to the maliciousness of a few wretches who feared the proximity of a man capable of shedding light on their conduct—the natural sciences in all fields would have obtained collections superior to any that had ever been made. Let the reader permit me at least to record here the regrets that naturalists owe him. This portion of his shipments has enabled us to get a more complete idea of most of the species that Mr. Hamilton-Buchanan[48] described in his handsome work on the fishes of the Ganges.

Dussumier, a merchant from Bordeaux who is enthusiastic about natural history and a young man who has already made several voyages in his own ships to China and India, has always taken care to bring back to us the more remarkable objects that he has found, and we owe to him several fishes that are interesting for their rarity and the singularity of their characteristics.[49] He even took the trouble to have some very elaborate paintings made at Canton of several beautiful Chinese species and has committed them to our care. Recently he sent us a valuable collection of fishes caught off the coast of Malabar and the Seychelles.

Ehrenberg, who has sampled the production of the Red Sea and the Nile with admirable discernment and perseverance, has been kind enough to send us his drawings and descriptions and give us his duplicates for the king's cabinet.[50] We scarcely know how to express our gratitude for such noble generosity. He has given us a means of clarifying most of the accounts, numerous but obscure, left by Forsskål on the Red Sea fishes.

We even have fishes from the Sea of Japan and Kamchatka, thanks to the generosity of Tilesius,[51] the learned companion of Captain von Krusenstern. Lichtenstein[52] has sent us all those collected during the same expedition by Langsdorff,[53] and given by him to the cabinet of Berlin, as well as all those that Pallas procured previously and that he described in his Russian zoography. In fact Temminck, without hesitation, has just placed at our disposal a large collection of fishes from those distant places, which has been received by the Royal Museum of the Netherlands.

While these generous friends of science have been thus depositing about us the fishes of the most distant countries, there are others who have been taking pleasure in procuring for us the fishes of Europe. In addition to the collections made by Delalande, Delaroche, and myself on the Mediterranean coasts [fig. 55], Risso has sent us his most interesting species from Nice, and the drawings he has had made of them from nature, without which we would not have a good idea of their colors.[54] Bonelli has also sent us fishes, and has lent us some of the rarer ones from the Museum of Turin.[55] But we have a particularly superb collection, numerous and well preserved, owing to the disinterested zeal of Savigny who, during a year's journey in Italy, ceaselessly sought all fishes that appeared in the various markets and who even went out on the boats several times to take the fish that the fishermen neglected.[56] He thus procured for the king's cabinet almost four hundred species, all of the highest standard and in a perfect state of perservation. Happy this ingenious observer would have been, had the state of his health permitted him to know that naturalists were enjoying the fruit of his efforts. We hasten to point out here, at least, the reasons for his deserving the gratitude of naturalists.

Bibron, an employee of the Muséum National d'Histoire Naturelle, Paris, went to Sicily and gathered several more species that had escaped Savigny;[57] Dr. Leach procured for us some from Malta;[58] Admiral de Rigny, during the noble expedition he commanded in the [Greek] archipelago,[59] took care to have the parrotfish *(scare)* caught for us, a fish famous among the ancients but never seen by modern man except Aldrovandi. At this moment we are awaiting some products from the area of the archipelago, where Dr. Bailly has promised us to look after the interests of ichthyology during his sojourn in Greece.[60] Adding to these numerous collections those that Geoffroy Saint-Hilaire has made on the Nile and on the coast of Egypt, we dare to congratulate ourselves for lacking nothing that would clarify what has been written about the fishes of the Mediterranean from earliest times.

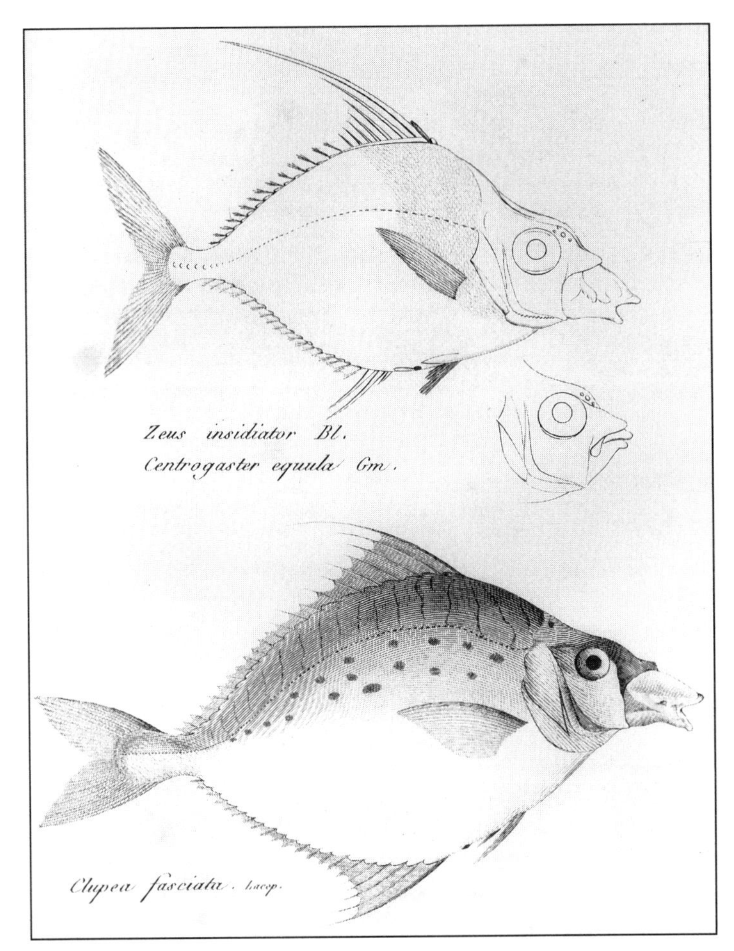

FIG. 55. Ponyfishes, genus *Leiognathus*, from the Mediterranean Sea. After Cuvier, *Mém. Mus. Nat. Hist. Nat.* (Paris), 1815f, vol. 1, pl. 23. Photograph by Stanley A. Shockey, Classroom Support Services, University of Washington, Seattle.

Also, at Marseilles we have in Roux, curator of that city's museum, a correspondent full of enlightenment and zeal who is willing to give us any information we ask of him[61] and who even proposes, when he has finished his ornithology of Provence, to publish colored drawings of the beautiful fishes of that coast, which are still little known and incorrectly represented.[62]

The fishes of our coasts have been studied with no less zeal. D'Orbigny, a correspondent of the museum of La Rochelle, has sent us all the species of the Bay of Biscay,[63] thus enabling us to comment on the treatise that Cornide published on the fishes of Galicia.[64] He has provided us the rudiments of the natural history of the albacore *(Germon)*, so interesting and quite forgotten by most naturalists.

At Brest, Garnot, a naval engineer, is willing not only to send us fishes, but also to commit to paper as accurately as possible their natural colors while the fishes are fresh.[65]

Baillon, a correspondent of the museum at Abbeville, whose name is well known to naturalists from the discoveries he and his father have made about birds, is no less enthusiastic and discerning in studying the fishes of the English Channel.[66] We are indebted to him for remarkable new species of such genera as *Pleuronectes (pleuronectes)*, of which it is almost inconceivable there should remain any more to discover on our coasts.

The late Noël de la Morinière, who perished in Norway during a journey to study fishing in the northern seas, collected several interesting fishes from those places.[67] We have also received others from Reinhardt, professor at Copenhagen,[68] who had been asked for them by my colleague Brongniart[69] on our behalf.

We would particularly like to receive European freshwater fishes, which are usually quite neglected in cabinets. We ourselves have intensively studied those of the Seine and the other rivers of the environs of Paris. My collaborator, Valenciennes,[70] went to Anvers and Dordrecht for the purpose of finding the so-called whitefish or hautin *(triptéronote* or *hautin)*, so poorly reproduced by Rondelet, which is none other than the lake whitefish *(lavaret)* [fig. 56].

Hammer, professor at Strasbourg, procured for us the fishes of the Rhine and the rivers that flow down from the Vosges.[71] De Candolle, the famous botanist,[72] along with Major, curator of the cabinet of Geneva,[73] and with the help of several Swiss naturalists, took pains to procure for us the fishes of Lake Leman and other lakes in Switzerland and Savoy. This has given us a means of disentangling the natural history of several species of trout *(truite)* and grayling *(ombre)*, poorly explained by Bloch and his correspondents. He even included some fishes from the lakes of Lombardy, several of which have also been given us by Bosc[74] and Savigny.[75]

Fishes from Lake Trasimeno have been sent to us by Canali, learned professor at Perugia.[76] Bredin, director of the veterinary school at Lyons, has given us the zingel *(apron)* from the Rhône.[77] Several interesting fishes from the Danube have been sent to us, superbly prepared by the labors of von Ritter Schreibers, the renowned director of the cabinet of natural history of Vienna.[78] Lichtenstein, learned professor at Berlin, sent us some from Brandenburg.[79] Thienemann, of Dresden, famous for his travels in the north, sent us many from Saxony.[80] Nitsch made us a collection at Halle.[81]

Des poiſſons au bec pointu,é du Hautin.

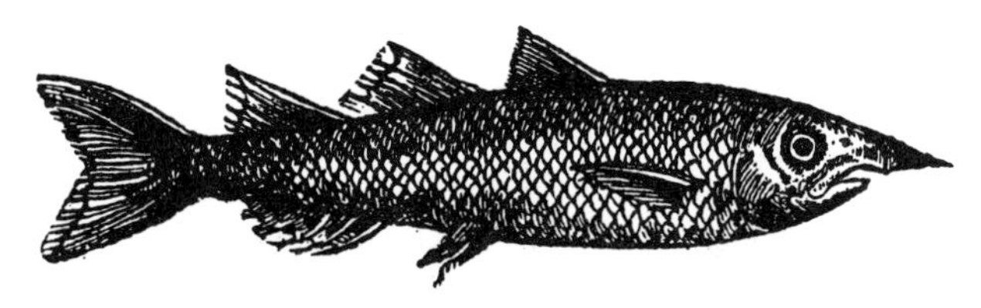

Chap. XVII.

IL i a en diuers lieux des poiſſons qui ont le bec fort pointu, qu'on peut nommer *Piſces Oxyrynchos*, poiſſons au bec pointu. Comme il en i a au Lac de Caſpie de huit coudées, que les Caſpiens vendent deſechés,é ſalés. D'iceux auſsi la greſſe ſeparée,en font de la farine,des entrailles cuittes ilz en font de la colle. Il i a vn autre *Oxyrynchus* au Nile, que les peſcheurs ſe gardent bien de prendre,l'aians en grande veneration. Il i a vn autre *Oxyrynchus* en la mer rouge, qui ha la bouche fort longue, les ieux reluiſans comme or, des merques palles au dos, les premieres æles noires,celles du dos blanches,la queüe longue é verte, departi par le millieu d'vne ligne dorée. Nous ne baillons les pourtraits de ces poiſſons que nous n'auons point veu,mais bien de celui que nous auons veu à Anuers,qui s'appelle Hautin,qui ha le bec long,menu, fort pointu,mol é noir. Il eſt couuert d'ecailles moiennes. Il ha trois æles au dos.autant au ventre que le Barbeau.

FIG. 56. Guillaume Rondelet's "hautin," from his *Histoire entière des poissons,* 1558, pt. 2, p. 141. T. W. Pietsch collection.

We are now particularly knowledgeable about the fishes of Germany and are able to verify all of Bloch's species because of Valenciennes's recent sojourn in Berlin, during which, through the good offices of the great von Humboldt, he was allowed to collect every species raised there, in large and fine samples, even from the pools belonging to the king.

We have received fishes even from the Don and the Phasis through the good offices of Gamba, French consul to Georgia;[82] and the shipment that Lichtenstein kindly sent us of Pallas's fishes, given to the cabinet of Berlin by Rudolphi,[83] has given us much information on the species of

Russia. But most of all, we are extremely grateful for the gracious attention that Her Imperial Highness the grand duchess Hélène has deigned to show us in sending handsome samples of the most remarkable fishes of that empire, along with their common names. May the reader permit us here to express our respectful gratitude for this proof, given by such a distinguished princess, of enlightened love for the sciences![84]

The great nautical expeditions commissioned by the late king have completed this long series of acquisitions that began with the expedition of Baudin. De Freycinet[85] and Duperrey,[86] on their voyages round the world, have had fishes collected from all the seas traversed, according to the instructions they received through the zeal for science that animates the Ministry of the Marine, and they have been well supported in this research—de Freycinet by Quoy[87] and Gaimard,[88] and Duperrey by Garnot[89] and Lesson.[90] The accounts of their voyages included drawings and descriptions of the more remarkable new species they discovered; but they have brought back many others that, although not new to science, still are of great interest for our work, either because we are thus enabled to describe them better than our predecessors have done, or because of anatomical and other peculiarities they exhibit. Moreover, it is thus that we have fishes from New Zealand, New Guinea, the Mariannas, the Sandwich Islands, Tierra del Fuego, and the south of Brazil. On other occasions we have even received fishes from the river Plata, and especially Buenos Aires; and we are expecting handsome collections from New Guinea, where Quoy and Gaimard, who have already accomplished such great work during the voyage of de Freycinet, are sailing with d'Urville. Inspired with a new zeal, and fortified by experience, they cannot help but obtain good results again.[91]

As for ourselves, our only remaining wish is that the work we have undertaken will not be found too unworthy of the illustrious writers whose work we seek to continue, or of the help and encouragement we have received from such a great number of friends and protectors of natural history. We are content to hope that in its turn it will take its place among the works that have advanced the cause of science. We bend our efforts toward that goal.

NOTES

1. To complete this history of ichthyology, we felt it necessary to repeat here the account of our work as it has been announced in our prospectus [to the *Histoire naturelle des poissons;* see Cuvier 1827b, c].

2. [Cuvier], *Tableau élémentaire de l'histoire naturelle des animaux,* Paris, 1798, one volume in octavo.

3. [Cuvier], *Leçons d'anatomie comparée,* Paris, 1800–1805, five volumes in octavo.

4. [Cuvier 1815a, c, d, e, f.]

5. [Péron, see chap. 13, n. 33.]

6. [Delaroche (see chap. 12, n. 17) brought back about one hundred species of fishes from Ibize, of which thirty were new to science.]

7. [Pierre Antoine Delalande, onetime naturalist-aide to Etienne Geoffroy Saint-Hilaire, was born at Versailles in 1787. Having traveled widely and returned to France with rich collections of natural objects, he died in 1823 of fever while editing the description of his travels. He brought back fishes from Toulon and Marseilles, Brazil, Cape Verde, and the Cape of Good Hope.]

8. [Cuvier], *The Animal Kingdom,* a classification of animals intended to be used as a basis for zoology and as an introduction to comparative anatomy, Paris, 1816, four volumes in octavo. [Although often dated 1817, even by Cuvier himself, there is ample evidence that all four volumes were available in November 1816; see Cowan 1969; Whitehead 1967; Roux 1976.] A second edition [was published in 1829, Paris, five volumes in octavo; the part on fishes is found in 2:122–406].

9. [Péron and Lesueur, see chap. 13, nn. 32, 33.]

10. [Delalande, see n. 7 above.]

11. [Augustin François César Provençal de Saint-Hilaire (Auguste de Saint-Hilaire, no known relation to Etienne Geoffroy Saint-Hilaire), born at Orléans in 1779, died near Sonnely (Loiret) in 1853. Accompanying Delalande aboard the *Hermione,* he traveled to Brazil, where from 1816 to 1822 he amassed large collections of plants and animals, including fifty-eight fishes that he donated to the Muséum National d'Histoire Naturelle, Paris.]

12. [Alexander Philipp Maximilian, prince of Wied-Neuwied, one of the great explorer-naturalists and ethnologists of the nineteenth century, was born at Neuwied, along the Rhine near Koblenz, Prussia, in 1782 and died there in 1867. He explored Brazil from 1815 to 1817 and North America from 1838 to 1843, bringing back large collections; he donated fishes from Brazil. For more on Wied-Neuwied, see Adler 1989, 22–23.]

13. [Spix, see chap. 15, n. 13.]

14. [Louis Claude Marie Richard, born at Versailles in 1754, died at Paris in 1821, traveled to Cayenne in 1781, Brazil in 1785, and numerous islands of the Antilles from 1786 to 1788. He sent back fishes from Martinique.]

15. [Le Blond, see chap. 12, n. 4.]

16. [Pierre Antoine Poiteau, born at Amblémy near Soissons (Aisne) in 1766, died at Paris in 1854, was sent on a mission to Haiti, where he was appointed chief gardener at the new botanical garden. Returning to France in 1815, he first worked at the garden at Versailles but later, in 1818, was appointed head of agriculture in Guiana. During his stay in the Antilles and Guiana, he collected plants and animals for the Muséum, donating in particular fishes from Cayenne.]

17. [Jean Baptiste Louis Claude Théodore Leschenault, sometimes called Leschenault de la Tour, born at Chalon-sur-Saône in 1773, went as botanist on the voyage to the southern lands on the *Géographe,* commanded by Nicolas Baudin (see chap. 13, n. 43). After exploring Timor, Java, and Ceylon and traveling to the

United States, he returned to France in 1807, only to depart again in 1816, this time for India, where he became director of the botanical gardens at Pondicherry. Returning again to France in 1822, he embarked on a third voyage, this time to South America, with Doumerc (see n. 18 below), where in Cayenne, Surinam, and the area around Rio de Janeiro and Bahia he made large collections for the Muséum. He died in Paris in 1826. For more on Leschenault, see Mearns and Mearns 1988, 231–35.]

18. [Adolphe Jacques Louis Doumerc, born in Hamburg in 1802, died at Paris in 1868, joined Leschenault in a voyage to South America (see n. 17 above).]

19. [Auguste Plée (also spelled Pley), born at Paris in 1786, was a voyager-naturalist who in 1820 sailed to the Antilles, where he visited the islands of Martinique, Guadeloupe, St. Thomas, St. Lucia, St. Barthelemy, Puerto Rico, and Cuba. He later traveled to Quebec and explored the coasts of Colombia and Venezuela. He died on Martinique in 1825, on the very day he was to return to France.]

20. [Pierre François Lefort, born at Mers (Somme) in 1767, was stationed at Fort Royal on Martinique as medical officer first class, physician to the king, from 1814 to 1817 and as second, then first, chief physician from 1818 to 1826. Knight of the Legion of Honor in 1820, he died at Amiens in 1843.]

21. [Mathieu Justinien Achard was a navy pharmacist stationed on Martinique. He donated fishes regularly from Martinique and Guadaloupe.]

22. [Alexandre Ricord, born in Baltimore, Maryland, in 1798, graduated as doctor of medicine at Paris in 1824, took lessons from Cuvier, traveled as correspondent of the Muséum (1827) and became naval surgeon at Santo Domingo. He died sometime after 1838.]

23. [Felipe Poey y Aloy, born at Havana, Cuba, in 1799, traveled to Paris to take courses from 1826 to 1832 under Cuvier, who entrusted him with the task of studying the collections of fishes from Cuba. Returning to Cuba, he was charged with the creation of a natural history museum in Havana, and in 1842 he became professor of zoology and comparative anatomy. By 1873 he was dean of the faculty of philosophy, sciences, and letters. Over a thirty-year period, beginning in the early 1850s, he published numerous papers on fishes, perhaps the most important being "Synopsis piscium Cubensium," part (2:279–468) of his much larger *Repertorio fisico-natural de la isla de Cuba,* in two volumes, 1865–68. He died at Havana in 1891.]

24. [Mocigno, biographical information unavailable.]

25. [Parra, see chap. 9, n. 41.]

26. [Humboldt (see chap. 15, n. 58) on *Eremophilus* and *Astroblepus,* two new genera (1811a); on a new species of the genus *Pimelodus* (1811b) in his] collected observations on zoology and comparative anatomy, Paris, vol. 1.

27. [Bosc, see chap. 12, n. 5.]

28. [Jacques Gérard Milbert, born at Paris in 1766, sailed as artist aboard the *Géographe* on Baudin's (see chap. 13, n. 43) great voyage around the world (1800–1804) but disembarked at Mauritius in 1801 because of illness. He returned to France in 1804. In 1815 he went to North America, where he stayed until 1822. Landscape painter and engraver as well as naturalist, he was named correspondent of the Muséum in 1820. He died at Paris in 1840.]

29. [Lesueur, see chap. 13, n. 32.]

30. [James Ellsworth DeKay, naturalist, 1792–1851, was later to publish an important faunal description of New York, 1842.]

31. [Mitchill, see chap. 13, n. 30.]

32. [Auguste Jean Marie Bachelot de La Pilaye, naturalist, born at Fougères in 1786, died at Marseilles in 1856.]

33. They are described [by Sir John Richardson] in an appendix [to Sir John Franklin's (1786–1847; see Mearns and Mearns 1988, 160–67) description] of this voyage [see Richardson 1823]. [Richardson, born at Dumfries, Scotland, in 1787, was naturalist and naval surgeon on the Arctic expeditions of William Edward Parry (1819–22) and Franklin (1825–27), charged with the search for the Northwest Passage between the Atlantic and Pacific oceans. Later he joined an expedition (1848–49) led by James Clark Ross (1800–1862) in search of Franklin, who had disappeared in 1845 (see Mearns and Mearns 1988, 307–15). Richardson died at Grasmere, England, in 1865.]

34. [Baron Jacques François Roger, a lawyer and governor of the French settlements in Senegal, thoroughly explored that country in 1826 with the naturalists Georges Samuel Perrottet (born at Vully, Switzerland, in 1793, died at Pondicherry in 1870) and François René Mathias Leprieur (born at Saint-Dié, Vosges, in 1799, died at Cayenne in 1870).]

35. [Etienne Geoffroy Saint-Hilaire; see chap. 13, n. 18.]

36. [Delalande, see n. 7 above.]

37. [Armand Jean Baptiste Louis Marceschaux (also spelled Maréchaux or Marcescheau), consul of France at Turin, traveled in 1826 to the south of the regency of Tunis.]

38. [Sonnerat, see chap. 9, n. 6.]

39. [Leschenault, see n. 17 above.]

40. [Mathieu, artillery officer and collector of crustaceans, mollusks, sea urchins, fishes, and birds, stationed on Mauritius; he probably returned to France between 1795 and 1800 (see Bauchot, Daget, and Bauchot 1990, 107.]

41. [Pierre Médard Diard, a student of Cuvier, born at Saint-Laurent in 1794. In 1817, with Alfred Duvaucel (see n. 42 below), Cuvier's stepson, he left for India, where they made collections for the Muséum. Later, in 1819, he went to Cochin China to collect plants on his own, ascended the Mekong, visited the ruins at Angkor, then went to Malacca. At the end of 1824 he settled in Batavia, where he collected for the Leiden museum. He died on Java in 1863, accidentally poisoned by the arsenic he used for scientific purposes.]

42. [Alfred Duvaucel, son from Madame Cuvier's first marriage, a voyager-naturalist of the Muséum, born in 1793. Given a scientific mission to India in 1817, he left France with Diard (see n. 41 above). After a brief stay at Calcutta and Chandannagar, where he established a botanical garden, he left in 1818 for Sumatra and Java, where he gathered collections that were later seized by the English East India Company at the behest of Sir Stanford Raffles. Much provoked by this incident, he returned to Calcutta, then departed again for Sumatra, where he succeeded in making other collections for Cuvier. Returning to India, he collected fishes from the Ganges and its tributaries and from Assam and Nepal. He died at Madras in 1825 of a fever contracted during his expedition.]

43. [Coenraad Jacob Temminck, born at Amsterdam in 1778, died at Leiden in 1858, was director of Leiden's royal cabinet of natural history from 1820 to 1858. He donated fishes from Vienna as well as the fishes collected by Kuhl and van Hasselt (see nn. 44 and 45 below). For more on Temminck, see Mearns and Mearns 1988, 372–76.]

44. [Heinrich Kuhl, a German naturalist, born at Hanau an der Main in 1797, was appointed in 1820 to the Commission for Natural Sciences of the East Indies and left that same year with his friend van Hasselt (see n. 45 below) for the Dutch East Indies. In Java he collected a great number of plants, animals, and minerals and sent to Holland many observations that were later published in the scientific journals. But the climate was fatal to him and he died in 1821 at Bogor, Java. For more on Kuhl, see Roberts 1993.]

45. [Jan Coenraad van Hasselt, born at Doesburg, Holland, in 1797, traveled in 1820 to the Dutch East Indies with Kuhl (see n. 44 above) to study the fauna and flora. Upon the death of Kuhl in 1821, he continued his exploration of Java and sent very large collections to the museum at Leiden, along with precise and detailed descriptions. He died of dysentery at Bogor in 1823. For more on van Hasselt, see Roberts 1993.]

46. [Gaspar Georg Carl Reinwardt was born in 1773 at Lutteringhausen im Bergischen, Germany. He was named professor of natural history at Harderwijk in 1800, director of Louis Napoléon's menagerie and director of the king's cabinet at Haarlem in 1808, and professor of chemistry, pharmacology, and natural history at Amsterdam in 1810. In 1815 he went to the East Indies, having been put in charge of organizing education, medical service, agriculture, industry, and scientific research in the Dutch colonies. In 1817, he created a botanical garden at Bogor, Java, and became its first director. Returning to Europe in 1822, he was named professor at the University of Leiden, where he died in 1854.]

47. [Duvaucel, see n. 42 above.]

48. [On Hamilton, see chap. 13, n. 39.]

49. [Jean Jacques Dussumier, ship owner and merchant at Bordeaux, born in 1792, made at least eleven commercial voyages between 1816 and 1840 aboard his own vessels, the *Buffon* and the *Georges Cuvier,* in the Indian Ocean and as far as the China seas (see Laissus 1973). In each port of call he made valuable zoological collections, which he gave to the Muséum National d'Histoire Naturelle, Paris, along with descriptive notes and drawings made from nature. He was named correspondent of the Muséum in 1827, knight of the Legion of Honor in 1831, and officer in 1841. After living in Paris near the Muséum for several years, he returned to Bordeaux, where he died in 1883.]

50. [Christian Gottfried Ehrenberg was born at Delitzsch in 1795 and died at Berlin in 1876.] By order of the Royal Academy of Sciences of Berlin [and with support from Alexander von Humboldt] he, along with [Friedrich Wilhelm] Hemprich [1796–1825], made an expedition to Libya, Egypt, Nubia, Arabia, and the western coast of Abyssinia from 1820 to 1825 that produced observations of the greatest interest to all branches of natural science [see Mearns and Mearns 1988, 183–87]. See the report by von Humboldt on this subject, Berlin, 1826, in quarto.

51. [Tilesius, see chap. 13, n. 42.]

52. [Martin Heinrich Karl Lichtenstein, a doctor of medicine, was born at Hamburg in 1780 and died at sea between Korfor and Kiel in 1857. He became professor of natural history at the University of Berlin and in 1813 was made director of the zoological museum. He made numerous fishes available to Cuvier and Valenciennes during their visits to Berlin. For more on Lichtenstein, see Mearns and Mearns 1988, 240–43.]

53. [Baron Georg Heinrich von Langsdorff, a German physician and naturalist, born at Wollstein, Hesse, in 1774, participated in von Krusenstern's voyage (see chap. 13, n. 41), joining the *Nadjedjeda* at Copenhagen and stopping at the Canaries, Brazil, and by way of Cape Horn, the Marquesas, Hawaiian Islands, and Kamchatka. Although the ship continued on to the China Seas, the Strait of Sunda, Cape of Good Hope, Saint Helena, and Scotland, finally arriving at Kronstadt on 9 August 1806, Langsdorff left the expedition in July 1804 at Saint-Pierre and Saint-Paul, returning to Europe by way of Siberia. Later he was named Russian consul-general to Brazil, arriving there in 1813. From 1813 to 1820 he lived at Rio de Janeiro. While leading a Russian expedition to the Amazon basin from 1821 to 1829, he fell ill and lost his mind; transported back to Europe, he died at Fribourg in 1852, never having recovered his reason.]

54. [Risso, see chap. 12, n. 19.]

55. [Bonelli, see chap. 13, n. 7.]

56. [Marie Jules César Lelorgne de Savigny, naturalist, born at Provins in 1777, joined the Egyptian campaign in 1798 as an assistant to Etienne Geoffroy Saint-Hilaire. An eye disease contracted in the sands of Africa and aggravated by his close work rendered him almost blind as early as 1829. He died at Gally near Versailles in 1851.]

57. [Gabriel Bibron (also sometimes spelled Biberon), born in 1806, was the son of an employee at the Jardin des Plantes, who from an early age devoted himself to the study of natural history. In about 1820 he was sent by the Muséum on scientific missions to Italy and Sicily, then to England and Holland. He became a member of the Philomatic Society of Paris in 1840 and later professor of natural history at the municipal college of Turgot. He died of tuberculosis in 1848 at the age of forty-two. For more on Bibron, see Adler 1989, 32–33.]

58. [William Elford Leach, born at Plymouth in 1790, graduated as surgeon at London in 1809 and physician at Edinburgh in 1811 but abandoned medicine to devote himself to natural history. In 1816 he was named curator of natural history at the British Museum but was forced to resign this position in 1822 because of poor eyesight. He left for Italy in 1826 and stayed there until his death at Piedmont in 1836. He wrote primarily on crustaceans and insects. For more on Leach, see Mearns and Mearns 1988, 223–26.]

59. [Henri Gauthier, comte de Rigny, a naval officer, was born at Toul in 1782. In 1822 he commanded the French naval forces in the Levantine seas; in 1825 he commanded the Anglo-Franco-Russian squadron sent to Greece; and in 1827 he was victor at Navarino against the Turco-Egyptian squadron of Ibrahim Pasha. Named vice admiral in 1827, maritime prefect of Toulon in 1829, and minister of foreign affairs in 1834, he died at Paris in 1835.]

60. [Etienne Marin Bailly, see chap. 15, n. 42.]

61. [Jean Louis Florent Polydore Roux, born at Marseilles in 1792, was appointed curator at the museum of Marseilles at the time of its creation in 1819 and conducted research on the zoology of the area, particularly on ornithology. In 1839 he traveled in the Orient, went up the Nile as far as Thebes, and then went to Bombay, where he died in 1833.]

62. [Roux's *Ornithologie provençale* appeared in 1825–30, in two volumes, but his proposed work on the fishes of that coast was never completed.]

63. [Charles Marie Dessalines d'Orbigny, naturalist and doctor of medicine, born at sea in 1770, helped to found and develop the museum of natural history of La Rochelle. He died at La Rochelle in 1856.]

64. [Cornide de Saavedra, see chap. 9, n. 40.]

65. [Prosper Garnot, born at Brest in 1794, joined the naval corps of surgeons in 1811 and defended his doctoral thesis in medicine in 1822. That same year, he embarked as assistant surgeon on the *Coquille*, and with his adjutant René Primevère Lesson (see n. 90 below), he was put in charge of zoological observations during that ship's voyage around the world under the command of Louis Isidore Duperrey (see chap. 13, n. 45). Upon his return he, together with Lesson, wrote the zoological part of the voyage (see Lesson and Garnot 1826–31). In 1828 he was appointed chief surgeon in Martinique, but afflicted with an incurable hepatitis, he was allowed to retire in 1833. He died at Paris in 1838.]

66. [Louis Antoine François Baillon, born at Montreuil-sur-Mer in 1778, died at Abbeville in 1851, was a botanist and zoologist who kept a cabinet of natural history at Abbeville and who donated fishes from Abbeville, Norway, Malaga, Buenos Aires, and La Plata.]

67. [Simon Barthélémy Joseph Noël de la Morinière, traveler and ichthyologist, born at Dieppe in 1765, became inspector general of marine fisheries of France. In 1819 he was put in charge of a voyage of exploration that was to go beyond North Cape. It was during this mission that he died at Trondheim in 1822.]

68. [Johannes Christopher Hagemann Reinhardt, 1776–1845, sent fishes from the North Sea and Greenland.]

69. [Alexandre Brongniart, mineralist and geologist, born at Paris in 1770 and died there in 1847, became professor at the Sorbonne in 1808 and professor of mineralogy at the Muséum in 1822. Traveling widely throughout Europe, particularly in Sweden, Norway, the Apennines, and Greece, he brought back fishes from Norway.]

70. [Valenciennes, see chap. 13, n. 10.]

71. [Frédéric Louis Hammer (b. 1762), became professor of natural history at the Ecole Centrale du Haut-Rhin at Colmar in 1798, professor of botany and natural history at the school of pharmacology at Strasbourg in 1805, and director of the experimental garden of the Society of Sciences and Agriculture at Strasbourg in 1806. He died at Ingershof in 1837.]

72. [Augustin Pyrame de Candolle was born at Geneva in 1778 and died there in 1841. Doctor of medicine in 1804, he was appointed professor of botany at Montpellier and director of the botanical garden of that city. Later he returned to Geneva and there received a chair in natural history and the directorship of a

botanical garden. Author of numerous works, he exerted considerable influence on botanical studies, of which Geneva became the center for many years. For more on Candolle, see Nelson 1978.]

73. [I have been unable to identify Major, curator of the cabinet of Geneva.]

74. [Bosc, see chap. 12, n. 5.]

75. [Savigny, see n. 56 above.]

76. [Luigi Canali, Italian scholar, born at Perugia in 1759 and died there in 1814, devoted his life to the study of natural sciences.]

77. [Claude Julien Bredin, born at Alfort in 1776, died at Nice in 1854, devoted himself to the study of nature, philosophy, and literature. He donated fishes from the Rhône and its tributaries.]

78. [Carl Franz Anton von Ritter Schreibers, born at Pressburg in 1775, became director of the imperial museum of Vienna in 1805. He died at Vienna in 1852.]

79. [Lichtenstein, see n. 52 above.]

80. [Friedrich August Ludwig Thienemann, German zoologist, 1793–1858, became inspector of the royal cabinet of Dresden in 1825. He sent fishes from the Elbe.]

81. [Christian Ludwig Nitsch, born at Beucha near Grimma in Saxony in 1782, studied the natural sciences at the University of Wittenberg, where he became professor of zoology and botany in 1808. From 1816 until his death in 1837, he was professor of zoology at the University of Halle.]

82. [Jacques François Gamba was born at Dunkirk in 1763. In 1817 he left for Ukraine, visited Georgia, returned to France, and left again for Odessa, from where he sent objects of natural history to the Muséum. He died in Georgia in 1833.]

83. [Rudolphi, see chap. 15, n. 57.]

84. [Her Imperial Highness the grand duchess Hélène is Helena Pavlovna, grand duchess of Württemberg. Born at Stuttgart in 1807, the daughter of Paul, duke of Württemberg (a naturalist and traveler who amassed immense collections of natural history at his castle of Mergentheim), she married in 1824 the grand duke Michel, brother of the Russian czar Nicolas I. She died at St. Petersburg in 1873.]

85. [Louis de Freycinet; see chap. 13, n. 44.]

86. [Duperrey, see chap. 13, n. 45.]

87. [Jean René Constant Quoy, born at Maillé in 1790, began medical studies in 1807 at the Rochefort naval school and sailed as auxiliary surgeon on a cruise to the Antilles (1808–9). After receiving the degree of doctor of medicine at Montpellier in 1814, he carried out a cruise to Réunion (1814–15) as surgeon-major and later participated, with Joseph Paul Gaimard as second surgeon, in the circumnavigation of the *Uranie* under the command of Louis de Freycinet (1817–29), during which time he was in charge of geological observations. Again in the company of Gaimard, he served as surgeon-major on the *Astrolabe* under the command of Jules Sébastian César Dumont d'Urville (1826–29). An excellent observer, an artist of great talent, and one of the greatest voyager-naturalists of his time, he amassed large collections of plants and animals. He died at Rochefort in 1869.]

88. [Joseph Paul Gaimard, born at Saint-Zacharie, Var, in 1796, took part, with Surgeon-Major Jean René Constant Quoy, in the expedition of the *Uranie,* commanded by Louis de Freycinet (1817–20) and the expedition of the *Astrolabe,* commanded by Dumont d'Urville (1826–29). After a journey to Poland, Prussia, Austria, and Russia (1831–32), he participated in two voyages aboard the *Recherche* to the coasts of Iceland and Greenland (1835–36). As president of the Scientific Commission of the North, he made yet another voyage to Scandinavia, Lapland, Spitsbergen, and the Faroes (1837), and two cruises to Iceland (1838 and 1839). Staying in Paris from 1839 to 1848 to direct the Scientific Commission of Polar Sea Expeditions, he wrote the account of his voyage. He died at Paris in 1858 in near poverty and had to be buried at government expense.]

89. [Garnot, see n. 65 above.]

90. [René Primevère Lesson, born at Rochefort in 1794, participated as naval physician-pharmacist in the circumnavigation of the *Coquille* under the command of Louis Isidore Duperrey (1822–25). In 1829 he was put in charge of a course in botany at the Rochefort School of Naval Medicine and was promoted to first pharmacist-in-chief for the navy in 1835. Nominated correspondent of the Royal Academy of Sciences, Paris, in 1833 and of the Royal Academy of Medicine in 1847, he died at Rochefort in 1849. A formidable collector of plants and animals and an excellent artist, he left a large number of paintings of fishes that he collected himself.]

91. At this very moment, Quoy and Gaimard have sent to the Muséum National d'Histoire Naturelle, Paris, from Port Jackson 270 fishes, among many other objects, from various areas of the Indian Ocean and the South Pacific.

APPENDIX
LITERATURE CITED
ILLUSTRATION CREDITS
INDEX

APPENDIX: PHILOSOPHY OF ANIMAL CLASSIFICATION, BY GEORGE CUVIER

The differences in external and internal organs appropriate for characterizing fishes are not only obvious but also numerous; in fact there are few classes of animals in which it is as easy to recognize natural genera and families and to classify the species.[1] Upon the slightest inspection [fig. 57], anyone can perceive the relationships that connect the herrings *(harengs)*, for example, to the shads *(aloses)*, anchovies *(anchois)*, tarpons *(mégalopes)*, ten-pounders *(élopes)*, and wolf herrings *(chirocentres)* and those that connect the freshwater eels *(anguilles)* to the moray eels *(murènes)*, swamp eels *(synbranches)*, and snake eels *(cécilies)*. No less striking are the similarities among the innumerable tribes of minnows *(cyprins)*, those of the catfishes *(silures)*, the salmon *(salmones)*, and the mackerels *(scombres)* and their kin. But to arrange these genera and families in some sort of order, it would be necessary to fix on a small number of important characters, which would result in a few large divisions that, without breaking up natural relationships, would be precise enough to leave no doubt about the place of each species. And this is what no one has yet been able to do in a sufficiently detailed manner.

For example, the numerous unique characters of the Chondropterygi, or fishes with cartilaginous skeletons (or to speak more precisely, those with a grained periosteum), are too obvious not to have been noticed by all methodical minds. Therefore all ichthyologists have placed these fishes in an order apart, but nearly all of them have altered the correctness of this division by including fishes that are similar only in the softness of the skeleton.

However, the latter fishes ought not to be thrown back indiscriminately into the crowd. There are indeed some, such as the anglerfish *(baudroie)* and the lumpfish *(lump)*, that, except for this softness, are no different from ordinary fishes and cannot always be separated from them; but there are also those that have particular characters of integuments, teeth,

This appendix is a translation of Cuvier's final chapter of volume 1 of the *Histoire naturelle des poissons.*

FIG. 57. Georges Cuvier (1769–1832). From a painting by Van Brae, 1798. Courtesy of M. Ducreux and the Bibliothèque Centrale, Muséum National d'Histoire Naturelle, Paris.

FIG. 58. Porcupinefishes, genus *Diodon*. After Cuvier, *Mém. Mus. Nat. Hist. Nat.* (Paris), 1818a, vol. 4, pl. 6. Photograph by Stanley A. Shockey, Classroom Support Services, University of Washington, Seattle.

and especially the disposition of the skeleton in the head. The pufferfishes *(tétrodons)*, the porcupine fishes *(diodons)*, the boxfishes *(coffres)*, and even the triggerfishes *(balistes)* are of this number [fig. 58], as are the pipefishes and their allies *(syngnathes)*, whose gills have distinctive characters of great importance. The remarkable external appearance of these genera has caused many naturalists to separate them from the others, but generally we have not been very fortunate in discovering their true characters.

Artedi, for example, not only included them [tetraodontiform fishes, i.e., *Balistes* and *Ostracion*] with the anglerfishes *(baudroies)* and the lumpfishes *(lumps)* in the order Branchiostegi, but also based that whole order on a false supposition, that these fishes do not have rays in the branchial membrane,[2] whereas they all have them, and Artedi himself described those of the lumpfish *(lump)*.[3]

Linnaeus, in the tenth edition of his *Systema naturae*, after placing the Chondropterygii with the reptiles, included, according to a scheme just as ill founded, the anglerfishes *(baudroies)* with the Chondropterygii. After including with Artedi's Branchiostegi the elephantfishes *(mormyres)* and pipefishes *(syngnathes)* and giving them all the characters of lacking not only branchiostegal rays but also opercula, which for several of these fishes is contrary to the simplest observation, he united in his twelfth edition the Chondropterygii and the Branchiostegi into one order of reptiles (Amphibia nantes), based on the character, again entirely opposite to the truth, of possessing both gills and lungs.

Gmelin reestablished Artedi's two orders but still attributed to the Branchiostegi the absence of branchiostegal rays.[4] Goüan characterized them only as having incomplete gills, a description that is vague and highly contestable in nearly all the genera.[5] Pennant reunited them with the Chondropterygii under the common name of cartilaginous fishes,[6] a denomination adopted by Lacepède,[7] the inappropriateness of which we have already seen. In fact it is not good in either a positive or a negative sense. It cannot in any way be said that the skeleton of the triggerfish *(balistes)* is cartilaginous [fig. 59], and among the number of fishes that Pennant and others following him have included among the bony forms there are some, such as the leptocephalus *(leptocéphale)*, that have hardly a sign of a skeleton.

And so, first, I have had to select from among these somewhat anomalous fishes those that deviate enough from the ordinary fish type to merit being separated from them, and then to find in them distinct characters capable of being clearly explained in words.

This examination has convinced me that it was wrong to remove from the great mass of ordinary fishes the anglerfishes *(baudroies)*, lumpfishes *(lumps)*, snipefishes *(centrisques)*, elephantfishes *(mormyres)*, and scabbardfishes *(macrorhynques)*,[8] which are essentially no different from ordinary fishes. But I have recognized that the pipefishes and their allies *(syngnathes)*, having such singular form and constitution, could be distinguished by their gills shaped like tufts, hidden under an opercle having only a small opening near the nape to let water out; and that the porcupinefishes *(diodons)*, pufferfishes *(tétrodons)*, boxfishes *(coffres)*, and triggerfishes *(balistes)*, apart from whatever incompleteness there is in the skeleton, and the strangeness of their shape, have jaws and a generally

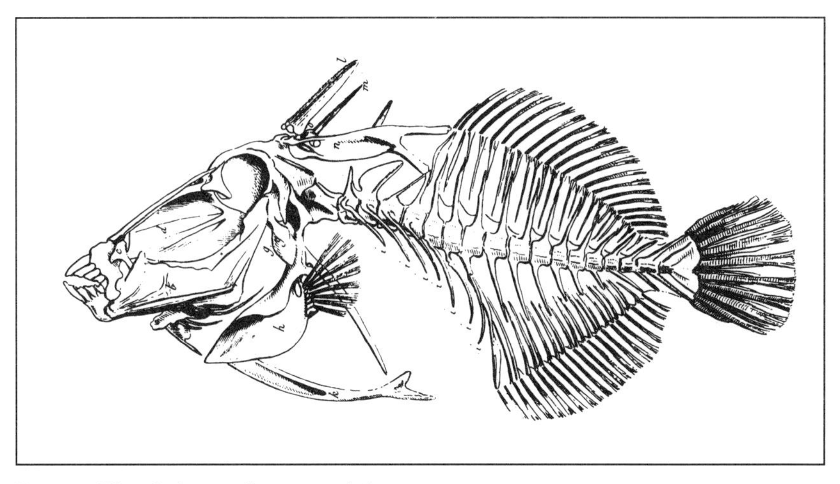

FIG. 59. The skeleton of a triggerfish. From Cuvier's *Leçons d'anatomie comparée,* 1800–1805, vol. 5, pl. 5.

complete skeleton in the head arranged slightly differently than in the ordinary fish; that the upper jaw and their palatal bones are articulated between themselves and with the vomer by means of immobile joints, which allow them much less freedom for opening and closing the mouth. Probably also related to this circumstance is the limited movement allowed their branchial apparatus by the skin that closely covers it and has kept some naturalists from observing that it is furnished with opercles and branchiostegal rays, as in all fishes.

But once these families are separated, there remain the nine-tenths of fishes among which the first observed distinction is that of fishes with soft fins, or with rays that are bifurcated and segmented, and fishes with spiny fins, of which some rays are small pointed bones without bifurcation or segmentation, or as Artedi has called the two types of fishes, the Malacopterygii and the Acanthopterygii. Unfortunately this division is still very general, but even to apply it one must make exception for the first rays of the dorsal fin or the pectoral fins in certain minnows *(cyprins)* and catfishes *(silures),* which have strong, solid spines, although it is true that the spines in these genera are formed by the agglutination of a multitude of small joints, the traces of which can be seen.

And there are other exceptions, or at least seemingly so, in certain fishes of the family of wrasses *(labres)* and of blennies *(blennies),* whose spines are so small or weak and few in number that they do not appear to have any; but except for these small irregularities, if this division does not lead far enough, at least it does not mislead and does not disunite fishes that nature has brought together.

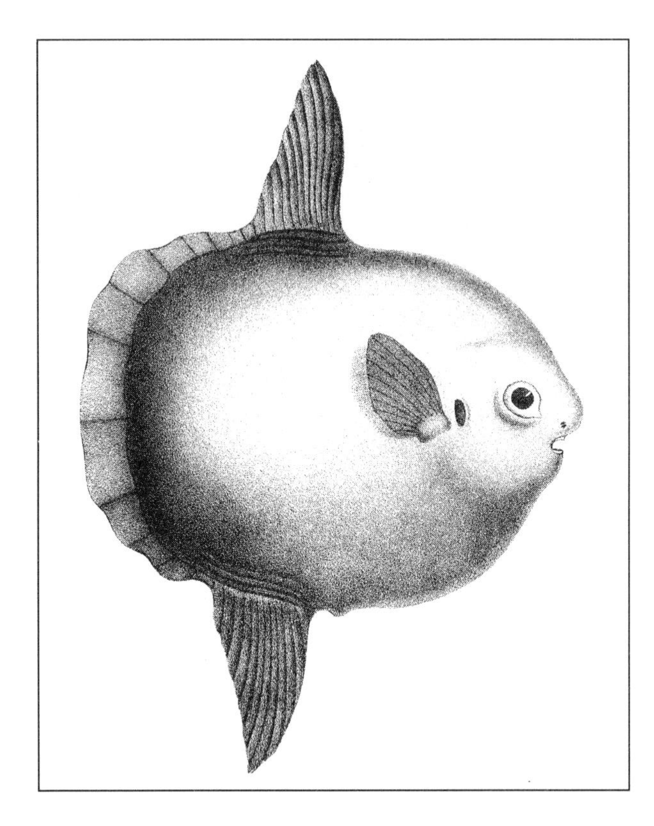

FIG. 60.
An ocean sunfish, *Mola mola*.
From Everhard Home's *Lectures on Comparative Anatomy*, 1814–28, vol. 6, pl. 50. Courtesy of Colleen M. Weum and the Health Sciences Library and Information Center. Photograph by Patricia L. McGiffert, Health Sciences Photography, University of Washington, Seattle.

The same cannot be said of the distinctions that naturalists have sought to establish on other principles, or of the subdivisions that those who have adopted the overall division according to spines have attempted to introduce into the two branches. Thus the general shape of the body and the absence of pelvic fins used by Ray, characters that for him took precedence over the presence or absence of spines, obliged him to group together the freshwater eel *(anguille)*, the freshwater burbot *(lote)* and the goby *(gobie)*, the pipefish *(syngnathe)*, the swordfish *(xiphias)*, and the mola or ocean sunfish *(poisson lune)* [fig. 60].[9]

Linnaeus, in his tenth edition, ignoring the distinction based on spines, was the first to conceive of dividing ordinary fishes into Apodes, Jugulares, Thoracici, and Abdominales according to whether they lacked pelvic fins or had the pelvics attached in front of the pectorals, under the pectorals, or farther behind [see table 8]. Thus he considered himself obliged to put the swordfish *(xiphias)*, the cutlassfish *(trichiure)*, and the butterfish *(stromatée)* together with the freshwater eel *(anguille)* and the knifefish *(gymnote)*, to place the codfishes *(gades)* between the weevers *(vives)* and blennies *(blennies)*, the flounders *(pleuronectes)* between the dories *(zeus)* and butterflyfishes *(chétodons)*, and the surgeonfishes or rabbitfishes *(teuthis* or *amphacanthes)* between the sheatfishes *(silures)* and the suckermouth armored catfishes *(loricaires)*.

Goüan, in combining the two methods and dividing each of Artedi's branches according to Linnaeus's four orders, avoids some unnatural groupings, yet he still places the swordfish *(xiphias)* and the cutlassfish *(trichiure)* far from the mackerels *(scombres)*; he also commits positive errors in making the cuskeel *(donzelle)* and the catfish *(silure)* acanthopterygians and the butterfish *(stromateus)* a malacopterygian.[10]

Lacepède takes up Pennant's characters and divides fishes into cartilaginous and bony forms [see table 10]. Each of these branches is subdivided, without regard for spines, according to the presence or absence of the opercle and the branchiostegal membrane. The final subdivisions are based on the relative positions of the pelvic and pectoral fins, a quite regular distribution that gives thirty-two a priori orders [fig. 61], but fifteen of them cannot be filled because the fishes relating to them have not been found in nature, and some seem to have been filled only because of the erroneous belief that the opercle or membrane is lacking in fishes that in fact have them, such as the elephantfishes *(mormyres)*, moray eels *(murènes)*, and swamp eels *(synbranches)*.

This classification, apart from the displacement of the anglerfishes *(baudroies)* and the lumpfishes *(lumps)*, and the mixing of malacopterygians with acanthopterygians that also occurred in Linnaeus's classification, would have the disadvantage of placing the moray eels *(murénes)* and the swamp eels *(synbranches)* far from the freshwater eels *(anguilles)* that resemble them so strongly, if, as regards this detail of its distribution, the classification were not founded, as we have just said, on characters that have no real existence. Nonetheless, Duméril has retained these orders in his own classification [fig. 62], which is basically that of Lacepède, subdivided according to body shape and other details, in order to bring together as much as possible the natural families; but the introduction of characters taken from the pelvic fins prevents the achieving of this goal [see table 11]. Thus one sees the anglerfishes *(baudroies)* with the triggerfishes *(balistes)* and chimaeras *(chimères)*, and the codfishes *(gades)* with the weevers *(vives)* and the stargazers *(uranoscopes)* [fig. 63]. A single family [the *péroptères*] unites the snake eels *(cécilies* and *ophisures)* and swamp eels *(monoptères)*, which are [better placed next to the] freshwater eels *(anguilles)*; the featherback *(notoptère)*, which is really a herring *(hareng)*; the cutlassfishes *(trichiures)*, which are related to the mackerels *(scombres)*, and so on [fig. 64].

These causes have led Rafinesque and Risso to similar conclusions in the combinations they attempted to make of Pennant's and Lacepède's classifications, either between the methods or with natural families [compare tables 12, 14, and 18].

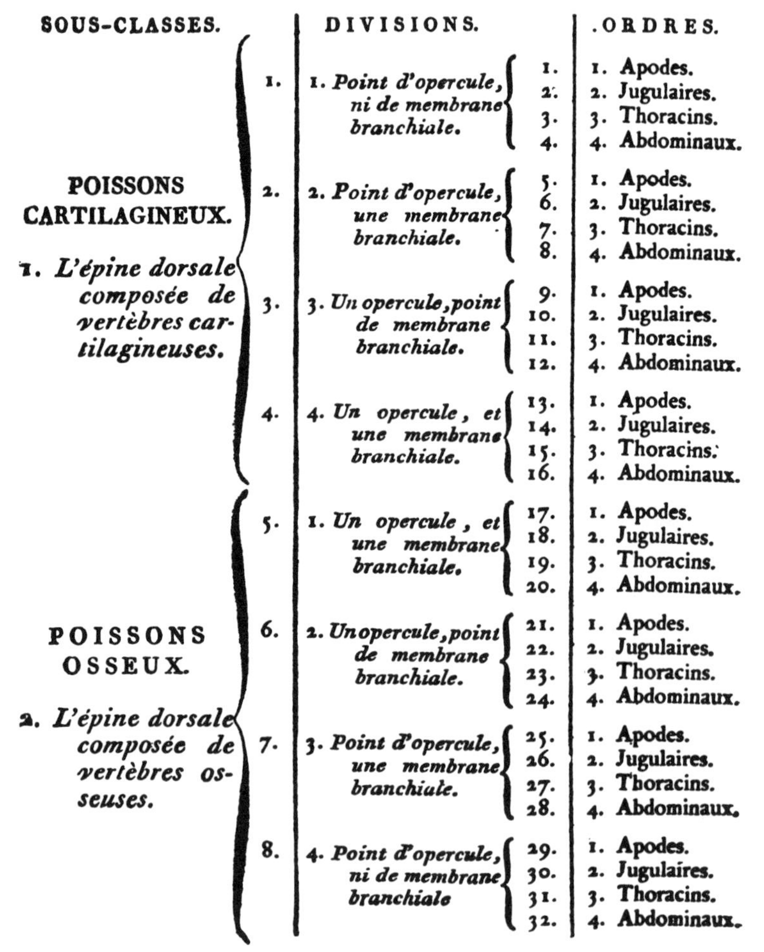

TABLE GÉNÉRALE DES POISSONS.

POISSONS.

Le sang rouge, des vertèbres, des branchies au lieu de poumons.

SOUS-CLASSES.		DIVISIONS.		ORDRES.
POISSONS CARTILAGINEUX. 1. *L'épine dorsale composée de vertèbres cartilagineuses.*	1.	1. *Point d'opercule, ni de membrane branchiale.*	1. 2. 3. 4.	1. Apodes. 2. Jugulaires. 3. Thoracins. 4. Abdominaux.
	2.	2. *Point d'opercule, une membrane branchiale.*	5. 6. 7. 8.	1. Apodes. 2. Jugulaires. 3. Thoracins. 4. Abdominaux.
	3.	3. *Un opercule, point de membrane branchiale.*	9. 10. 11. 12.	1. Apodes. 2. Jugulaires. 3. Thoracins. 4. Abdominaux.
	4.	4. *Un opercule, et une membrane branchiale.*	13. 14. 15. 16.	1. Apodes. 2. Jugulaires. 3. Thoracins. 4. Abdominaux.
POISSONS OSSEUX. 2. *L'épine dorsale composée de vertèbres osseuses.*	5.	1. *Un opercule, et une membrane branchiale.*	17. 18. 19. 20.	1. Apodes. 2. Jugulaires. 3. Thoracins. 4. Abdominaux.
	6.	2. *Un opercule, point de membrane branchiale.*	21. 22. 23. 24.	1. Apodes. 2. Jugulaires. 3. Thoracins. 4. Abdominaux.
	7.	3. *Point d'opercule, une membrane branchiale.*	25. 26. 27. 28.	1. Apodes. 2. Jugulaires. 3. Thoracins. 4. Abdominaux.
	8.	4. *Point d'opercule, ni de membrane branchiale*	29. 30. 31. 32.	1. Apodes. 2. Jugulaires. 3. Thoracins. 4. Abdominaux.

FIG. 61. Bernard Germain Etienne de la Ville, the comte de Lacepède's classification of fishes, as illustrated in his *Histoire naturelle des poissons,* 1798–1803, vol. 5.

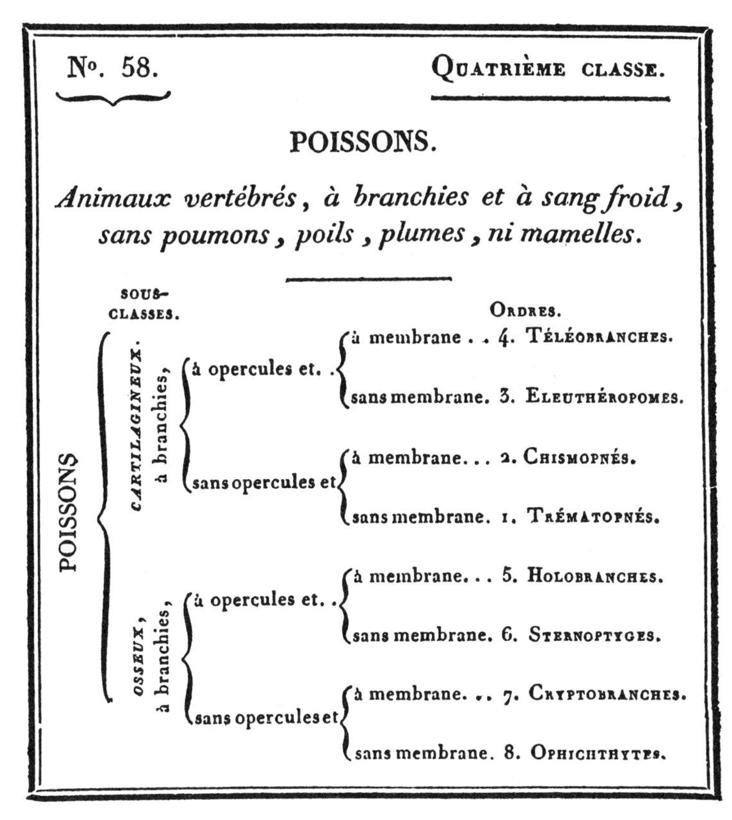

FIG. 62. The classification of André Marie Constant Duméril, from his *Zoologie analytique, ou Méthode naturelle de classification des animaux,* 1806, p. 97.

I do not see that one has been any more fortunate in these attempts in Germany. Goldfuss, making no other changes to Linnaeus's division than to combine the Jugulares with the Thoracici and the Branchiostegi with the Chondropterygii, has deprived himself of any way of arranging the families in the order of their similarities [see table 17]. The lumpfishes *(cycloptères)* and the anglerfishes *(baudroies)* will never go, as he has placed them, between the lampreys *(lamproies)* and sharks *(squales);* one will never be able rationally to place, as he did, the cutlassfish *(trichiure)* with the freshwater eels *(anguilles)* and far from the scabbardfish *(lepidopus),* which resembles it at almost every point; the longfin herring *(gnathobolus),* which is a true herring *(hareng),* can never remain with the butterfish *(stromateus),* which is almost a butterflyfish *(chétodon).*

The author himself [Cuvier] thought he was obliged to break his own rule for the swordfish *(xiphias),* which he leaves with the mackerels

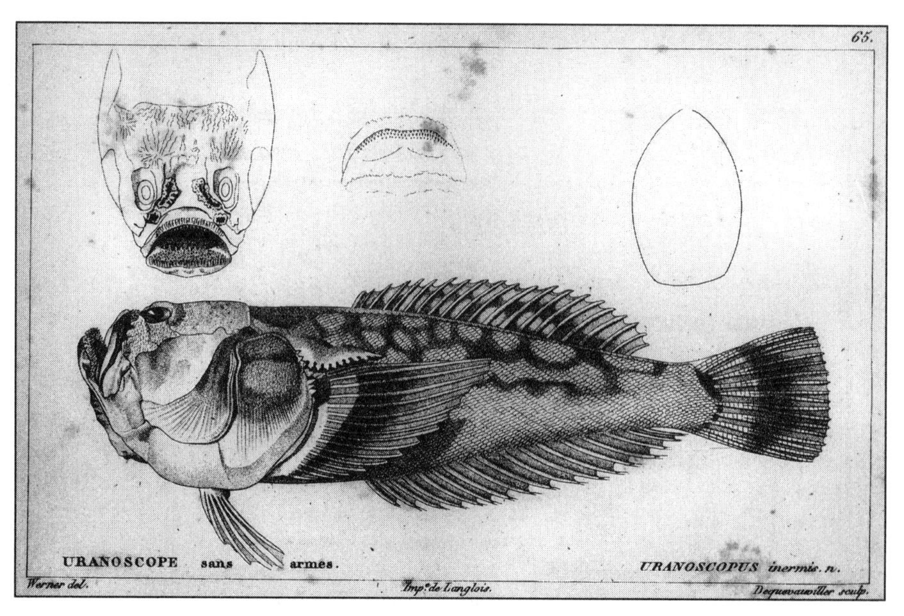

FIG. 63. A stargazer, *Ichthyscopus inermis.* After Cuvier and Valenciennes, *Histoire naturelle des poissons,* 1828–49, vol. 3, pl. 65. Courtesy of Leslie Overstreet and the Special Collections Department, Smithsonian Institution Libraries, Washington, D.C.

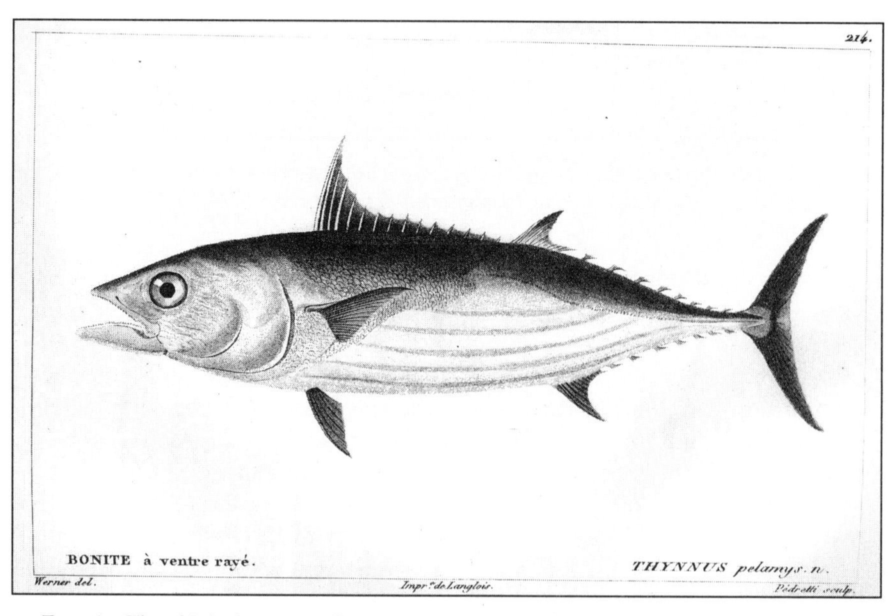

FIG. 64. The skipjack tuna or bonito, *Katsuwonus pelamis.* After Cuvier and Valenciennes, *Histoire naturelle des poissons,* 1828–49, vol. 8, pl. 214. Courtesy of Leslie Overstreet and the Special Collections Department, Smithsonian Institution Libraries, Washington, D.C.

(scombres) among the Subbrachiens, although it is most assuredly a member of the Apodes.[11]

Oken had more freedom in arranging his families, because he gave to his grand orders—his fish fishes, amphibian fishes, bird fishes, and mammal fishes—almost indeterminate characteristics [see tables 19–23]; and yet, because he used the position of the pelvic fins in his subdivisions, one sees him placing the herrings *(clupées)* between the mullets *(muges)* and the rabbitfishes *(amphacanthes, buro)*, the codfishes *(gades)* near the sticklebacks *(gasterostes)*, and the swordfish *(xiphias)* near the wolffishes *(anarrhiques)*, while leaving the spiny eels *(rhinchobdelles)* and ribbonfishes *(bogmares)* in the same family as the freshwater eels *(anguilles)*.

After studying fishes for almost forty years, not according to the authors but the fishes themselves, their skeletons and internal organs, and after dissecting several hundred species, I am convinced of the necessity of never mixing an acanthopterygian with fishes of other families; I have learned that the acanthopterygians, which make up three-fourths of all known fishes, are also the type that nature has been the most careful with, the type that she has kept most nearly the same in all the variations in detail to which she has submitted it.

All other characters should be used only in subordination to this one and without ever contradicting it, but the extreme constancy of the general plan and the predominant influence of this regulating character has made it very difficult to give to fishes in which it exists precise and perceptible applications of subordinate characters; therefore the different families of acanthopterygians thus overlap one another, so that it is not known where one begins and the other ends. For example, the Percidae *(famille des perches)* [fig. 65], which is distinguished from the Sciaenidae *(sciènes)* by its palatal teeth, is composed of a rather large but otherwise entirely natural group, part possessing teeth and part not.

The same happens in the otherwise well characterized family of fishes with protected cheeks [mailed-cheek or scorpaeniform fishes]: most of its genera are related to the perches *(perches)*, and the rest to the drums *(sciènes)* because of the palatal teeth.

There are observable overlappings of some genera of the Sciaenidae *(famille des sciènes)* with those of the Chaetodontidae *(chétodons)* on account of the scales that more or less cover their vertical fins [fig. 66]; on the other hand, one is obliged to bring together the Sparidae *(famille des spares)* and several genera of the Sciaenidae *(sciènes)* that have no trace of these scales.

Overlappings no less noticeable combine certain genera of porgies *(spares)*, such as [members of the centracanthid genus] *Spicara (picarels)* and the mojarras *(gerres)*, with other genera such as the ponyfishes *(équules)*,

FIG. 65. The yellow perch, *Perca flavescens.* After Cuvier and Valenciennes, *Histoire naturelle des poissons,* 1828–49, vol. 2, pl. 9. Courtesy of Leslie Overstreet and the Special Collections Department, Smithsonian Institution Libraries, Washington, D.C.

FIG. 66. The raccoon butterflyfish, *Chaetodon lunula.* After Cuvier and Valenciennes, *Histoire naturelle des poissons,* 1828–49, vol. 7, pl. 173. Courtesy of Leslie Overstreet and the Special Collections Department, Smithsonian Institution Libraries, Washington, D.C.

which cannot be separated from the dory *(zeus),* which leads to the Scombridae *(famille des scombres),* which in turn blends by such imperceptible nuances into those fishes that are extremely elongate and narrow, which we call ribbonfishes *(taenioïdes),* that it is almost impossible to say where one can draw the line separating the two.

What recourse remains for naturalists desiring to describe organisms according to their true relationships except to acknowledge that the acanthopterygian fishes—which make up the old genera of perches *(perches),* drums *(sciènes),* porgies *(spares),* butterflyfishes *(chétodons),* dories *(zeus),* and mackerels *(scombres),* including the bandfishes *(cépoles)* and other ribbon-shaped fishes—compose, despite their innumerable species, one single natural family, in which nuances can be defined and incipient groups or slight differences can be perceived but in which it is impossible to determine perfectly clear boundaries that do not overlap each other at any point?

It is not entirely the same with the anglerfishes *(baudroies),* the toadfishes *(batrachus),* the gobies *(gobies),* the blennies *(blennies),* and especially the wrasses *(labres).* Their characters are precise enough and, though in part internal, easy enough to assign and discern: in the first of these groups [the anglerfishes], the small restricted gill opening and the pectoral fins, with the base lengthened in the shape of an arm; in the second group [the toadfishes], similar pectoral fins joined to pelvic fins that consist of three rays; the flexible dorsal-fin spines in the third and fourth groups [the gobies and blennies]; the fleshy lips of the fifth [the wrasses]; the total absence of cecal appendages in nearly all these groups. While the characters mentioned above separate them from the other acanthopterygians, the last character relates them to the catfishes *(silures)* and minnows *(cyprins),* families that begin the order of malacopterygians [fig. 67]; the malacopterygians themselves, as we have said, are related to the acanthopterygians by the spiny form of some of their rays.

The families of the malacopterygians offer more differences and traits that are easier to recognize, and several of them are as natural as if they were subject to fixed limits, so much does each one of them, while clearly separate from the others, keep internally a great resemblance in details. This fixedness is so perceptible that most of the natural families we shall establish for this part of the class have already been discerned by Artedi and presented under the names of genera. His catfishes *(silures),* minnows *(cyprins),* salmons *(saumons),* herrings *(clupées),* and pikes *(brochets)* can remain together: it is not even inappropriate to distribute them according to the presence and position of the pelvic fins, for this character, slight as it is, does not vary in any of them. But I have found it impossible to preserve the distinction of the Jugulares, the Thoracici, and the Abdominales in the terms used by Linnaeus.[12] In fact it is of little

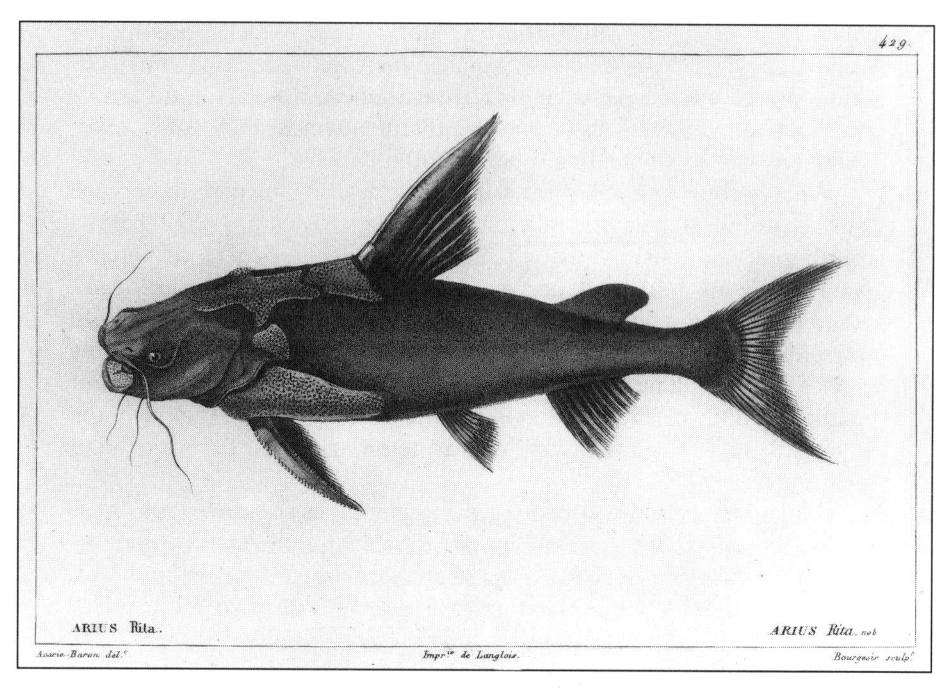

FIG. 67. A bagrid catfish of the genus *Rita*. After Cuvier and Valenciennes, *Histoire naturelle des poissons*, 1828–49, vol. 15, pl. 429. Courtesy of Melanie Wisner, Roger E. Stoddard, and the Houghton Library, Harvard University, Cambridge.

importance whether the pelvic fins are situated outside, slightly forward, a little behind the pectoral, or exactly below it; the important circumstance, based on the very structure of the fish, is knowing whether the pelvis is attached to the bones of the shoulder or simply suspended in the flesh of the belly. Thus I have coined the term "Subbrachien" to designate fishes of the first category, regardless of the location of their pelvic fins, which depends only upon the greater or shorter length of the bones of the pelvis; and I have left the name Abdominales for those of the second category. The Apodes are naturally found to be malacopterygians without pelvic fins.

We shall thus begin this natural history of fishes with the acanthopterygians, which actually constitute scarcely more than one immense family. We shall place after them the diverse families of the malacopterygians in the order in which they appear to us to be the most closely related to the acanthopterygians; but I would not wish it to be believed that they are related along only one line and in a single series. Although the abdominal malacopterygians can be arranged thus, and can even begin with those among them having spiny rays, neither the Apodes nor the Subbrachiens are to come next after them.

The codfishes *(gades),* for example, are related as closely as any member of the Abdominales to certain acanthopterygians, and there would be no reason to place them after the Abdominales if one wishes to indicate their rank in nature. If we speak in terms of a series, it is only because facts put into books can be expressed only one after the other.

The same observation applies to the other fishes, those having a fixed upper jaw, those with puff-shaped gills, and especially to the large and important family of Chondropterygii, with which we shall end this natural history.[13]

The vanity of these systems that tend to arrange organisms in a straight line is easily seen in the Chondropterygii. Several of these genera—for example, those of the rays *(raies)* and sharks *(squales)*—are egregiously complex in some of their sense organs and in their organs of generation and are more developed in some of their parts than are birds. Other genera, which one reaches by obvious stages—for example, the lampreys *(lamproies)* and the ammocoetes *(ammocètes)*—are on the other hand so simplified that one might be authorized to consider them an overlap of fishes with segmented worms *(vers articulés);* most certainly the ammocoetes *(ammocètes)* at least do not have a skeleton, and their whole muscular apparatus has only tendinous or membranous supports.

Therefore let it not be thought that, because we shall be placing one genus or family before another, we actually consider it superior or more nearly perfect in the system of organisms. Only he can make that claim who pursues the wildly fanciful project of ranking organisms in a straight line, and this is an approach we renounced long ago. The more progress we make in our study of nature, the more we are convinced that this is one of the most erroneous ideas that has ever been held in natural history, and the more we have recognized the need to consider each organism or each group of organisms by itself, according to the role it plays on account of its characteristic qualities and its organization, and not to exclude any of its relationships, any of the ties that bind it to organisms nearest to and farthest from it.

Once the naturalist adopts this point of view, the difficulties vanish and things fall into place for him. Our present systematic classifications see only the nearest relationships: they only wish to place one organism between two others, and they are constantly at fault. The true classification sees each organism in the midst of the others; it shows all the radiations or lineages by which organisms are linked more or less closely in this immense network that makes up organized nature; and only the true classification gives us great and true ideas of nature, worthy of this nature and its author—but ten or twenty radiating lineages would not often suffice to explain these innumerable relationships.

Therefore we inform our readers, once and for all, that it is in the descriptions themselves, which we shall give, that one must seek the proper idea of the degrees of organization, and not in the positions we are obliged to assign to the species.[14] And yet we are far from claiming that relationships do not exist, that no classification is possible, and that one must not form groups of species and define them.

Buffon was entirely correct when he asserted that absolute characters and clear distinctions between genera do not always exist, that there is no way of lining them up without coercion in our methodical framework; but this great man went too far when he rejected all relationships, when he objected to any ordering based on the similarities of organisms.[15]

These relationships are so real, our mind is drawn to them by a propensity so necessary, that ordinary folk as well as naturalists have always had their genera.

We shall therefore bring together what nature brings together, without forcing organisms into our groupings if nature has not placed them there; and we shall not hesitate—after describing all the species that may be arranged into a well-defined genus and all the genera that may compose a well-circumscribed family—to omit one or more isolated species or one or more genera not connected to the others in a natural way. We prefer a frank recognition of these sorts of so-called irregularities rather than to introduce errors by placing these anomalous species and genera in a series that does not embrace their characters.

Our list of fishes, based on these principles, can be distributed into families approximately as indicated in the following table [table 24]. Unable to assign to each family one unequivocal and exclusive character, we indicate them for the moment by names derived from the most widely known genus of each, the genus that may be regarded as the type from which it is easiest to get an idea of the family. At the beginning of each family will be found a more extensive list of its characters, as well as the combinations according to which we are subdividing the family and that lead us to the different genera that compose the family.

NOTES

1. [As a basis for a comparative anatomy of all fishes, Cuvier (1828c) provides in volume 1 of the *Histoire naturelle des poissons* (288–551) an anatomical description of the European perch (*Perca fluviatilis* Linnaeus, 1758), "Idée générale de la nature et de l'organisation des poissons" (General idea of the nature and organization of fishes) that includes chapters on external morphology, osteology, myology, nervous system, sense organs, organs of digestion (from mastication to excretion, including the circulatory and respiratory systems), and reproductive system.]

2. At least this is how one explains the definition Artedi [1738] gives for the Branchiostegi [in his *Genera piscium*]: "branchiis ossibus destitutis" [opposite p. 1]; "branchiis, uti praecedentes, nulla ossicula gerunt" (85).

3. [Artedi 1738], *Genera piscium* (61): "Membrana Branchiostega ossicula sex gracilia et teretia utrinque continet."

4. [Johann Friedrich Gmelin (see chap. 10, n. 15), in the thirteenth edition of Linnaeus's *Systema naturae*, vol. 1, pt. 3, pp. 1126–1516, 1789 (for a full citation of Gmelin's edition of the *Systema naturae*, see Linnaeus, 1788–93).]

5. [Goüan (see chap. 8, n. 39), in his *Historia piscium*, 1770.]

6. [Pennant (see chap. 8, n. 38), in his *British Zoology*, vol. 3, 1769.]

7. [*Poissons cartilagineux*, Lacepède; see, e.g., his première table méthodique, in the *Histoire naturelle des poissons*, 1798–1803, vol. 1, a large foldout preceding p. 1.]

8. The *Macrorhinque argenté* of Lacepède (1798–1803, 2:76) or *Cheval marin argenté (Syngnathe argenté)* of Bonnaterre (1788, 32), is none other than a scabbardfish *(lépidope)* incompletely described by Osbeck (1771, 107).

9. [Ray (see chap. 6, n. 1), in his *Synopsis methodica piscium*, 1713.]

10. [Goüan (see chap. 8, n. 39), *Historia piscium*, 1770.]

11. [Despite the absence of the pelvic girdle and fins in the swordfish *(Xiphias)*, Cuvier places this fish together with the billfishes *(Tétraptures, Makaira, Voiliers)* in a tribe of their own ("Tribu espadons, ou Scombéroïdes à museau en forme de dard ou d'épée"), among other scombroids; see Cuvier 1831, 255.]

12. Those who have combined the Thoracici and the Jugulares have done so only on the basis of my *Règne animal* [see Cuvier 1816, vol. 2].

13. [Cuvier intended to describe all known fishes in the *Histoire naturelle des poissons*, but a number of important groups were left out, including the Chondropterygii or chondrichthyan fishes. Before he died in 1832 he left instructions for Valenciennes (see chap. 13, n. 10) to continue the work, leaving to him all the appropriate notes, manuscripts, and library necessary for the job (Monod 1963, 25). But as pointed out by a number of authors (e.g., Günther 1880, 18; Jordan 1902, 440; Monod 1963, 25), Valenciennes did not complete the work as originally conceived. For reasons unknown, publication ceased after volume 22, the last volume appearing in December 1849 (Bailey 1951), even though Valenciennes lived for another sixteen years. In addition to groups then unknown to science—deep-sea fishes, fishes of the Antarctic, Lake Baikal, Great Lakes of Africa, and so on—there are some obvious omissions, including the "Ganoides," the Chondrichthyes, Anguilliformes, Gadiformes, Pleuronectiformes, and Tetraodontiformes. Substantial evidence that additional volumes were planned has survived. Valenciennes himself seems to have considered the twenty-two volumes a first series and believed that a second series dealing with the taxa not yet covered was to follow (Valenciennes 1849, vi; Monod 1963, 25). See Pietsch 1985, 59–61; Bauchot, Daget, and Bauchot 1990, 7–17.]

14. I make this observation because one author [whom we have not been able to identify] thought himself clever in noticing that the lampreys *(lamproies)* are not near the reptiles, and that consequently I had been wrong to place them immediately after this class in my *Règne animal* [Cuvier 1816, vol. 2]; but because

this time I shall place them at the other end of the class of fishes, one should not conclude that I wish to place them after all the others.

15. [Buffon's ideas about plant and animal relationships are developed in the first discourse *(Premier discours)* of vol. 1 of his *Histoire naturelle, générale et particulière,* 1749–1804.]

TABLE 24. [The classification of Cuvier and Valenciennes adopted in the twenty-two volumes of their *Histoire naturelle des poissons* (see Cuvier 1828b, 572–73.]

POISSONS		
OSSEUX [bony fishes]		
Gills comblike or laminate		
Upper jaw free		
ACANTHOPTÉRYGIENS		
Percoïdes	Chétodonoïdes	Lophioïdes
Polynèmes	Scombéroïdes	Gobioïdes
Mulles	Muges	Labroïdes
Joues cuirassées	Branchies labyrinthiques	
Sciènoïdes		
Sparoïdes		
MALACOPTÉRYGIENS		
Abdominaux		
Cyprinoïdes	Salmonoïdes	Lucioïdes
Siluroïdes	Clupéoïdes	
Subbrachiens		
Gadoïdes	Pleuronectes	Discoboles
Apodes		
Murénoïdes		
Upper jaw fixed		
Sclérodermes	Gymnodontes	
Gills tufted		
Lophobranches		
CARTILAGINEUX or CHONDROPTÉRYGIENS [cartilaginous fishes]		
Sturioniens	Plagiostomes	Cyclostomes

LITERATURE CITED

Abildgaard, P. C. 1792. Kurze anatomische Beschreibung des Säugers *(Myxine glutinosa* Linn.). *Schr. Berl. Ges. Naturf. Fr.* 10:193–200, pl. 4.

Adanson, M. 1757. *Histoire naturelle du Sénégal. Coquillages. Avec la relation abrégée d'un voyage fait en ce pays, pendant les années 1749, 50, 51, 52 et 53.* Paris: Claude-Jean-Baptiste Bauche.

Adelmann, H. B. 1942. *The embryological treatises of Hieronymus Fabricius of Aquapendente: The formation of the egg and of the chick* [De formatione ovi et pulli], *The formed fetus* [De formato foetu]. A facsimile edition, with an introduction, a translation, and a commentary. Ithaca, N.Y.: Cornell University Press.

Adler, K. 1989. *Contributions to the history of herpetology.* Contributions to herpetology no. 5. Oxford, Ohio: Society for the Study of Amphibians and Reptiles.

Albertus Magnus. 1651. *Beati Alberti Magni, Ratisbonensis episcopi, ordinis praedicatorum, de animalibus lib. XXVI.* vol. 6. Lyons: Claudii Prost, Petrus et Claudius Rigaud, Hieronymus Delagarde, Joan. Ant. Huguetan.

Aldrovandi, U. 1613. *De piscibus libri V, et de cetis liber unus.* Bononia: Bellagambam.

———. 1623. *De piscibus libri V, et de cetis liber unus.* Frankfurt: I. N. Stoltzenbergeri et I. Treudelij.

———. 1629. *De piscibus libri V, et de cetis liber unus.* Frankfurt: Matthaeus Kempfferus.

———. 1638. *De piscibus libri V, et de cetis liber unus.* Bononia: Nicolaus Thebaldinus.

———. 1640. *De piscibus libri V, et de cetis liber unus.* Frankfurt: Caspari Rôtelij et Ioanis Treudelii.

———. 1644. *De piscibus libri V, et de cetis liber unus.* Bononia: Nicolaus Thebaldinus. [Title page dated 1638, colophon at rear of volume dated 1644.]

Allamand, J. N. S. 1755. Kort verhaal van de uitwerkzelen, welke een Americaanse vis veroorzaakt op de geenen, die hemaanraaken. *Verh. Holl. Maatsch. Wetensch. Haarlem,* 2:372–79.

Allen, B. 1698. Of the manner of the generation of eels. *Phil. Trans. Roy. Soc. Lond.* 19:664–66.

Allen, E. G. 1951. The history of American ornithology before Audubon. *Trans. Amer. Phil. Soc.*, n.s., 41 (3): 386–591.

Allgayer, R. 1991. Un Naturaliste et son oeuvre oubliés, Léonard Baldner (1612–1694). *Rev. Franç. Aquar.*, suppl. 1, 1–12.

Allinson, F. G. 1905. *Lucian: Selected writings.* Boston: Ginn.

Alpinus, P. 1645. *De medicina Aegyptiorum, libri quatuor, et Jacobi Bonti in Indiis archiatri, de medicina Indorum.* Editio ultima. Paris: Nicolaus Redelichuysen.

Amsterdammers, Collegium Privatum. 1667. *Observationes anatomicae selectiores collegii privati Amstelodamensis.* Amsterdam: Caspar Commelin.

———. 1673. *Observationum anatomicarum collegii privati Amstelodamensis, pars altera, in quibus praecipue de piscium pancreate ejusque succo agitur, figuris elegantioribus illustrata.* Amsterdam: Caspar Commelin.

Anderson, J. 1746. *Nachrichten von Island, Grönland und der Strasse Davis, zum wahren nussen der Wissenschaften und der Handlung.* Hamburg: Georg Christian Grund.

———. 1750. *Histoire naturelle de l'Islande, du Groenland, du Détroit de Davis, et d'autres pays situés sous le nord, traduite de l'Allemand de M. Anderson, de l'Académie Impériale, bourg-mestre en chef de la ville de Hambourg.* Par M** [Gottfried Sellius], de l'Académie Impériale, et de la Société Royale de Londres. Paris: Sebastien Jorry.

Anderson, W. 1777. An account of some poisonous fish in the South Seas. In a letter to Sir John Pringle, Bart. P.R.S. from Mr. William Anderson, late surgeon's mate on board His Majesty's ship the Resolution, now surgeon of that ship. *Phil. Trans. Roy. Soc. Lond.* 66:544–74.

Andre, W. 1784. A description of the teeth of the *Anarrhichas lupus* Linnaei, and of those of the *Chaetodon nigricans* of the same author; to which is added, an attempt to prove that the teeth of cartilaginous fishes are perpetually renewed. *Phil. Trans. Roy. Soc. Lond.* 74:274–82, pls. 11–13.

Appel, T. A. 1973. Lacepède. In *Dictionary of Scientific Biography,* 7:546–48. New York: Scribner's.

———. 1976. Valenciennes, Achille. In *Dictionary of Scientific Biography,* 13:554–55. New York: Scribner's.

Arderon, W. 1750. Extract of a letter from Mr. William Arderon, F.R.S. to Mr. Henry Baker, F.R.S. concerning the hearing of fish. *Phil. Trans. Roy. Soc. Lond.* 45 (486): 149–55.

Arendt, E. 1822. *De capitis ossei esocis lucci structura singulari. Dissertatio inauguralis zootomica.* Regiomonti: Typis Academicis Hartungianis.

Argillander, A. 1753. Rön om gjädd-leken. *Kongl. Svenska Vetensk. Acad. Handling.* (Stockholm) 14:74–77.

Arnault de Nobleville, L. D., and F. Salerne. 1756–57. *Suite de la matière medicale, Règne animal.* 6 vols. Paris: Desaint et Saillant, G. Cavelier, et Le Prieur. Vol. 1, 1756; vol. 2, 1756; vol. 3, 1756; vol. 4, 1757; vol. 5, 1757; vol. 6, 1757.

Arsakis, A. 1813. *Commentatio de piscium cerebro et medulla spinali.* Inaug. diss. Halle: Hendelianis.

———. 1836. *Commentatio de piscium cerebro et medulla spinali, scripta auspiciis et ductu Joannis Frederici Meckelii denuo edita fragmentis de eadem re additis ab Gustavo Guilielmo Minter.* Leipzig: Robertus Friesius.

Artedi, P. 1738. *Ichthyologia, sive Opera omnia piscibus scilicet: Bibliotheca ichthyologica. Philosophia ichthyologica. Genera piscium. Synonymia specierum. Descriptiones specierum. Omnia in hoc genere perfectiora, quam antea ulla. Posthuma vindicavit, recognovit, coaptavit et edidit Carolus Linnaeus, Med. Doct. et Ac. Imper. N.C.* Leiden: Wishoff.

Ascanius, P. 1767. *Icones rerum naturalium, ou figures enluminées d'histoire naturelle.* Pt. 1. Copenhagen and Geneva: Claude Philibert. [Pt. 1, an oblong folio, was subsequently reprinted, the text entirely altered to make it uniform with the succeeding parts, which appeared under a modified title; see Ascanius 1772–1806.]

———. 1772–1806. *Icones rerum naturalium, ou Figures enluminées d'histoire naturelle du Nord.* Copenhagen: Claude Philibert. Pt. 1 [2d ed.; see Ascanius, 1767], 1806; pt. 2, 1772; pt. 3, 1775; pt. 4, 1777; pt. 5, 1805.

Ash, H. B. 1941. *Lucius Junius Moderatus Columella on agriculture, with a recension of the text and an English translation.* Vol. 1. Cambridge, Mass.: Harvard University Press. [For vols. 2 and 3, see Forster and Heffner 1954-55.]

Autenrieth, J. H. F. von. 1800. Bemerkungen über den Bau der Scholle *Pleuronectes platessa* L. insbesondere, und den Bau der Fische, hauptsächlich ihres Scelets, im Allgemeinen. *Arch. Zool. Zoot., Wiedemann* 1 (2): 47–103.

Baehrens, E. 1879–83. *Poetae Latini minores.* Leipzig: B. G. Teubner. Vol. 1, 1879; vol. 2, 1880; vol. 3, 1881; vol. 4, 1882; vol. 5, 1883.

Bailey, R. M. 1951. The authorship and names proposed in Cuvier and Valenciennes' *Histoire naturelle des poissons. Copeia* 1951 (3): 249–51.

Bailly, E. M. 1824. Description des filets pêcheurs de la baudroie. *Ann. Sci. Nat.* (Paris) 2:323–32.

Bakker, G. 1822. *Osteographia piscium; gadi praesertim aeglefini, comparati cum lampride guttato, specie rariori, icones accedunt forma majore, aere ac lapide expressae* (text). *Icones, ad illustrandam Gerbrandi Bakker piscium osteographiam* (atlas). Gröningen: W. van Boekeren.

Baldinger, E. G. 1802. Sur l'âge d'un brochet. *Neues Med. Physik. J.* (Marburg) 5 (1): 29.

Baldwin, S. A. 1986. *John Ray (1627–1705): Essex naturalist.* Witham, Essex: Baldwin Books.

Balme, D. M. 1970. Aristotle and the beginnings of zoology. *J. Soc. Bibliogr. Nat. Hist.* 5 (4): 272–85.

Bancroft, E. 1769a. *Naturgeschichte von Guiana in Süd-Amerika. Worinn von der natürlichen Beschaffenheit und den vornehmsten Naturproducten des Landes, ingleichen der Religion, Sitten und Gebräuchen verschiedener Stämme der wilden Landes-Einwohner, Nachricht ertheilet wird.* Frankfurt and Leipzig: J. Dobslen.

———. 1769b. *An essay on the natural history of Guiana, in South America. Containing a description of many curious productions in the animal and vegetable systems of that country. Together with an account of the religion, manners, and customs of several tribes of its Indian inhabitants. Interpreted with a variety of literary and medical observations. In several letters from a gentleman of the medical faculty, during his residence in that country.* London: T. Becket and P. A. De Hondt.

Barrington, D. 1774. Of the gillaroe trout: A letter from the Hon. Daines Barrington, to the Rev. Dr. Horsley. *Phil. Trans. Roy. Soc. Lond.* 64:116–20.

Bartholin, C. 1632. *Institutiones anatomicae, corporis humaniutriusq.* Goslaroia: Nicolaus Dunckerus. Impensis Johannis Hallervordius Bibliopolae Rostochiensis.

———. 1645. *De unicornu observationes novae.* Padua: Cribellianis.

———. 1678. *De unicornu observationes novae.* Secunda editione auctiores et emendatiores editae à filio Casparo Bartholino. Amsterdam: Henr. Wetstenius.

Bartholin, C., II. 1704. *De glossopetris disputatio physica.* Copenhagen: Reg. Majest. et Universit.

Bartholin, E. 1669. *Experimenta crystalli Islandici disdiaclastici quibus mira et insolita refractio detegitur.* Copenhagen: Danielis Paullus.

Bartholin, T. 1647. *De luce animalium libri III.* Leiden: Franciscus Hackius.

———. 1661. *Historiarum anatomicarum et medicarum rariorum centuriae V. et VI., accessit viri clarissimi Joannis Rhodii mantissa anatomica.* Copenhagen: Petrus Hauboldus.

———. 1669. *De luce hominum et brutorum libri III, novis rationibus, et raris historiis secundum illustrati.* Copenhagen: Petrus Hauboldus.

———., ed. 1673–80. *Acta medica et philosophica Hafniensia, cum aeneis figuris.* 5 vols.

Copenhagen: Petrus Hauboldus. Vol. 1, 1673; vol. 2, 1675; vols. 3 and 4, 1677; vol. 5, 1680.

Basore, J. W. 1928. *Seneca, Moral essays.* Vol. 1. New York: G. P. Putnam's Sons.

Baster, J. 1759–65. *Opuscula subseciva, observationes miscellaneas de animalculis et plantis quibusdam marinis, eorumque ovariis et seminibus continentia.* 2 vols. in 1. Haarlem: Joannem Bosch. Vol. 1, pt. 1, 1759; pt. 2, 1760; pt. 3, 1761. Vol. 2, pt. 1, 1762; pt. 2, 1765; pt. 3, 1765.

———. 1762. Verhandeling over de bekleedselen van de huid der dieren in 't Algemeen, en byzonder over de schubben der vissen. *Verh. Holl. Maatsch. Wetensch. Haarlem* 6 (2): 746–66, pl. 15.

Battarra, G. A. 1771. Observationes zootomicae. *Atti Accad. Sci. Siena Fisio-Crit.* 4: 353–56, pl. 1.

Bauchot, M. L. 1969. Les Poissons de la collection de Broussonet au Muséum National d'Histoire Naturelle de Paris. *Bull. Mus. Nat. Hist. Nat.* (Paris), ser. 2, 41 (1): 125–43.

Bauchot, M. L., J. Daget, and R. Bauchot. 1990. L'ichtyologie en France au début du XIXe siècle: *L'Histoire naturelle des poissons* de Cuvier et Valenciennes. *Bull. Mus. Nat. Hist. Nat.* (Paris), ser. 4, 12, sec. A, no. 1, suppl., pp. 3–142.

Baudrier, H. L. 1964. *Bibliographie Lyonnaise. Recherches sur les imprimeurs, libraires, relieurs et fondeurs de lettres de Lyon au XVIe siècle. Publiées et continées par J. Baudrier* Vol. 2. Paris: F. de Nobele.

Beckmann, J. 1769. Beitrag zur Naturgeschichte, des sogenannten Fiecks in Fischen. *Hannov. Mag.* 1769:527–28, 665–72.

———. 1786. *Aristotelis liber De mirabilibus auscultationibus explicatus a Joanne Beckmann.* Göttingen: Abraham Vandenhoek.

Belin de Ballu, J. N. 1786. *Oppiani poemata de venationes et piscatione cum interpretatione Latina et Scholiis.* Vol. 1. *Cynegetica.* Strasbourg: Bibliopolius Academicus.

Bell, W. 1793. Description of a species of *Chaetodon*, called, by the Malays, Ecan bonna. *Phil. Trans. Roy. Soc. Lond.* 83:7–9, pls. 5–6.

Belon, P. 1551. *L'Histoire naturelle des estranges poissons marins, avec la vraie peincture et description du daulphin, et de plusieurs autres son espèce, observée par Pierre Belon du Mans.* Paris: Regnaud Chaudière.

———. 1553a. *De aquatilibus, libri duo, cum conibus ad vivam ipsorum effigiem, quoad eius fieri potuit, expressis.* Paris: Carolus Stephanus.

———. 1553b. *Les Observations de plusieurs singularitez et choses mémorables, trouvées en Grèce, Asie, Judée, Egypte, Arabie, et autres pays estranges, redigées en trois livres, par Pierre Belon du Mans.* Paris: Gilles Corrozet.

———. 1554. *Les Observations de plusieurs singularitez et choses mémorables, trouvées en Grèce, Asie, Judée, Egypte, Arabie, et autres pays estranges, redigées en trois livres, par Pierre Belon du Mans.* Reveuz de nouveau et augmentez de figures. Paris: Guillaume Cavellat.

———. 1555a. *La Nature et diversité des poissons, avec leurs pourtraicts, representez au plus près du naturel.* Paris: Charles Estienne.

———. 1555b. *Observations de plusieurs singularitez et choses mémorables, trouvées en Grèce, Asie, Judée, Egypte, Arabie, et autres pays estranges, redigées en trois livres, par Pierre Belon du Mans.* Paris: Benoist Preuost pour Gilles Corrozet et Guillaume Cavellat.

Berkeley, E., and D. S. Berkeley. 1969. *Dr. Alexander Garden of Charles Town.* Chapel Hill: University of North Carolina Press.

Bertin, L. 1950. Les Poissons en herbier et le système ichtyologique de Michel Adanson. *Mém. Mus. Nat. Hist. Nat.* (Paris), ser. A, Zool., 1 (1): 1–45.

Beyers, C. J. 1977. Valentijn, François. In *Dictionary of South African Biography*, 3:796–97.

Biot, J. B. 1807. Mémoire sur la nature de l'air contenu dans la vessie natatoire des poissons. *Mém. Phys. Chim. Soc. Arcueil* (Paris) 1:252–81, 2:487–91.

Birkholz, J. C. 1770. *Ökonomische Beschreibung aller arten Fische welche in den Gewässern der Churmark gefunden werden.* Berlin and Stralsund: Gottlieb August Lange.

Blainville, H. M. D. de. 1810. Note sur plusieurs espèces de squale, confondues sous le nom de *Squalus maximus* de Linnée. *J. Physique* (Paris) 71:248–59, pl. 2.

———. 1811. Mémoire sur le squale pélerin. *Ann. Mus. Hist. Nat.* (Paris) 18:88–135, pl. 6.

———. 1816. Prodrome d'une nouvelle distribution systématique du règne animal. *J. Physique* (Paris) 83:244–67.

———. 1818. Note sur le *Stylephorus chordatus* de Shaw. *J. Physique* (Paris) 87:68–71, figs. 1–4.

———. 1822. *De l'organisation des animaux, ou Principes d'anatomie comparée.* Paris: F. G. Levrault.

Blasius, G. 1681. *Anatome animalium, terrestrium variorum, volatilium, aquatilium, serpentum, insectorum, ovorumque, structuram naturalem ex veterum, recentiorum, propriisque observationibus proponens, figuris variis illustrata.* Amsterdam: Joannnis à Someren, Henricus et Vidua Theodorus Boom.

Bloch, M. E. 1779. Naturgeschichte der Maräne. *Beschäft. Berl. Ges. Naturf. Fr.* 4:60–94, pl. 4.

———. 1780. Oeconomische Naturgeschichte der Fische in den preussischen Staaten, besonders der Märkischen und Pommerschen Provinzen. *Schr. Berl. Ges. Naturf. Fr.* 1:231–96.

———. 1782. *Abhandlung von der Erzeugung der Eingeweidewürmer und den Mitteln wider dieselben.* Berlin: Siegismund Friedrich Hesse.

———. 1782–84. *Oeconomische Naturgeschichte der Fische Deutschlands.* Berlin: Auf Kosten des Verfassers und in Commission bei dem Buchhändler Hr. Hesse. Vol. 1, 1782. Auf Kosten des Verfassers und in Commission in der Buchhandlung der Realschule. Vol. 2, 1783; vol. 3, 1784. [Index in Bloch, 1785–95, with numeration from vol. 1 to vol. 12.]

———. 1783–85. *Oeconomische Naturgeschichte der Fische Deutschlands.* Berlin: Auf Kosten des Verfassers, und in Commission in der Buchhandlung der Realschule. Vol. 1, 1783; vol. 2, 1784; vol. 3, 1785.

———. 1785. Von den vermeinten doppelten Zeugungsgliedern der Rochen und Haye. *Schr. Berl. Ges. Naturf. Fr.* 6:377–93, pl. 9.

———. 1785–88. *Ichthyologie, ou Histoire naturelle, générale et particulière des poissons. Avec des figures enluminées, dessinées d'après nature.* Berlin: Chez l'auteur, et François de La Garde. Pt. 1, 1785; pt. 2, 1785; pt. 3, 1786. Chez l'Auteur, et Didot le jeune et White et Fils. Pt. 4, 1787; pt. 5, 1787; pt. 6, 1788.

———. 1785–95. *Naturgeschichte der ausländischen Fische.* Berlin: Auf Kosten des Verfassers, und in Commission in der Buchhandlung der Realschule. Vol. 1, 1785; vol. 2, 1786; vol. 3, 1787. Bey den Königl. Akademischen Kunsthändlern J. Morino und Comp. Vol. 4, 1790; vol. 5, 1791; vol. 6, 1792; vol. 7, 1793; vol. 8, 1794; vol. 9, 1795.

———. 1785–97. *Ichthyologie, ou Histoire naturelle générale et particulière des poissons. Avec des figures enluminées, dessinées d'après nature.* Trans. from the German by Laveaux. Berlin: François de La Garde. Vol. 1, 1785; vol. 2, 1785; vol. 3, 1786; vol. 4, 1787; vol. 5, 1787; vol. 6, 1788; vol. 7, 1797; vol. 8, 1797; vol. 9, 1797; vol. 10, 1797; vol. 11, 1797; vol. 12, 1797.

———. 1786–1787. *Naturgeschichte der ausländischen Fische.* Berlin: Auf Kosten des Verfassers, und in Commission in der Buchhandlung der Realschule. Vol. 1, 1786; vol. 2, 1787.

———. 1788a. Abhandlung von den vermeinten männliche Gliedern des Dornhayes. *Schr. Berl. Ges. Naturf. Fr.* 8 (2): 9–15, pl. 2.

———. 1788b. Pleuronectarum duplex species descripta a Marco Elieser Bloch. *Nova Acta Acad. Sci. Impér. Petro.* 3:139–43, pl. 4.

———. 1788c. Beskrivelse over tvende nye Aborrer fra Indien. *Nye Saml. K. Dansk. Vidensk. Skr.* 3:383–85, pls. 1–2.

———. 1789. Två Utländska fiskar. *Kongl. Svenska Vetensk. Acad. Nya Handling.* (Stockholm) 10:234–36, pl. 7.

———. 1792a. Bemerkungen zu obiger Abhandlung des Herrn Abildgaard über Ansauger (*Myxine glutinosa* Lin.). *Schr. Berl. Ges. Naturf. Fr.* 10:244–51.

———. 1792b. Beschreibung zweyer neuen Fische. *Schr. Berl. Ges. Naturf. Fr.* 10:422–24, pl. 9.

———. 1801. *Histoire naturelle des poissons, avec les figures dessinées d'après nature par Bloch. Ouvrage classé par ordres, genres et espèces, d'après le système de Linné; avec les caractères génériques; par René Richard Castel, auteur du poème des plantes.* 10 vols. Paris: Deterville.

Bloch, M. E., and J. G. Schneider. 1801. *M. E. Blochii . . . Systema ichthyologiae iconibus CX illustratum. Post obitum auctoris opus inchoatum absolvit, correxit, interpolavit Jo. Gottlob Schneider, Saxo.* Vol. 1, text; vol. 2, *Icones CX systema ichthyologiae M. E. Blochii illustrantes,* Berlin: Sanderiano, 110 pls.

Boccone, P. [S.] 1674. *Recherches et observations naturelles.* Amsterdam: Jean Jansson à Waesberge.

——— 1697. *Museo di fisica e di esperienze variato, e decorato di osservazioni naturali, note medicinali, e ragionamenti secondo i principii de' moderni.* Venice: Baptista Zuccato.

Bochart, S. 1712. Geographia sacra, seu Phaleg. In *Samuelis Bocharti opera omnia. Hoc est Phaleg, Chanaan, et Hierozoicon,* 1:vi–318. Leiden: Cornelius Boutesteyn et Samuel Luchtmans; Trajecti ad Rhenum: Guilielmus vande Water.

Boeseman, M. 1970. The vicissitudes and dispersal of Albertus Seba's zoological specimens. *Zool. Meded.* (Leiden) 44 (13): 177–206, pls. 1–4.

Boinet, A., ed. 1914. *Catalogue général des manuscrits des bibliothèques publiques de France.* Vol. 2. *Muséum d'Histoire Naturelle, Ecole des Mines, Ecole des Ponts-et-Chaussées, Ecole Polytechnique.* Paris: Plon-Nourrit.

Boissard, J. J. 1599. *Icones quinquaginta virorum illustrium, doctrina et eruditione praestantium . . . cum eorum vitis descriptis a Ian. Iac. Boissardo. Omnia recens in aes artificiose incisa, et demum foras data.* Pt. 4. Frankfort am Main: Theodori de Bry.

Bojanus, L. H. 1818. Versuch einer Deutung der Knocken im Kopfe der Fische. *Isis* 2 (3): 498–510, pl. 7.

———. 1819–21. *Anatomie testudines Europaeae.* Vilnae: Impensis auctoris, Typis Josephus Zawadzki, Typographi Universitatis.

———. 1821a. Gehörknocken im Fische. *Isis* 8 (3): 272–77, pl. 4.

———. 1821b. Abermals ein Wort zur Deutung der Kopfknocken. *Isis* 9 (12): 1145–58, pl. 8.

Bonelli, F. A. 1820. Description d'une nouvelle espèce de poisson de la Méditerranée appartenant au genre Trachyptère avec des observations sur les caractères de ce même genre. *Mem. R. Accad. Sci. Torino* 24:485–94, pl. 9.

Bonnaterre, P. J. 1788. *Tableau encyclopédique et méthodique des trois règnes de la nature: Ichthyologie.* Paris: Panckoucke.

Bontius, J. 1658. Historiae naturalis et medicae indiae orientalis libri sex. I. De conservanda valetudine. II. Methodus medendi. III. Observationes e cadaveribus. IV. Notae in garciam ab orta. V. Historia animalium. VI. Historia plantarum. In quorum librorum penultimo, naturae animalium, avium, et piscium: In ultimo autem, arborum et plantarum species mirae, europaeis incognitae, ac ad vivum delineatae, explicantur. Commentarii, quos auctor, morte in indiis praeventus, indigestos reliquit, a Gulielmo Pisone, in ordinem redacti et illustrati, atque annotationibus et additionibus rerum et iconum necessariis adaucti. In *De Indiae utriusque re naturali et medica libri quatuordecim, quorum contenta pagina sequens exhibet,* by G. Piso, 1–150. Amsterdam: L. en D. Elsevier.

Boogaart, E. van den, H. R. Hoetink, and P. J. P. Whitehead, eds. 1979. *Johan Maurits van Nassau-Siegen, 1604–1679, a humanist prince in Europe and Brazil.* The Hague: Johan Maurits van Nassau Stichting.

Borelli, G. A. 1680–81. *De motu animalium, Io. Alphonsi Borelli Neapolitani Matheseos professoris opus posthumum.* Rome: Angeli Bernabo. Vol. 1, 1680; vol. 2, 1681.

———. 1743. *De motu animalium. Ed. nova, a plurimis mendis repurgata, ac dissertationibus physico-mechanicis de motu musculorum, et de effervescentia, et fermentatione . . . Joh. Bernoulli aucta.* 2 vols. in 1. The Hague: Petrum Gosse.

Borlase, W. 1758. *The natural history of Cornwall, the air, climate, waters, rivers, lakes, sea and tides; of the stones, semimetals, metals, tin, and the manner of mining; the constitution of the stannaries; iron, copper, silver, lead, and gold, found in Cornwall. Vegetables, rare birds, fishes, shells, reptiles, and quadrupeds; of the inhabitants, their manners, customs, plays or interludes, exercises, and festivals; the Cornish language, trade, tenures, and arts.* Oxford: W. Jackson.

Bornbusch, A. H. 1989. Lacépède and Cuvier: A comparative case study of goals and methods in late eighteenth- and early nineteenth-century fish classification. *J. Hist. Biol.* 22 (1): 141–61.

Borrichius, O. 1675. Aci marini anatome. In *Acta medica et philosophica Hafniensia,* ed. T. Bartholin, 2:149–51. Copenhagen: Petrus Hauboldus.

Bosman, W. 1705. *Voyage en Guinée, contenant une description nouvelle et très-exacte de cette côte où l'on trouve & où l'on trafique l'or, les dents d'éléphant, et les esclaves.* Utrecht: Antoine Schouten.

Bougainville, L. A. de. 1771. *Voyage autour du monde, par la frégate du roi la Boudeuse, et la flûte l'Etoile; en 1766, 1767, 1768 et 1769.* Paris: Saillant et Nyon.

———. 1772. *Voyage autour du monde, par la frégate du roi la Boudeuse, et la flûte l'Etoile; en 1766, 1767, 1768 & 1769.* 2d ed., enl. 2 vols. in 1. Paris: Saillant et Nyon.

Boulenger, G. A. 1907. *Zoology of Egypt: The fishes of the Nile.* London: Hugh Rees for the Egyptian government.

Boyer, F. 1971. Le transfert à Paris des collections du Stathouder (1795). *Ann. Hist. Rév. Franç.* 43 (205): 389–404.

Brewer, D. J., and R. F. Friedman. 1989. *The natural history of Egypt.* Vol. 2. *Fish and fishing in ancient Egypt.* Warminster, Eng.: Aris and Phillips.

Brisson, M. J. 1756. *Le Règne animal, divisé en IX classes.* Paris: Jean-Baptiste Bauche.
———. 1760. *Ornithologie, ou Méthode contenant la division des oiseaux en ordres, sections, genres, espèces et leurs variétés.* 6 vols. Paris: Jean-Baptiste Bauche.
Broch, A. J. 1929. *Galen, On the natural faculties, with an English translation by Arthur John Brock, M.D.* Cambridge: Harvard University Press.
Broussonet, P. M. A. 1782. *Ichthyologia sistems piscium descriptiones et icones. Decas I.* London: Petr. Elmsly.
———. 1784. Mémoire sur les différentes espèces de chiens de mer. *Hist. Mém. Acad. Roy. Sci.* (Paris) 1780:641–80.
———. 1785a. Extrait d'un mémoire sur les différentes espèces de chiens de mer. *J. Physique* (Paris) 26:51–66, 120–31.
———. 1785b. Mémoire sur le trembleur, espèce peu connue de poisson électrique. *J. Physique* (Paris) 27:139–43.
———. 1785c. Mémoire sur le trembleur, espèce peu connue de poisson électrique. *Hist. Mém. Acad. Roy. Sci.* (Paris) 1782:692–98, pl. 17.
———. 1787a. Observations sur les écailles de plusieurs espèces de poissons qu'on croit communément dépourvues de ces parties. *J. Physique* (Paris) 31:12–19.
———. 1787b. Mémoire pour servir à l'histoire de la respiration des poissons. *J. Physique* (Paris) 31:289–304.
———. 1788a. Observations sur le loup marin. *Hist. Mém. Acad. Roy. Sci.* (Paris) 1785:161–69, pl. 3.
———. 1788b. Observations sur les vaisseaux spermatiques des poissons épineux. *Hist. Mém. Acad. Roy. Sci.* (Paris) 1785:170–73.
———. 1788c. Mémoire pour servir à l'histoire de la respiration des poissons. *Hist. Mém. Acad. Roy. Sci.* (Paris) 1785:174–96.
———. 1788d. Mémoire sur le voilier, espèce de poisson peu connue, qui se trouve dans les Mers des Indes. *Hist. Mém. Acad. Roy. Sci.* (Paris) 1786:450–55, pl. 10.
———. 1788e. Observations sur la régénération de quelques parties du corps des poissons. *Hist. Mém. Acad. Roy. Sci.* (Paris) 1786:684–88.
———. 1789. Mémoire sur la régénération de quelques parties du corps des poissons. *J. Physique* (Paris) 35:62–65.
Brown, T. 1779. A description of the *Exocoetus volitans,* or flying fish. *Phil. Trans. Roy. Soc. Lond.* 68:791–800, pl. 12.
Brown, Y. 1988. *Japanese book illustration.* London: British Library.
Browne, P. 1756. *The civil and natural history of Jamaica.* London: Printed for the author.
Brückmann, F. E. 1734. Communicata. *Commerc. Litter. Med. Sci. Nat.* (Nuremburg) 39 (September): 305, pl. 9, figs. 4–6.
Bruggen, A. C. Van, and F. F. J. M. Pieters. 1990. Notes on a drawing of Indian elephants in red crayon by Petrus Camper (1786) in the archives of the Rijksmuseum van Natuurlijke Historie. *Zool. Meded.* (Leiden) 63 (19): 255–66.
Bruin, C. De. 1700. *Voyage au Levant, c'est à dire dans les principaux endroits de l'Asie Mineure dans les isles de Chio, de Rhodes, de Chypre etc., de même que dans les plus considérables villes d'Egypte, de Syrie, et de la Terre Sainte.* Delft: Henri de Kroonevelt.
———. 1714. *Cornelis de Bruins reizen over Moskovie, door Persie en Indie: Verrykt met driehondert kunstplaten, vertoonende de beroemste lantschappen en steden, ook de byzondere dragten, beesten, gewassen en planten, die daar gevonden worden: Voor al derzelver oudheden, en wel voornamentlyk heel uitvoerig, die van het heerlyke en van oudts de geheele werrelt door befaemde hof van Persepolis, by den Persianen Tchilminar genaemt.* Amsterdam: Wetstein, Oosterwyk, en Gaete.

———. 1718. *Voyages de Corneille Le Brun par la Moscovie, en Perse, et aux Indes Orientales*. 2 vols. Amsterdam: Wetstein.

Brünnich, M. T. 1768. *Ichthyologia Massiliensis, sistens piscium descriptiones eorumque apud incolas nomina, accedunt spolia maris Adriatici*. Copenhagen and Leipzig: Rothii Viduam et Proft.

Buffon, G. L. L. de. 1749–1804. *Histoire naturelle, générale et particulière, avec la description du cabinet du Roy* [by Buffon, L. J. M. Daubenton, P. Guéneau de Montbeillard, G. L. C. A. Bexon, and B. G. E. Lacepède.] 44 vols. Paris: Imprimerie Royale: Vol. 1, 1749; vol. 2, 1749; vol. 3, 1749; vol. 4, 1753; vol. 5, 1755; vol. 6, 1756; vol. 7, 1758; vol. 8, 1760; vol. 9, 1761; vol. 10, 1763; vol. 11, 1764; vol. 12, 1764; vol. 13, 1765; vol. 14, 1766; vol. 15, 1767 Supplément, 1774–89, 7 vols.; Oiseaux, 1770–83, 9 vols.; Minéraux, 1783–88, 5 vols.; Quadrupèdes ovipares et des serpens [see Lacepède 1788–89], 2 vols.; Poissons [see Lacepède 1798–1803], 5 vols.; Cétacés [see Lacepède 1804b], 1 vol.

Butler, H. E., and A. S. Owen. 1914. *Apvlei apologia sive Pro se de magia liber, with introduction and commentary*. Oxford: Clarendon Press.

Butterworth, G. W. 1919. *Clement of Alexandria, with an English translation by G. W. Butterworth*. New York: Putnam's Sons.

Cailliaud, F. 1823. *Voyage à Méroé, au fleuve blanc, au delà de Fâzoql dans le midi du royaume de Sennâr, à Syouah et dans cinq autres oasis; fait dans les années 1819, 1820, 1821 et 1822*. 2 vols. in 1. Paris: Imprimerie de Rignoux. [Plates only; for text, see Cailliaud 1826–27.]

———. 1826–27. *Voyage à Méroé, au Fleuve Blanc, au delà de Fâzoql dans le midi du royaume de Sennâr, à Syouah et dans cinq autres oasis; fait dans les années 1819, 1820, 1821 et 1822*. Paris: Imprimerie Impériale. Vol. 1, 1826; vol. 2, 1826; vol. 3, 1826; vol. 4, 1827. [Text only; for plates, see Cailliaud 1823.]

Cain, A. J. 1990. *Constantine Samuel Rafinesque Schmaltz on classification: A translation of early works by Rafinesque with introduction and notes*. Tryonia no. 20. Philadelphia: Department of Malacology, Academy of Natural Sciences.

Call, R. E. 1895. *The life and writings of Rafinesque*. Prepared for the Filson Club and read at its meeting, Monday, 2 April 1894. Filson Club Publications no. 10. Louisville, Ky.: John P. Morton for the Filson Club.

———. 1899. *Ichthyologia Ohiensis, or Natural history of the fishes inhabiting the river Ohio and its tributary streams by C. S. Rafinesque. A verbatim et literatim reprint of the original, with a sketch of the life, the ichthyologic work, and the ichthyologic bibliography of Rafinesque*. Cleveland: Burrows Brothers.

Campbell, R. 1969. *Letters from a Stoic: Epistulae morales ad Lucilium*. Harmondsworth: Penguin.

Camper, P. 1762. Verhandeling over het gehoor der geschubde visschen. *Verh. Holl. Maatsch. Wetensch. Haarlem* 7:79–117, pls. 1–2.

———. 1774. Mémoire sur l'organe de l'ouïe des poissons. *Mém. Math. Phys. Acad. Sci.* (Paris) 6:177–97, pls. 1–3.

———. 1784. Abhandlung über das Gehör der schuppichten Fische. In his *Sämmtliche kleinere Schriften die Arzney-Wundarzneykunst und Naturgeschichte betreffend, mit vielen neuen Zusätzen und Vermehrungen des Verfassers bereichert, von J. F. M. Herbell*, vol. 1, pt. 2, pp. 1–31, pls. 1–2. Leipzig: Siegfried Lebrecht Crusius.

———. 1784–90. *Herrn Peter Campers . . . sämmtliche kleinere Schriften die Arzney-Wundarzneykunst und Naturgeschichte betreffend, mit vielen neuen Zusätzen und Vermehrungen des Verfassers bereichert, von J. F. M. Herbell*. 6 pts. in 3 vols. Leipzig:

Siegfried Lebrecht Crusius. Vol. 1, pt. 1, 1784; pt. 2, 1784. Vol. 2, pt. 1, 1785; pt. 2, 1787. Vol. 3, pt. 1, 1788; pt. 2, 1790.
———. 1787a. Abhandlung über des Gehörorgan der Fische. In his *Sämmtliche kleinere Schriften die Arzney-Wundarzneykunst und Naturgeschichte betreffend, mit vielen neuen Zusätzen und Vermehrungen des Verfassers bereichert, von J. F. M. Herbell,* vol. 2, pt. 2, pp. 1–34, pls. 1–3. Leipzig: Siegfried Lebrecht Crusius.
———. 1787b. Bemerkungen über die Klasse derienigen Fische, die vom Ritter Linné schwimmende Amphibien genannt werden. *Schr. Berl. Ges. Naturf. Fr.* 7:197–218.
———. 1802. *Description anatomique d'un éléphant mâle.* Paris: H. J. Jansen.
———. 1820. *Observations anatomiques sur la structure intérieure et le squelette de plusieurs espèces de cétacés.* Paris: Gabriel Dufour.
Camus, A. G. 1783. *Histoire des animaux d'Aristote, avec la traduction françoise.* 2 vols. Paris: Veuve Desaint.
Carr, D. J. 1983. *Sydney Parkinson, artist of Cook's "Endeavour" voyage.* Canberra: Nova Pacifica in association with the British Museum (Natural History) and the Australian National University Press.
Carter, H. B. 1987. *Sir Joseph Banks (1743–1820): A guide to biographical and bibliographical sources.* Winchester: St. Paul's Bibliographies, in association with the British Museum.
———. 1988. *Sir Joseph Banks, 1743–1820.* London: British Museum (Natural History).
Carus, C. G. 1818. *Lehrbuch der Zootomy, mit stäter hinsicht auf Physiologie ausgearbeitet, und durch zwanzig Kupferntaflen erläutert.* Leipzig: Gerhard Fleischer dem Jüngern.
Casserius, J. 1600. *De vocis auditusque organis historia anatomica.* Ferrara: Victorius Baldinus.
———. 1610. *Pentaestheseion, hoc est de quinque sensibus liber.* Venice: Nicolaus Misserinus.
Catesby, M. 1731–43. *The natural history of Carolina, Florida and the Bahama Islands, containing the figures of birds, beasts, fishes, serpents, insects, and plants, particularly, the forest-trees, shrubs, and other plants, not hitherto described, or very incorrectly figured by authors, together with their descriptions in English and French, to which are added observations on the air, soil, and waters, with remarks upon agriculture, grain, pulse, roots, etc. To the whole is prefixed a new and correct map of the countries treated of.* 2 vols. London: W. Innys, at the expense of the author.
———. 1749–70. *Sammlung verschiedener ausländischer und seltener Vögel, worinnen ein jeder dererselben nicht nur auf das genaueste beschrieben, sondern auch in einer richitigen und sauber illuminirfen Abbildung vorgestellet wird von Johann Michael Seligmann.* 7 pts. in 3 vols. Nuremberg: Johann Joseph Fleischmann. Vol. 1, pt. 1, 1749; pt. 2, 1751. Vol. 2, pt. 3, 1753; pt. 4, 1755. Vol. 3, pt. 5, 1759; pt. 6, 1764; pt. 7, 1770.
———. 1771. *The natural history of Carolina, Florida and the Bahama Islands, containing the figures of birds, beasts, fishes, serpents, insects, and plants.* 2 vols. London: Benjamin White.
———. 1974. *The natural history of Carolina, Florida and the Bahama Islands: Containing two hundred and twenty figures of birds, beasts, fishes, serpents, insects and plants by the late Mark Catesby, F.R.S., with an introduction by George Frick and notes by Joseph Ewan.* Savannah, Ga.: Beehive Press.
Cavolini, F. 1787. *Memoria sulla generazione dei pesci e dei granchi.* Naples.

———. 1792. *Abhandlung über die Erzeugung der Fische und der Krebse.* Aus dem Italiänischen übersetzt, mit Anmerkungen herausgegeben von E. A. W. Zimmermann. Berlin: Vossischen Buchhandlung.

Cetti, F. 1774–78. *Storia naturale di Sardegna.* Sassari: Giuseppe Piattoli. Vol. 1, *I quadrupedi di Sardegna,* 1774; vol. 2, *Gli uccelli di Sardegna,* 1776; vol. 3, *Appendice alla storia naturale dei quadrupedi di Sardegna,* 1777; vol. 4, *Anfibi e pesci di Sardegna,* 1778.

Charlevoix, P. F. X. de. 1736. *Histoire et description générale du Japon.* 2 vols. Paris: Pierre-François Giffart.

Cheselden, W. 1733. *Osteography, or The anatomy of the bones.* London. Unpaged folio, 56 pls.

Chevalier, A. 1934. *Michel Adanson: Voyageur, naturaliste et philosophe.* Paris: Editions Larose.

Chevreul, M. E. 1811a. Expériences chimiques sur le cartilage du *Squalus peregrinus. Ann. Mus. Hist. Nat.* (Paris), 18:136–53, 154.

———. 1811b. Sur la liqueur contenue dans les cavités intervertébrales du *Squalus peregrinus. Ann. Mus. Hist. Nat.* (Paris) 18:154–55.

Churchill, A., and J. Churchill. 1732. Mr. John Nieuhoff's remarkable voyage & travells into ye best provinces of ye West and East Indies. In their *A collection of voyages and travels, some now first printed from original manuscripts, others now first published in English. In six volumes. With a general preface, giving an account of the progress of navigation, from its first beginning. Illustrated with a great number of useful maps and cuts, curiously engraven,* 2:1–326. London: J. Walthoe.

Clarke, J. C. 1910. *Physical science in the time of Nero, being a translation of the Quaestiones naturales of Seneca.* London: Macmillan.

Cleevely, R. J. 1978. Some background to the life and publications of Colonel George Montagu (1753–1815). *J. Soc. Bibliogr. Nat. Hist.* 8 (4): 445–80.

Cloquet, H. 1823. *Encyclopédie méthodique: Système anatomique.* Vol. 1. *Dictionnaire raisonné des termes d'anatomie et de physiologie.* Paris: Agasse.

———. 1830. *Encyclopédie méthodique: Système anatomique.* Vol. 4. *Reptiles, poissons, mollusques, crustacés, annélides, arachnides, insectes, radiaires.* Paris: Agasse.

Clusius, C. 1605. *Exoticorum libri decem: Quibus animalium, plantarum, aromatum, aliorumque peregrinorum fructuum historiae describuntur: Item Petri Bellonie observationes, eodem Carolo Clusio interprete.* Antwerp: Plantin.

Coiter, V. 1566. *De ossibus, et cartilaginibus humani corporis.* Bologna: Joannem Rossius.

———. 1575. *Lectiones Gabrielis Fallopii de partibus similaribus humani corporis, ex diversis exemplari . . . his accessere diversorum animalium sceletorum explicationes iconibus artificiosis, et genuinis illustratae.* Nuremberg: Theodoricus Gerlachius.

Coker, J. 1732. *A survey of Dorsetshire, containing the antiquities and natural history of that county.* London: J. Wilcox.

Cole, F. J. 1938. *Observations anatomicae selectiores Amstelodamensium, 1667–1673.* Edited with an introduction by F. J. Cole, F.R.S. Reading: University of Reading.

———. 1949. *A history of comparative anatomy, from Aristotle to the eighteenth century.* London: Macmillan.

Coleman, W. 1964. *Georges Cuvier zoologist: A study in the history of evolution theory.* Cambridge: Harvard University Press.

Collins, S. 1685. *A system of anatomy, treating of the body of man, beasts, birds, fish, insects, and plants, illustrated with many schemes, consisting of variety of elegant figures, drawn from the life, and engraven in seventy four folio copper-plates.* 2 vols. London: Thomas Newcomb.

Columna, F. 1606. Aquatilium, et terrestrium aliquot animalium, aliarumque, naturalium rerum observationes. In his *Minus cognitarum stirpium aliquot, ac etiam rariorum nostro coelo orientium*, i–lxxiii. Rome: Guilielmus Facciottus.

———. 1752. De glossopetris dissertatio. In *De corporibus marinus lapidescentibus quae defossa reperiuntur auctore Augustino Scilla, addita dissertatione Fabii Columnae De glossopetris edition altera emendatior*, by A. Scilla, 75–84. Rome: Venantius Monaldinus.

Commelin, H. 1597. *Athenaei Deipnosophista rum libri XV.* 2 vols. Heidelberg: Hieronymus Commelin.

Comparetti, A. 1789. *Observationes anatomicae de aure interna comparata.* Padua: S. Bartholomaeus.

Configliachi, P. 1809. *Sull'analisi dell'aria contenuta nella vescica natatoria dei pesci.* Pavia: Capelli.

Cook, J. 1774. Relation d'un voyage fait autour du monde, dans les années 1769, 1770 et 1771, par le lieutenant Jacques Cook, commandant le vaisseau du roi l'Endeavour. In *Relation des voyages entrepris par ordre de Sa Majesté Britannique, pour faire des découvertes dans l'hémisphère méridional, et successivement exécutés par le Commodore Byron, le Capitaine Carteret, le Capitaine Wallis et le Captaine Cook, dans les vaisseaux le Dauphin, le Swallow et l'Endeavour*, by J. Hawkesworth, vols. 2–4 Lausanne and Neuchatel: Société Typographique.

———. 1777. *A voyage towards the South Pole, and round the world, performed in His Majesty's ships the Resolution and Adventure, in the years 1772, 1773, 1774, and 1775.* 2 vols. London: W. Strahan and T. Cadell.

———. 1778. *Voyage dans l'hémisphère austral, et autour du monde, fait sur les vaisseaux de roi, l'Adventure, et la Résolution, en 1772, 1773, 1774 & 1775. Ecrit par Jacques Cook, commandant de la Résolution; dans lequel on a inséré la relation du Capitaine Furneaux, et celle de MM. Forster. traduit de l'Anglais.* 5 vols. Paris: Hôtel de Thou.

———. 1785. *Troisième voyage de Cook, ou Voyage à l'Océan Pacifique, ordonné par le roi d'Angleterre.* 4 vols. Paris: Hôtel de Thou.

Cornide de Saavedra, J. A. 1788. *Ensayo de una historia de los peces y otras producciones marinas de la costa de Galicia, arreglado al sistema del caballero Cárlos Linneo. Con un tratado de las diversas pescas, y de las redes y aparejos con que se practican.* Madrid: Benito Cano.

Cowan, C. F. 1969. Cuvier's *Règne animal*, first edition. *J. Soc. Bibliogr. Nat. Hist.* 5 (3): 219.

Cranz, D. 1765–70. *Historie von Grönland enthaltend die Beschreibung des Landes und der Einwohner etc. insbesondere die Geschichte der dortigen Mission der evangelischen Brüder zu Neu-Herrnhut und Lichtenfels.* 2 vols. Barby: Heinrich Detlef Ebers.

———. 1767. *The history of Greenland: Containing a description of the country, and its inhabitants.* 2 vols. London: Brethren's Society for the Furtherance of the Gospel among the Heathen.

Cupani, F. 1713. *[Panphyton Siculum, sive Historа naturalis de animalibus, stirpibus et fossilibus quae in Sicilia, vel in circuitu ejus inveniuntur. Opus posthumum admodum Rev. Patris Francisci Cupani.]* Panormi: Typographia Regia Antonini Epiro. [Handwritten title page + 168 uncolored engravings of plants, without text—these must be the proofs Cuvier speaks of.]

Cuvier, G. 1798. *Tableau élémentaire de l'histoire naturelle des animaux.* Paris: Baudouin.

———. 1800–1805. *Leçons d'anatomie comparée.* Paris: Baudouin. Vol. 1, 1800; vol. 2, 1800; vol. 3, 1805; vol. 4, 1805; vol. 5, 1805.

———. 1809. Rapport fait à la classe des sciences physiques et mathématiques, sur le mémoire de M. Delaroche, relatif à la vessie aérienne des poissons. *Ann. Mus. Hist. Nat.* (Paris) 14:165–83.

———. 1812. Sur la composition de la tête osseuse dans les animaux vertébrés. *Ann. Mus. Hist. Nat.* (Paris) 19:123–28.

———. 1813. Note sur un poisson peu connu, pêché récemment dans le golfe de Gênes (le *Lophote cepedien,* Giorna.). *Ann. Mus. Hist. Nat.* (Paris) 20:393–400, pl. 17.

———. 1815a. Notice sur un poisson célèbre, et cependant presque inconnu des auteurs systématiques, appelé sur nos côtes de l'océan, aigle ou maigre, et sur celles de la Méditerranée, umbra, fegaro et poisson royal; avec une description abrégée de sa vessie natatoire. *Mém. Mus. Nat. Hist. Nat.* (Paris) 1:1–21, pls. 1–3.

———. 1815b. Mémoire sur la composition de la mâchoire supérieure des poissons, et sur le parti que l'on peut en tirer pour la distribution méthodique de ces animaux. *Mém. Mus. Nat. Hist. Nat.* (Paris) 1:102–32.

———. 1815c. Observations et recherches critiques sur différens poissons de la Méditerranée, et à leur occasion sur des poissons d'autres mers, plus ou moins liés avec eux. *Mém. Mus. Nat. Hist. Nat.* (Paris) 1:226–41, pl. 11.

———. 1815d. Suite des observations et recherches critiques sur différens poissons de la Méditerranée, et à leur occasion sur des poissons d'autres mers, plus ou moins liés avec eux. *Mém. Mus. Nat. Hist. Nat.* (Paris) 1:312–30, pl. 16.

———. 1815e. Suite des observations et recherches critiques sur différens poissons de la Méditerranée, et à leur occasion sur des poissons d'autres mers, plus ou moins liés avec eux. *Mém. Mus. Nat. Hist. Nat.* (Paris) 1:353–63.

———. 1815f. Suite des observations et recherches critiques sur différens poissons de la Méditerranée, et à leur occasion sur des poissons d'autres mers, plus ou moins liés avec eux. *Mém. Mus. Nat. Hist. Nat.* (Paris) 1:451–66, pl. 23.

———. 1816. *Le Règne animal distribué d'après son organisation, pour servir de base à l'histoire naturelle des animaux et d'introduction à l'anatomie comparée.* Paris: Deterville. Vol. 1, L'Introduction, les mammifères et les oiseaux; vol. 2, Les Reptiles, les poissons, les mollusques et les annélides; vol. 3, Les Crustacés, Les arachnides et les insectés; vol. 4, Les Zoophytes, les tables, et les planches.

———. 1817. Sur le genre *Chironectes* Cuv. (*Antennarius.* Commers.). *Mém. Mus. Nat. Hist. Nat.* (Paris) 3:418–35, pls. 16–18.

———. 1818a. Sur les diodons, vulgairement orbes-épineux. *Mém. Mus. Nat. Hist. Nat.* (Paris) 4:121–38, pls. 6–7.

———. 1818b. Sur les poissons du sous-genre mylètes. *Mém. Mus. Nat. Hist. Nat.* (Paris) 4:444–56, pls. 21–22.

———. 1819a. Eloge historique de Pierre-Simon Pallas, lu le 5 janvier 1813. In *Recueil des éloges historiques lus dans les séances publiques de l'Institut Royal de France,* 2:109–56. Paris and Strasbourg: Prince et Levrault.

———. 1819b. Sur les poissons du sous-genre *Hydrocyn,* sur deux nouvelles espèces de *Chalceus,* sur trois nouvelles espèces de *Serrasalmes,* et sur l'*Argentina glossodonta* de Forskahl, qui est l'*Albula gonorhynchus* de Bloch. *Mém. Mus. Nat. Hist. Nat.* (Paris) 5:351–79, pls. 26–28.

———. 1827a. Eloge historique de M. Le C.te de Lacépède, lu le 5 juin 1826. In *Recueil des éloges historiques lus dans les séances publiques de l'Institut Royal de France,* 3:281–335. Paris and Strasbourg, Levrault.

———. [1827b]. *Histoire naturelle des poissons, ouvrage contenant plus de cinq mille espèces de ces animaux, décrites d'après nature et distribuées conformément à leurs rapports d'organisation, avec des observations sur leur anatomie et des recherches critiques sur leur nomenclature ancienne et moderne; par M. le Baron Cuvier . . . et par M. Valenciennes. . . . Prospectus rédigé par M. le baron Cuvier, Librairie de F.-G. Levrault, Paris et Strasbourg.* Strasbourg: Levrault.

———. 1827c. *Histoire naturelle des poissons, ouvrage contenant plus de cinq mille espèces de ces animaux, décrites d'après nature et distribuées conformément à leurs rapports d'organisation, avec des observations sur leur anatomie et des recherches critiques sur leur nomenclature ancienne et moderne; par M. le baron Cuvier . . . et par M. Valenciennes. . . . Prospectus rédigé par M. le baron Cuvier.* Paris: Mallet-Bachelier.

———. 1828a. Tableau historique des progrès de l'ichtyologie, depuis son origine jusqu'à nos jours. In *Histoire naturelle des poissons,* by G. Cuvier and A. Valenciennes, 1:1–270. Paris: Levrault.

———. 1828b. Livre premier. Distribution méthodique des poissons en familles naturelles et en divisions plus élevées. In *Histoire naturelle des poissons,* by G. Cuvier and A. Valenciennes, Paris: Levrault.

———. 1828c. Livre deuxième. Idée générale de la nature et de l'organisation des poissons. In *Histoire naturelle des poissons,* by G. Cuvier and A. Valenciennes, 1:288–551 Paris: Levrault.

———. 1829. *Le Règne animal distribué d'après son organisation, pour servir de base à l'histoire naturelle des animaux et d'introduction à l'anatomie comparée.* New ed. rev. and enl. 5 vols. Paris: Deterville.

———. 1831. Tribu espadons, ou Scombéroïdes a museau en forme de dard ou d'épée. In *Histoire naturelle des poissons,* by G. Cuvier and A. Valenciennes, 8:254–309. Paris and Strasbourg: Levrault.

Cuvier, G., and A. Valenciennes. 1828. *Histoire naturelle des poissons.* Vol. 1. Paris and Strasbourg: F. G. Levrault.

Daget, J. 1994. Achille Valenciennes, zoologiste complet. *Cybium,* 18 (2), suppl.:103–39.

Dale, S. 1699. An account of a very large eel, lately caught at Maldon in Essex; with some considerations about the generation of eels. *Phil. Trans. Roy. Soc. Lond.* 20:90–97.

Danelius, E., and H. Steinitz. 1967. The fishes and other aquatic animals on the Punt-reliefs at Dier el-Bahri. *J. Egypt. Archeol.* 53:15–24.

Darton, F. J. H. 1924. *The Golden ass of Lucius Apuleius.* In the translation by William Adlington, edited, with an introduction, by F. J. Harvey Darton. London: Privately printed for the Navarre Society.

Daubenton, L. J. M. 1787. *Encyclopédie méthodique. Histoire naturelle.*Vol. 3. *Contenant les poissons.* Paris and Liège: Panckoucke et Plomteux.

Dean, B. 1923. *Bibliography of fishes.* Vol. 3. New York: Published by the American Museum of Natural History.

DeKay, J. E. 1842. Zoology of New York, or The New York fauna, comprising detailed descriptions of all the animals hitherto observed within the state borders. Class V. Fishes. *Nat. Hist. N.Y. Geol. Survey* (Albany) 1 (3–4): 1–415.

Delaroche, F. E. 1809a. Suite du mémoire sur les espèces de poissons observées a Iviça. *Ann. Mus. Hist. Nat.* (Paris) 13:313–61, pls. 20–25.

———. 1809b. Observations sur la vessie aérienne des poissons. *Ann. Mus. Hist. Nat.* (Paris) 14:184–217.

Description of Egypt. 1809–30. *Description de l'Egypte, ou Recueil des observations et des recherches qui ont été faites en Egypte pendant l'expédition de l'armée française.* Edited by E. F. Jomard, with a historical preface by M. Fourier. Text, 9 vols.; atlas, 11 vols. Paris: Imprimerie Impériale.

———. 1821–29. *Description de l'Egypte, ou Recueil des observations et des recherches qui ont été faites en Egypte pendant l'expédition de l'armée française.* Text only, 24 vols. Paris: Panckoucke.

Desmarest, A. G. 1823. Première décade ichthyologique, ou Description complète de dix espèces de poissons nouvelles, ou imparfaitement connues, habitant la mer qui baigne les côtes de l'île de Cuba. *Mém. Soc. Linn.* (Paris) 2 (2): 163–210, pls. 16–18.

Desmoulins, L. A., and F. Magendie. 1825. *Anatomie des systèmes nerveux des animaux à vertèbres, appliquée à la physiologie et à la zoologie. Ouvrage dont la partie physiologique est faite conjointement avec F. Magendie.* Text, 2 vols., and atlas. Paris: Méquignon-Marvis.

Diment, J. A., and A. Wheeler. 1984. Catalogue of the natural history manuscripts and letters by Daniel Solander (1733–1782), or attributed to him, in British collections. *Arch. Nat. Hist.* 11 (3): 457–88.

Dobson, J. 1969. *John Hunter.* Edinburgh and London: E. and S. Livingston.

Dodd, J. S. 1752. *An essay towards a natural history of the herring.* London: T. Vincent.

Doellinger, I. 1811. Ueber den eigentlichen Bau des Fischherzens. *Ann. Wetterau. Gesell. Natur.* (Frankfurt) 2 (2): 311–14, pl. 13, figs. 1–4.

Driesch, A. Von den. 1983. Some archaeozoological remarks on fishes in ancient Egypt. In *Animals and archaeology*, vol. 2, *Shell middens, fishes and birds*, ed. C. Grigson and J. Clutton-Brock, 87–110. B.A.R. International Series 163. Oxford: B.A.R.

———. 1986. Fische im alten Aegypten—eine osteoarchaeologische Untersuchung. *Doc. Nat., Forsch. Naturwiss.* (München), 34:1–33.

Duhamel du Monceau, H. L., and L. H. de La Marre. 1769–82. *Traité général des pêsches, et histoire des poissons qu'elles fournissent, tant pour la subsistance des hommes, que pour plusieurs autres usages qui ont rapport aux arts et au commerce.* 2 pts., 13 secs. Paris: Saillant et Nyon, et Desaint. Pt. 1, sec. 1, 1769; sec. 2, 1772; sec. 3, 1772. Pt. 2, sec. 1, 1772; sec. 2, 1773; sec. 3, 1772; sec. 4, 1777; sec. 5, 1778; sec. 6, 1778; sec. 7, 1779; sec. 8, 1779; sec. 9, 1781; sec. 10, 1782.

Duméril, A. M. C. 1804. *Traité élémentaire d'histoire naturelle.* Paris: Deterville.

———. 1806. *Zoologie analytique, ou Méthode naturelle de classification des animaux, rendue plus facile à l'aide de tableaux synoptiques.* Paris: Allais.

———. 1807a. *Traité élémentaire d'histoire naturelle.* 2d ed. 2 vols. Paris: Deterville.

———. 1807b. *Mémoires de zoologie et d'anatomie comparée.* Paris: J. B. Sajou.

———. 1812. *Dissertation sur les poissons qui se rapprochent le plus des animaux sans vertèbres; thèse soutenue publiquement dans l'amphithéâtre de la Faculté des Sciences de Paris, en présence des juges du concours, le 28 mars 1812.* Paris: Imprimerie de Didot Jeune.

du Tertre, J. B. 1654. *Histoire générale, des isles de S. Christophe, de la Guadeloupe, de la Martinique, et autres dans l'Amérique. Où l'on verra l'establissement des colonies françoises, dans ces isles; leurs guerres civiles et estrangères, et tout ce qui se passe dans les voyages et retours des Indes. Comme aussi plusieurs belles particularitez des Antilles de l'Amérique; une description générale de l'isle de la Guadeloupe: De tous ses minéraux, de ses pierreries, de ses rivières, fontaines et estangs, et de toutes ses plantes. De plus, la description de tous les animaux de la mer, de l'air, et de la terre: et un traité fort ample *des moeurs des sauvages du pays, de l'estat de la colonie françoise, et des esclaves, tant mores que sauvages.* Paris: Jacques Langlois et Emmanuel Langlois.

———. 1667–71. *Histoire générale des Antilles habitées par les François. Divisée en deux tomes, et enrichy de cartes et de figures. Contenant tout ce qui s'est passé dans l'establissement des colonies françoises.* Paris: Jolly. Vol. 1, 1667; vol. 2, 1667; vol. 3, 1671; vol. 4, 1671.

du Val, G. 1629. *Aristotelis opera omnia quae extant, Graecè et Latinè.* 4 pts. in 2 vols. Paris: Antonii Stephani, Typographi Regii, apud Societatem Graecarum Editionum.

Duvau, A. 1823. Plumier. In *Biographie universelle, ancienne et moderne, ou Histoire, par ordre alphabétique, de la vie publique et privée de tous les hommes qui se sont fait remarquer par leurs écrits, leurs actions, leurs talents, leurs vertus ou leurs crimes,* 35:93–99. Paris: Michaud.

Duverney, J. G. 1743. Sur la circulation du sang des poissons qui ont des ouyes, et sur leur respiration. *Hist. Mém. Acad. Roy. Sci.* (Paris) 1701:226–41.

———. 1761. *Oeuvres anatomiques de M. Duverney.* 2 vols. Paris: Charles-Antoine Jombert.

———. 1789. Mémoire sur la circulation du sang des poissons qui ont des ouïes. In *Petri Artedi renovati,* by J. J. Walbaum, 1788–93, vol. 1, pt. 2, pp. 167–83. Grypeswaldia: Ant. Ferdin. Roese.

Eames, J. 1742. An account . . . of a book intituled, Jacobi Theodori Klein "Historiae piscium naturalis promovendae missus primus Gendani," 1740. Or the first number of an essay toward promoting the natural history of fishes, etc. *Phil. Trans. Roy. Soc. Lond.* 42:27–33.

Ebel, J. G. 1788. *Observationes neurologicae ex anatome comparata.* Frankfurt an der Oder: Traiecti ad Viadrum, e Typographeo Apitziano.

———. 1793. Observationes neurologicae ex anatome comparata. In *Scriptores neurologici minores selecti sive Opera minora ad anatomiam physiologiam et pathologiam nervorum spectantia,* by C. F. Ludwig, 3:148–61. Leipzig: Jo. Frid. Iunius, pls. 4–5.

Edmonds, J. M. 1929. *The characters of Theophrastus newly edited and translated.* New York: G. P. Putnam's Sons.

Edmundson, G. 1922. *History of Holland.* Cambridge: Cambridge University Press.

Edwards, G. 1743–51. *A natural history of uncommon birds, and of some other rare and undescribed animals, quadrupeds, reptiles, fishes, insects, etc.* 4 pts. in 2 vols. London: Printed for the author, at the College of Physicians, in Warwick-Lane. Vol. 1, pt. 1, 1743; pt. 2, 1747. Vol. 2, pt. 3, 1750; pt. 4, 1751.

———. 1758–64. *Gleanings of natural history, exhibiting figures of quadrupeds, birds, insects, plants, etc., most of which have not, till now, been either figured or described, with descriptions of seventy different subjects, designed, engraved, and coloured after nature, on fifty copper-plate prints.* London: Printed for the author, at the Royal College of Physicians, in Warwick-Lane. Vol. 1, 1758; vol. 2, 1760; vol. 3, 1764.

Edwards, P. I. 1976. Robert Brown (1773–1858) and the natural history of Matthew Flinders' voyage in H.M.S. *Investigator,* 1801–1805. *J. Soc. Bibliogr. Nat. Hist.* 7:385–407.

———. 1981. Sir Hans Sloane and his curious friends. In *History in the service of systematics,* ed. A. C. Wheeler and J. H. Price, 27–35. Papers from the conference to celebrate the centenary of the British Museum (Natural History) 13–16 April, 1981. Special Publication no. 1. London: Society for the Bibliography of Natural History.

Egede, H. 1741. *Det gamle Grønlands nye perlustration, eller naturel historie og beskrivelse over det gamle Grønlands situation, lust, temperament og beskassenhed.* Copenhagen: Johan Christoph Groth.

———. 1745. *A description of Greenland.* London: C. Hitch, S. Austen, and J. Jackson.

———. 1763. *Description et histoire naturelle du Groenland.* Copenhagen and Geneva: C. et A. Philibert.

Engel, H. 1937. The life of Albert Seba. *Svenska Linné-Sällsk. Årsskr.* 20:75–100.

———. 1961. The sale-catalogue of the cabinets of natural history of Albertus Seba (1752), a curious document from the period of the *Naturae curiosi. Bull. Res. Counc. Israel* 10B (1–3): 119–31.

———. 1986. *Hendrik Engel's alphabetical list of Dutch zoological cabinets and menageries.* 2d ed., enl. Ed. P. Smit, A. P. M. Sanders, and J. P. F. van der Veer. Amsterdam: Rodopi.

Erman, P. 1808. Untersuchungen über das Gas in der Schwimmblase der Fische, und über die Mitwirkung des Darmkanals zum Respirationsgeschäfte bei der Fischart *Cobitis fossilis* (Schlammpitzger). *Ann. Physik.* (Gilbert, Leipzig) 30:113–60.

Erxleben, J. C. P. 1776. Ueber den Nutzen der Schwimmblase bey den Fischen. In his *Physikalisch-chemische Abhandlungen,* 343–47. Leipzig: Weygandschen Buchhandlung.

Eschmeyer, W. N. 1990. *Catalog of the genera of recent fishes.* San Francisco: California Academy of Sciences.

Fabricius, H. 1600. *De formato foetu.* Venice: Franciscus Bolzetta.

Fabricius, J. C. 1779. *Reise nach Norwegen mit bemerkungen aus der Naturhistorie und Oeconomie.* Hamburg: Carl Ernst.

———. 1802. *Voyage en Norwège, avec des observations sur l'histoire naturelle et l'économie.* Trans. from the German by Jean-Chrétien Fabricius. Paris and Strasbourg: Levrault.

Fabricius, O. 1780. *Fauna Groenlandica, systematice sistens animalia Groenlandiae occidentalis hactenus indagata, quoad nomen specificum, triviale, vernaculumque; synonyma auctorum plurium, descriptionem, locum, victum, generationem, mores, usum, capturamque singuli, prout detegendi occasio fuit, maximaque, parte secundum proprias observationes.* Copenhagen and Leipzig: Joannis Gottlob Rothe.

Falck, J. P. 1785–86. *Beyträge zur topographischen Kenntniss des russischen Reichs.* 3 vols. in 2. St. Petersburg: Academie der Wissenschaften. Vol. 1, 1785; vol. 2, 1786; vol. 3, 1786.

Farrington, [—]. 1756. Some account of the charr-fish, as found in North-Wales. In a letter from the Rev. Mr. Farrington, of Dinas, near Caernarvon, to Mr. Thomas Collinson, of London. *Phil. Trans. Roy. Soc. Lond.* 49:210–12.

Faujas-Saint-Fond, B. 1806. Lettre adressée à M. de Lacepède, sur les poissons du golfe de la Spezzia et de la mer de Gênes. *Ann. Mus. Hist. Nat.* (Paris) 8:365–71.

Fenner, C. W. H. 1820. *De anatomia comparata et naturali philosophia commentatio sistens descriptionem et significationem cranii, encephali et nervorum encephali in piscibus.* Jena: Schmid et Schreiberi.

Ferguson, J. 1764. An account of a remarkable fish, taken in King-Road, near Bristol: In a letter from Mr. James Ferguson, to Thomas Birch, D.D. Secret. R.S. *Phil. Trans. Roy. Soc. Lond.* 53:170–72, pl. 13.

Fernholm, B., and A. C. Wheeler. 1983. Linnaean fish specimens in the Swedish Museum of Natural History, Stockholm. *Zool. J. Linn. Soc.* 78:199–286.

Ferris, [—]. 1782. Sur la génération des saumons. *J. Physique* (Paris) 20:321–22.
Feuillée, L. 1714. *Journal des observations physiques, mathématiques et botaniques, faites par l'ordre du roy sur les côtes orientales de l'Amérique méridionale, et dans les Indes Occidentales, depuis l'année 1707, jusques en 1712.* 2 vols. Paris: Pierre Giffart.
———. 1725. *Journal des observations physiques, mathématiques et botaniques, faites par l'ordre du roy sur les côtes orientales de l'Amérique méridionale, et aux Indes Occidentales. Et dans un autre voïage fait par le même ordre à la Nouvelle Espagne, et aux isles de l'Amérique.* Paris: Jean Mariette.
Fischer, J. B. 1778. *Versuch einer Naturgeschichte von Livland.* Leipzig: Johann Gottlob Immanuel Breitkopf.
———. 1791. *Versuch einer Naturgeschichte von Livland.* Königsberg: Friedrich Nicolovius.
Fischer von Waldheim, G. F. 1795. *Versuch über die Schwimmblase der Fische.* Leipzig: Christian Gottlieb Rabenhorst.
Fitzpatrick, T. J. 1982. *Fitzpatrick's Rafinesque: A sketch of his life with bibliography.* Rev. and enl. by Charles Boewe. Weston, Mass.: M and S Press.
Fohmann, V. 1827. *Das Saugadersystem der Wirbelthiere.* Vol. 1. *Das Saugadersystem der Fische.* Heidelberg and Leipzig: Verlage der Neuen Academischen Buchhandlung von Karl Groos.
Forsskål, P. 1775a. *Descriptiones animalium avium, amphibiorum, piscium, insectorum, vermium; quae in itinere orientali observavit Petrus Forskål, Prof. Haun.* Post mortem auctoris edidit Carsten Niebuhr. Copenhagen: Mölleri.
———. 1775b. *Flora Aegyptiaco-Arabica, sive Descriptones plantarum, quas per Aegyptum inferiorem et Arabiam felicem detexit, illustravit Petrus Forskål, Prof. Haun.* Post mortem auctoris Carsten Niebuhr, accedit tabula Arabiae felicis geographia-botanica. Copenhagen: Mölleri.
———. 1776. *Icones rerum naturalium, quas in itinere orientali depingi curavit Petrus Forskål, Prof. Haun.* Post mortem auctoris ad regis mandatum aeri incisas edidit Carsten Niebuhr. Copenhagen: Mölleri.
Forster, E. S., and E. H. Heffner, 1954–55. *Lucius Junius Moderatus Columella on agriculture, with a recension of the text and an English translation.* Cambridge: Harvard University Press. Vol. 2, 1954; vol. 3, 1955. [For vol. 1, see Ash 1941.]
Forster, J. G. A. 1778. *Geschichte der See-Reisen und Entdeckungen im Süd-Meer. Dr. Johann Reinhold Forster's und seines Sohnes Georg Forster's Reise um die Welt auf kosten der Grosbritannischen Regierung zu Erweiterung der Naturkenntniss unternommen und während den Jahren 1772 bis 1775 in dem vom Capitain J. Cook commandirten Schiffe the Resolution ausgeführt.* 2 vols. Berlin: Haude und Spenser.
Forster, J. R. 1771. *A catalogue of the animals of North America. Containing, an enumeration of the known quadrupeds, birds, reptiles, fish, insects, crustaceous and testaceous animals; many of which are new, and never described before. To which are added, short directions for collecting, preserving, and transporting, all kinds of natural history curiosities.* London: Benjamin White.
———. 1772. A letter from Mr. John Reinhold Forster, F.R.S. to the Hon. Daines Barrington, Vice-Pres. R.S. on the management of carp in Polish Prussia. *Phil. Trans. Roy. Soc. Lond.* 61:310–25.
———. 1773. An account of some curious fishes, sent from Hudson's Bay; by Mr. John Reinhold Forster, F.R.S. in a letter to Thomas Pennant, Esq.; F.R.S. *Phil. Trans. Roy. Soc. Lond.* 63:149–60, pl. 6.
———. 1778. *Observations made during a voyage round the world, on physical geography,*

natural history, and ethic philosophy, especially on 1. the earth and its strata, 2. water and the ocean, 3. the atmosphere, 4. the changes of the globe, 5. organic bodies, and 6. the human species. London: G. Robinson.

———. 1781. *Indische Zoologie, oder Systematische Beschreibungen seltener und unbekannter Thiere aus Indien. Zoologia Indica selecta Tabulis XV aeneis illustrata.* Halle: Johann Jacob Gebauer.

———. 1788. *Enchiridion historiae naturali inserviens, quo termini et delineationes ad avium, piscium, insectorum et plantarum adumbrationes intelligendas et concinnandas, secundum methodum systematis Linnaeani.* Halle: Hemmerde und Schwetschke.

———. 1790. An essay on India. In *Indian zoology,* 2d ed., by T. Pennant, 3–27. London: Printed by Henry Hughs for Robert Faulder.

———. 1795. *Indische Zoologie, in welcher zu finden sind I. Beschreibungen einiger seltenen in Kupfern vorgestellten Thiere; II. Bemerkungen über den Umfang und die Beschaffenheit des Himmelsstriches, des Bodens, und der Meere von Indien; letzlich III. Auch eine Indische Fauna, oder ein so viel möglich vollständiges Verzeichniss aller Thierarten von Indien.* Halle: Johann Jacob Gebauer.

———. 1799. *Manuel pour servir à l'histoire naturelle des oiseaux, des poissons, des insects et des plantes; où sont expliqués les termes employés dans leurs descriptions, et suivant la méthode de Linné; traduit du Latin de J. Reinhold Forster augmenté d'un mémoire de Murray sur la conchyliologie, traduit de la même langue, et de plusieurs additions considérables extraites des ouvrages des Cit. Lacépède, Jussieu, Lamarck, Cuvier, etc.* Par J. B. F. Léveillé, médecin de l'Ecole de Paris, etc. Paris: Villier.

Fourcroy, A. F. de, and L. N. Vauquelin. 1807. Extrait d'un mémoire ayant pour titre: Expériences chimiques pour servir à l'histoire de la laite des poissons. *Ann. Mus. Hist. Nat.* (Paris) 10:169–78.

Fournier, P. 1932. Aux Antilles, Charles Plumier, Minime. In *Contribution à l'histoire des sciences naturelle. Voyages et découvertes scientifiques des missionnaires naturalistes français à travers le monde pendant cinq siècles XVe à XXe siècles,* 53–59. Paris: Lechevalier.

Fowler, H. W., and F. G. Fowler. 1905. *The works of Lucian of Samosata, complete with exceptions specified in the preface.* Oxford: Clarendon Press.

Frängsmyr, T. 1983. *Linnaeus, the man and his work.* Berkeley and Los Angeles: University of California Press.

Franz, J. G. F. 1778–91. *Caii Plinii Secundi Naturalis historiae cum interpretatione et notis integris Johannis Harduini.* Leipzig: Guilielmi Gottlob Sommeri. Vol. 1, 1778; vol. 2, 1778; vol. 3, 1779; vol. 4, 1781; vol. 5, 1785; vol. 6, 1787; vol. 7, 1788; vol. 8, 1788; vol. 9, 1788; vol. 10, 1791.

Freycinet, L. C. de S. de. 1815. *Voyage de découvertes aux Terres Australes, exécuté sur les corvettes le Géographe, le Naturaliste, et la goëlette le Casuarina, pendant les années 1800, 1801, 1802, 1803 et 1804; sous le commandement du Capitaine de Vaisseau N. Baudin. Navigation et géographie.* Paris: Imprimerie Royale.

Frick, G. F., and R. P. Stearns. 1961. *Mark Catesby, the colonial Audubon.* Urbana: University of Illinois Press.

Gaillard, C. 1923. *Recherches sur les poissons représentés dans quelques tombeaux Egyptiens de l'ancien empire.* Mémoires publiés par les membres de l'Institut Français d'Archéologie Orientale du Caire sous la direction de M. George Foucart. Cairo: Imprimerie de l'Institut Français d'Archéologie Orientale.

Gamer-Wallert, I. 1970. Fische und Fischkulte im alten Aegypten. *Aegyptol. Abhandl.* (Wiesbaden), vol. 21.

Gaselee, S. 1915. *The golden ass, being the Metamorphosis of Lucius Apuleius.* Vol. 1. London: William Heinemann.

Gautier d'Agoty, J. 1752–55. *Observations sur l'histoire naturelle, sur la physique et sur la peinture, avec des planches imprimées en couleur.* Paris: Delaguette Vol. 1, 1752, pts. 1–3; vol. 2, 1752, pts. 4–6; vol. 3, 1753, pts. 7–9; vol. 4, 1754, pts. 10–12; vol. 5, 1755, pts. 13–15; vol. 6, 1755, pts. 16–18. Paris: Pissot, Lambert, et Cailleau. Vol. 8, 1757. [Title varies somewhat by volume.]

———. 1756. *Observations périodiques, sur la physique, l'histoire naturelle, et les beaux arts. Avec des planches imprimées couleur.* Par Monsieur Gautier, pensionnaire du roi, de l'Académie des Sciences et Belles Lettres de Dijon. Paris: Cailleau et Gautier.

———. 1757. *Observations périodiques, sur la physique, l'histoire naturelle et les arts, ou Journal de sciences et arts.* Par M. Toussaint, avocat du Parlement de Paris, de l'Académie Royale de Prusse. Avec des planches imprimées couleurs, par M. Gautier fils. Paris: Pissot, Lambert, et Cailleau.

Gaza, T. 1476. *Theodori: Graeci: Thessalonicensis: praefatio: in libros: De animalibus: Aristotelis: Philosophi: adxystum: quartum: maximum.* Venice: Johannem de Colonia.

Geer, C. de. 1750. Vid föregående rön har Herr de Geer gjort följande anmärkningar. *Kongl. Svenska Vetensk. Acad. Handling.* (Stockholm) 11:196–97.

Geoffroy, E. L. 1755. Premier Mémoire sur l'organe de l'ouie des reptiles, et de quelques poissons que l'on doit rapporter aux reptiles. *Mém. Math. Phys. Acad. Sci.* (Paris) 2:164–96, pls. 1–3.

Geoffroy Saint-Hilaire, E. 1802a. Histoire naturelle et description anatomique d'un nouveau genre de poisson du Nil, nommé polyptère. *Ann. Mus. Hist. Nat.* (Paris) 1:57–68, pl. 5.

———. 1802b. Description de l'achire barbu, espèce de pleuronecte indiquée par Gronou. *Ann. Mus. Hist. Nat.* (Paris) 1:152–55, pl. 11.

———. 1802c. Mémoire sur l'anatomie comparée des organes électriques de la raie torpille, du gymnote engourdissant, et du silure trembleur. *Ann. Mus. Hist. Nat.* (Paris) 1:392–407, pl. 26.

———. 1807a. Premier Mémoire sur les poissons, où l'on compare les pièces osseuses de leurs nageoires pectorales avec les os de l'extrémité antérieure des autres animaux à vertèbres. *Ann. Mus. Hist. Nat.* (Paris) 9:357–72, pl. 29.

———. 1807b. Second Mémoire sur les poissons. Considérations sur l'os furculaire, une des pièces de la nageoire pectorale. *Ann. Mus. Hist. Nat.* (Paris) 9:413–27.

———. 1807c. Troisième Mémoire sur les poissons, où l'on traite de leur sternum sous le point de vue de sa détermination et de ses formes générales. *Ann. Mus. Hist. Nat.* (Paris) 10:87–104, pl. 4.

———. 1807d. Considérations sur les pièces de la tête osseuse des animaux vertébrés, et particulièrement sur celles du crâne des oiseaux. *Ann. Mus. Hist. Nat.* (Paris) 10:342–65, pl. 27.

———. 1807e. Sur le sac branchial de la baudroie, et l'usage qu'elle en fait pour pêcher. *Ann. Mus. Hist. Nat.* (Paris) 10:480–81.

———. 1809a. De la synonymie des espèces du genre *Salmo* qui existent dans le Nil. *Ann. Mus. Hist. Nat.* (Paris) 14:460–66.

———. 1809b. Histoire naturelle des poissons du Nil. In *Description de l'Egypte, ou Recueil des observations et des recherches qui ont été faites en Egypte pendant l'expédition de l'armée française, publié par ordre de Sa Majesté l'empereur Napoléon le Grand.* Vol. 1, pt. 1, pp. 1–52. Paris: Imprimerie Impériale, Histoire Naturelle.

———. 1818. *Philosophie anatomique. Des organes respiratoires sous le rapport de la détermination et de l'idendité de leurs pièces osseuses, avec figures de 116 nouvelles préparations d'anatomie.* Paris: Méquignon-Marvis.

———. 1822. Considérations générales sur la vertèbre. *Mém. Mus. Hist. Nat.* (Paris) 9:89–119, pls. 5–7.

———. 1824a. De l'aile operculaire ou auriculaire des poissons, considérée comme un principal pivot, sur lequel doit rouler toute recherche de détermination des pièces composant le crâne des animaux; suivi de tableaux synoptiques donnant le nombre et expliquant la composition de ces pièces. *Mém. Mus. Hist. Nat.* (Paris) 11:420–44, pl. 21.

———. 1824b. Composition de la tête osseuse de l'homme et des animaux. *Ann. Sci. Nat.* (Paris) 3:173–92, 245–99, pl. 16.

———. 1825. Mémoire sur la structure et les usages de l'appareil olfactif dans les poissons, suivi de considérations sur l'olfaction des animaux qui odorent dans l'air. *Ann. Sci. Nat.* (Paris) 6:322–54, pls. 14–15.

Geoffroy Saint-Hilaire, I. 1827a. Suite de l'histoire naturelle des poissons du Nil. In *Description de l'Egypte, ou Recueil des observations et des recherches qui ont été faites en Egypte pendant l'expédition de l'armée française, publié par ordre de Sa Majesté l'empereur Napoléon le Grand,* vol. 1, pt. 1, pp. 265–310, pls. 6–17. Paris: Imprimerie Impériale, Histoire Naturelle.

———. 1827b. Histoire naturelle des poissons de la Mer Rouge et de la Méditerranée. In *Description de l'Egypte, ou Recueil des observations et des recherches qui ont été faites en Egypte pendant l'expédition de l'armée française, publié par ordre de Sa Majesté l'empereur Napoléon le Grand.* Vol. 1, pt. 1, pp. 311–40, pls. 18–27. Paris: Imprimerie Impériale, Histoire Naturelle.

Georgi, J. G. 1775. *Bemerkungen einer Reise im russischen Reich im Jahre 1772.* 2 vols. in 1. St. Petersburg: Academie der Wissenschaften.

———. 1797–1802. *Johann Gottlieb Georgi . . . geographisch-physikalische und naturhistorische Beschreibung des russischen Reichs zur Uebersicht bisheriger Kenntnisse von demselbe.* 13 pts. in 4 vols. Königsberg: Friedrich Nocolovius. Vol. 1, 1797. Vol. 2, pt. 1, 1798; pt. 2; pt. 3, 1799; pt. 4, 1799. Vol. 3, pt. 1, 1798; pt. 2, 1798; pt. 3, 1798; pt. 4, 1800; pt. 5, 1800; pt. 6, 1800; pt. 7, 1801. Vol. 4, 1802.

Gessner, C. 1551–87. *Historiae animalium.* 5 bks. in 3 folio vols. Zurich: Christoph Froschoverus. Vol. 1, bk. 1, De quadrupedibus viviparis, 1551; bk. 2, De quadrupedibus oviparis, 1554. Vol. 2, bk. 3, De avium natura, 1555. Vol. 3, bk. 4, De piscium et aquatilium animantium natura, 1558; bk. 5, De serpentium natura, 1587.

———. 1556. *Claudii Aeliani praenestini pontificis et sophistae, qui Romae sub Imperatore Antonino Pio vixit, Meliglossus aut Meliphthongus ab orationis suauitate cognominatus, opera, quae extant, omnia, Graecè Latinéque è regione.* Lyons: Gesneros.

———. 1560. *Nomenclator aquatilium animantium. Icones animalium aquatilium in mari et dulcibus aquis degentium, plusquam DCC. cum nomenclaturis singulorum Latinis, Grecis, Italicis, Hispanicis, Gallicis, Germanicis, Anglicis, alii'sq; interdum, per certos ordines digestae.* Zurich: Christoph. Froschoverus.

———. 1604. *Historiae animalium liber IV. Qui est de piscium et aquatilium animantium natura. Cum iconibus singulorum ad viuum expressis fere omnibus DCCXII.* 2d ed. Frankfurt: Andrea Cambierus.

———. 1620. *Historiae animalium liber IV. Qui est de piscium et aquatilium animantium natura.* Frankfurt: Henricus Lavrentius.

Gijzen, A. 1938. *'S Rijks Museum van Natuurlijke Historie 1820–1915*. Rotterdam: W. L. en J. Brusse's Uitgeversmaatschappij N.V.

Gill, T. N. 1872. Arrangement of the families of fishes, or classes Pisces, Marsipobranchii, and Leptocardii. *Smiths. Misc. Coll.* 247.

Gilles, P. 1533. *Ex Aeliani historia per Petrum Gyllium Latini facti, itemque ex Porphyrio, Heliodoro, Oppiano, tum eodem Gyllio luculentis accessionibus aucti libri XVI. De vi et natura animalium. Eiusdem Gyllii liber unus, De Gallicis et Latinis nominibus piscium Massiliensium.* Lyons: Seb. Gryphius.

———. 1535. *Ex Aeliani historia per Petrum Gyllium Latini facti, itemque ex Porphyrio, Heliodoro, Oppiano, tum eodem Gyllio luculentis accessionibus aucti libri XVI. De vi et natura animalium. Eiusdem Gyllii liber unus, De Gallicis et Latinis nominibus piscium Massiliensium.* Lyons: Seb. Gryphius.

Gillispie, C. C., and M. Dewachter. 1987. *Monuments of Egypt.* The Napoleonic edition. The complete archaeological plates from *La Description de l'Egypte*, edited with introduction and notes by C. C. Gillispie and M. Dewachter. 2 vols. Princeton: Princeton Architectural Press.

Giorna, M. E. 1809. Mémoire sur des poissons d'espèces nouvelles et de genres nouveaux. *Mem. R. Accad. Sci. Torino* 16:1–19, pls. 1–2.

Giovio, P. 1524. *Pauli Iovii Novocomensis medici De Romanis piscibus libellus ad Ludovicum Borbonium, Cardinalem amplissimum.* Rome: Minitius Calvus.

———. 1527. *Pauli Iovii Novocomensis, De piscibus marinis, lacustribus, fluviatilibus, item de testaceis ac salsamentis liber. De Romanis piscibus libellus ad Ludovicum Borbonium Cardinalem amplissimum.* Rome: Minitius Calvus.

———. 1560. *Libro di Mons. Paolo Giovio De' pesci Romani.* Tradotto in volgare da Carlo Zancaruolo. Venice: Gualtieri.

Gleditsch, J. G. 1766. Exposition abrégée d'une fécondation artificielle des truites et des saumons, qui est appuyée sur des expériences certaines, faites par un habile naturaliste. *Mém. Acad. Roy. Sci.* (Berlin) 1764:47–64.

Gmelin, J. F. 1789. Pisces. In *Caroli a Linné Systema naturae per regna tria naturae, secundum classes, ordines, genera, species, cum characteribus, differentiis, synonymis, locis.* Editio decima tertia, aucta, reformata. Cura Jo. Frid. Gmelin. Vol. 1, pt. 3, pp. 1126–1516. Leipzig: Georg. Emanuel Beer. [For a full citation of Gmelin's edition of *Systema naturae*, see Linnaeus 1788–93.]

Gmelin, J. G. 1747–49. *Flora Sibirica, sive Historia plantarum Sibiriae.* St. Petersburg: Typographia Academiae Scientiarum. Vol. 1, 1747; vol. 2, 1749. [For vols. 3–4, see S. G. Gmelin 1768–69.]

———. 1751–52. *Reise durch Sibirien, von dem Jahr 1733 bis 1743.* Göttingen: Abram Vandenhoecks. Vol. 1, 1751; vol. 2, 1752; vol. 3, 1752; vol. 4, 1752.

———. 1767. *Voyage en Sibérie, contenant la description des moeurs et usages des peuples de ce pays, le cours des rivières considérables, la situation des chaînes de montagnes, des grandes forêts, des mines, avec tous les faits d'histoire naturelle qui sont particuliers à cette contrée.* 2 vols. St. Petersburg: Desaint.

Gmelin, S. G. 1768–69. *Flora Sibirica, sive Historia plantarum Sibiriae.* Auctore D. Joanne Georgio Gmelin, editore D. Samuel Gottlieb Gmelin. St. Petersburg: Typographia Academiae Scientiarum. Vol. 3, 1768; vol. 4, 1769. [For vols. 1–2, see J. G. Gmelin 1747–49.]

———. 1770–74. *Reise durch Russland zur Untersuchung der drey Natur-Reiche.* St. Petersburg: Academie der Wissenschaften. Vol. 1, 1770; vol. 2, 1774; vol. 3, 1774. [For vol. 4, see Pallas 1784.]

Godolphin, F. R. B. 1942. *The Greek historians.* The complete and unabridged historical works of Herodotus, translated by G. Rawlinson, Thucydides, translated by B. Jowett, Xenophon, translated by H. G. Dakyns, Arrian, translated by E. J. Chinnock, edited, with an introduction, revisions and additional notes, by F. R. B. Godolphin. 2 vols. New York: Random House.

Goldfuss, G. A. 1820. *Handbuch der Zoologie.* 2 vols. Nuremburg: Johann Leonhard Schrag.

González, A. G. 1989. *Antonio Parra en la ciencia hispanoamericana del siglo XVIII.* Havana: Editorial Academia.

Gonzalez-Crussi, F. 1988. Anatomy and old lace: An eighteenth-century attitude toward death. *Sciences, N.Y. Acad. Sci.* 28 (1): 48–49.

Goode, G. B., and T. H. Bean. 1885. On the American fishes in the Linnaean collection. *Proc. U.S. Nat. Mus.* 8 (13): 193–208.

Goüan, A. 1770. *Historia piscium, sistens ipsorum anatomen externam, internam, atque genera in classes et ordines redacta.* Strasbourg: Amandi König.

Gould, S. J. 1993. The Razumovsky duet. *Nat. Hist.* (New York) 102 (10): 10–19.

Grant, W. 1752. Auszug aus Herrn Dr. W. Grant Schreiben an Herrn Hofrath von Haller von der Paarung und Fortpflanzung des Lachses. *Königl. Schwed. Akad. Wissenschaft. Abhandl.* 14:142–46.

Griffith, F. L., and P. E. Newberry. [1893]. *Archaeological survey of Egypt. El Bersheh. Part II. With appendix, plans and measurements of the tombs by G. Willoughby Fraser.* London: Egypt Exploration Fund.

Gronovius, A. 1744. *Aeliani, De natura animalium libri XVII.* 2 vols. London: Gulielmus Bowyer.

Gronovius, J. F. 1744a. Cottus ossiculo pinnae dorsalis primo longitudine corporis. *Acta Soc. Reg. Sci. Upsal.* 1740:121–23, pl. 8.

———. 1744b. A method of preparing specimens of fish, by drying their skins, as practised by John Frid. Gronovius, M.D. at Leyden. *Phil. Trans. Roy. Soc. Lond.* 42:57–58.

———. 1746a. Pisces Belgii seu piscium in Belgio natantium, et a se observatorum catalogus. *Acta Soc. Reg. Sci. Upsal.* 1741:67–76.

———. 1746b. Salmo oblongus, maxillae inferioris apice introrsum reflexo. *Acta Soc. Reg. Sci. Upsal.* 1741:85–90.

———. 1747. The figure of the Mustela fossilis; communicated from Dr. Gronovius at Leyden to Mr. Peter Collinson, F.R.S. *Phil. Trans. Roy. Soc. Lond.* 44 (483): 451, pl. 2, fig. 1.

———. 1748. Pisces Belgii descripti. *Acta Soc. Reg. Sci. Upsal.* 1742 (9): 79–107, pl. 3.

———. 1751. Pisces duo descripti a Jo. Fr. Gronovio. *Acta Soc. Reg. Sci. Upsal.* 1744 (15): 36–42, pl. 4.

Gronovius, L. T. 1754–56. *Museum ichthyologicum, sistens piscium indigenorum et quorumdam exoticorum, qui in museo Laurentii Theodori Gronovii, J.U.D. adservantur, descriptiones ordine systematico. Accedunt nonnullorum exoticorum piscium icones aeri incisae.* Leiden: Theodor Haak. Pt. 1, 1754; pt. 2, 1756.

———. 1760a. Animalium in Belgio habitantium centuria prima. *Acta Helvet.* 4:243–56, pl. 13.

———. 1760b. Centuria animalium secunda in Belgio a me observatorum. *Acta Helvet.* 4:256–70.

———. 1762a. Animalium Belgicorum, a Laur. Theod. Gronovio . . . observatorum centuria tertia. *Acta Helvet.* 5:120–38.

———. 1762b. Animalium Belgicorum centuria quarta. *Acta Helvet.* 5:138–53.

———. 1762c. Animalium Belgicorum a Laur. Theod. Gronovio observatorum centuria quinta. *Acta Helvet.* 5:353–82, pl. 5.

———. 1763–81. *Zoophylacium Gronovianum, exhibens animalia quadrupeda, amphibia, pisces, insecta, vermes, mollusca, testacea, et zoophytaa, quae in museo suo adservavit, examini subjecit, systematice disposuit atque descripsit.* 3 pts. in 1. Leiden: Theodor Haak et Samuel et Johannes Luchtmans. Pt. 1, 1763; pt. 2, 1764; pt. 3, 1781.

Gudger, E. W. 1912. George Marcgrave, the first student of American natural history. *Pop. Sci. Month.* 81:250–74.

———. 1924. The sources of the material for Hamilton-Buchanan's fishes of the Ganges, the fate of his collections, drawings and notes, and the use made of his data. *J. Proc. Asiatic Soc. Bengal.*, n.s., 29 (4): 121–36.

———. 1934. The five great naturalists of the sixteenth century, Belon, Rondelet, Salviani, Gesner and Aldrovandi: A chapter in the history of ichthyology. *Isis* 22 (1): 21–40.

Gueroult, P. C. B. 1802. *Histoire naturelle des animaux par Pline, traduction nouvelle.* 3 vols. Paris: Delance et Lesueur.

Güldenstädt, J. A. von. 1772. *Salmo leucichthys* et *Cyprinus chalcoides* descripti. *Novi Comment. Acad. Sci. Impér. Petro.* 16:531–47, pl. 16.

———. 1773. *Cyprinus capoeta* et *Cyprinus mursa. Novi Comment. Acad. Sci. Impér. Petro.* 17:507–20, pls. 8–9.

———. 1775. Acerina; piscis, ad percae genus pertinens, descriptus. *Novi Comment. Acad. Sci. Impér. Petro.* 19: 455–62, pl. 11.

———. 1787–91. See P. S. Pallas 1787–91.

Gunner, J. E. 1761–64. *Tronhiemske Samlinger, udgivne af philaletho.* Trondheim: J. E. Gunner. Vol. 1, 1761; vol. 2, 1762; vol. 3, 1763; vol. 4, 1764; vol. 5, 1764.

Günther, A. C. L. G. 1880. *An introduction to the study of fishes.* Edinburgh: Adam and Charles Black.

Gunther, R. T. 1934. *The Greek herbal of Dioscorides, illustrated by a Byzantine A.D. 512, Englished by John Goodyer A.D. 1655, edited and first printed A.D. 1933.* Oxford: Printed for the author at the University Press.

Haan, F. de. 1902. Rumphius en Valentijn als geschiedschrijvers van Ambon. In *Rumphius gedenkboek, 1702–1902*, ed. M. Greshoff, 17–25. Haarlem: Koloniaal Museum.

Haller, A. von. 1757. *Elementa physiologiae corporis humani.* Vol. 1. Lausanne: Marci-Michael Bousquet et Sociorum.

———. 1762. Addendum [on the brain of various fishes]. In his *Elementa physiologiae corporis humani*, 4:591–96. Lausanne: Marci-Michael Bousquet et Sociorum.

———. 1764. Mémoire sur les yeux de quelques poissons. *Hist. Mém. Acad. Roy. Sci.* (Paris) 1762:76–95.

———. 1766. De oculis piscium. *Gött. Anz. Gelehr. Sach.* 1766 (2): 1169–70.

———. 1768a. Verhandeling over de hersenen der vogelen en visschen. *Verh. Holl. Maatsch. Wetensch. Haarlem* 10 (2): 287–386.

———. 1768b. Cerebro avium et piscium. In his *Operum anatomici argumenti minorum*, 3:191–217. Lausannae: Francisci Grasset.

———. 1768c. Pisces, salmonum oculi. In his *Operum anatomici argumenti minorum*, 3:250–62. Lausannae: Francisci Grasset.

———. 1774. Liber VI. Animalium incisiones. In his *Bibliotheca anatomica, qua scripta ad anatomen et physiologiam facientia rerum initiis recensentur*, 1:362–709. Lyons: Orell, Gessner, Fuessli, et Socc.

Hamilton, F. [Buchanan]. 1822. *An account of the fishes found in the river Ganges and its branches.* Vol. 1, text; vol. 2, plates. Edinburgh: Archibald Constable.

Hammond, D. 1970. *News from New Cythera: A report of Bougainville's voyage, 1766–1769.* Minneapolis: University of Minnesota Press.

Hamy, E. T. 1906. Alexandre de Humboldt et le Muséum d'Histoire Naturelle. Etude historique publiée à l'occasion du centenaire du retour en Europe de Humboldt et Bonpland. *Nouv. Arch. Mus. Hist. Nat.* (Paris), ser. 4, 8:1–32.

Harduin, J. 1723. *Caii Plini Secundi Historiae naturalis libri XXXVII.* 2 vols. Paris: Antonii-Urban Coustelier.

Harmer, T. 1768. Remarks on the very different accounts that have been given of the fecundity of fishes, with fresh observations on that subject: By Mr. Thomas Harmer; communicated by Samuel Clark, Esq.; F.R.S. *Phil. Trans. Roy. Soc. Lond.* 57:280–92.

Hartmann, P. J. 1689. Anatome ventriculi piscis siluri. *Misc. Curio. Medico-Physica,* decur. 2, 7 (40): 80–83.

———. 1695. Descriptio anatomico-physica xiphiae, sive Gladii piscis. *Misc. Curio. Medico-Physica,* decur. 3, 2 (appendix): 1–22, pl. 1.

Harwood, J. 1827. On a newly discovered genus of serpentiform fishes. *Phil. Trans. Roy. Soc. Lond.* 118:49–57, pl. 7.

Hasselquist, F. 1757. *Iter Palaestinum eller resa til Heliga Landet, forrättåd ifrån år 1749 til 1752.* Stockholm: Lars Salvius.

———. 1762. *Reise nach Palästina in den Jahren von 1749 bis 1752.* Rostock: Johann Christian Koppe.

———. 1766. *Voyages and travels in the Levant; in the years 1749, 50, 51, 52. Containing observations in natural history, physick, agriculture, and commerce: Particularly on the Holy Land, and the natural history of the scriptures.* London: L. Davis and C. Reymers, printers to the Royal Society.

———. 1769. *Voyages dans le Levant, dans les années 1749, 50, 51 et 52. Contenant des observations sur l'histoire naturelle, la médecine, l'agriculture et le commerce, et particulièrement sur l'histoire naturelle de la Terre Sainte.* 2 pts. in 1. Paris: Delalain.

Haüy, R. J. 1801. *Traité de minéralogie.* 4 vols. Paris: Delance.

———. 1822. *Traité de cristallographie, suivi d'une application des principes de cette science à la détermination des espèces minérales, et d'une nouvelle méthode pour mettre les formes cristallines en projection.* 2 vols. Paris: Bachelier et Huzard.

Hawkesworth, J. 1773. *An account of the voyages undertaken by the order of His Present Majesty for making discoveries in the Southern Hemisphere, and successively performed by Commodore Byron, Captain Carteret, Captain Wallis, and Captain Cook, in the Dolphin, the Swallow, and the Endeavour: Drawn up from the journals which were kept by the several commanders, and from the papers of Joseph Banks, Esq.* 3 vols. London: W. Strahan and T. Cadell.

Hazen, A. T. 1939. Johnson's life of Frederic Ruysch. *Bull. Hist. Medicine* 7:324–34.

Hederström, H. 1759. Rön om fiskars ålder. *Kongl. Svenska Vetensk. Acad. Handling.* (Stockholm) 20:222–29.

Heiss, A. 1870. *Description générale des monnaies antiques de l'Espagne.* Paris: Imprimerie Nationale.

Hellant, A. 1745. Berättelse om laxens alstrande, i ackt tagit vid des fiskande. *Kongl. Svenska Vetensk. Acad. Handling.* (Stockholm) 6:267–79.

Hemming, F., and D. Noakes. 1958. *Official list of works approved as available for zoological nomenclature. First instalment: Names 1–38.* London: International Trust for Zoological Nomenclature.

Heniger, J. 1986. *Hendrik Adriaan van Reede tot Drakenstein (1636–1691) and Hortus Malabaricus: A contribution to the history of Dutch colonial botany.* Rotterdam: A. A. Balkema.

Henry, R. 1974. *Photius bibliothèque tome VII (codices 246–256).* Text established and trans. by René Henry. Vol. 7. Paris: Société d'Edition les Belles Lettres.

Herbell, J. F. M. 1784–90. *Herrn Peter Campers . . . Sämmtliche kleinere Schriften die Arzney-Wundarzneykunst und Naturgeschichte betreffend, mit vielen neuen Zusätzen und Vermehrungen des Verfassers bereichert, von J. F. M. Herbell.* 6 pts. in 3 vols. Leipzig: Siegfried Lebrecht Crusius. Vol. 1, pt. 1, 1784; pt. 2, 1784. Vol. 2, pt. 1, 1785; pt. 2, 1787. Vol. 3, pt. 1, 1788; pt. 2, 1790.

Hérissant, F. D. 1753. Recherches sur les usages du grand nombre de dents du *Canis carcharias. Hist. Mém. Acad. Roy. Sci.* (Paris) 1749:155–62, pls. 7–9.

Hermann, J. 1781. Schreiben an den Herausgeber über ein neues amerikanisches Fischgeschlecht, *Sternoptyx diaphana,* der durchsichtige Brust-falten-Fisch. *Naturforscher* (Halle) 16:8–36, pl. 1, figs. 1–2.

———. 1782. Verbesserungen und Zusass zur Beschreibung des Brustfaltenfisches im sechzehnten Stück. *Naturforscher* (Halle) 17:249–50.

———. 1783. *Tabula affinitatum animalium olim academico specimine edita nunc uberiore commentario illustrata cum annotationibus ad historiam naturalem animalium augendam facientibus.* Strasbourg: Joh. Georgius Treuttel.

———. 1804. *Observationes zoologicae quibus novae complures, aliaeque animalium species describuntur et illustrantur opus posthumum edidit Fridericus Ludovicus Hammer.* Strasbourg and Paris: Amandus Koenig.

Heusinger, C. F. von. 1826. Bemerkungen über das Gehörwerkzeug des *Mormyrus cyprinoides, Gastroblecus compressus* und *Pimelodus synodontis. Arch. Anat. Physiol.* (Meckel) 9 (3): 324–27, pl. 4, figs. 8–10.

Hewson, W. 1770. An account of the lymphatic system in fish. *Phil. Trans. Roy. Soc. Lond.* 59:204–15.

———. 1772. Histoire des vaisseaux lymphatiques dans les animaux amphibies. *J. Physique* (Paris), 1:350–53, 401–6.

Hill, J. 1748–52. *A general natural history: or, New and accurate descriptions of the animals, vegetables, and minerals, of the different parts of the world; with their virtues and uses, as far as hitherto certainly known in medicine and mechanics.* London: Thomas Osborne. Vol. 1, 1748; vol. 2, 1751; vol. 3, 1752.

Hoeven, J. van der. 1822. *Dissertatio philosophica inauguralis, de sceleto piscium.* Leiden: L. Herdingh.

Holthuis, L. B. 1969. Albertus Seba's "Locupletissimi rerum naturalium thesauri . . ." (1734–1765) and the "Planches de Seba" (1827–1831). *Zool. Meded.* (Leiden) 43 (19): 239–52, pls. 1–3.

Holthuis, L. B., and M. Boeseman. 1977. Notes on C. S. Rafinesque Schmaltz's (1810) *Caratteri di alcuni nuovi generi e nuove specie di animali e piante della Sicilia. J. Soc. Bibliogr. Nat. Hist.* 8 (3): 231–34.

Home, E. 1809. An anatomical account of the *Squalus maximus* (of Linnaeus), which in the structure of its stomach forms an intermediate link in the gradation of animals between the whale tribe and cartilaginous fishes. *Phil. Trans. Roy. Soc. Lond.* 99:206–20, pls. 6–9.

———. 1810. Description anatomique du *Squalus maximus* de Linnée. Qui, par la forme de son estomac, établit un passage entre les poissons cartilagineux et les cétacés. Extrait par H. de Blainville, D.-M. P. professor d'anatomie. *J. Physique* (Paris) 71:241–47.

———. 1814–28. *Lectures on comparative anatomy; in which are explained the preparations in the Hunterian collection.* London: W. Bulmer for G. and W. Nicol. Vol. 1, 1814; vol. 2, 1814; vol. 3, 1823; vol. 4, 1823; vol. 5, 1828; vol. 6, 1828.

Horner, F. B. 1987. *The French reconnaissance: Baudin in Australia, 1801–1803.* Carlton, Victoria: Melbourne University Press.

Horrebow, N. 1752. *Tilforladelige efterretninger om Island med et nyt landkort og 2 aars meteorologiske observationer.* Copenhagen: Arue-Herre og Konge.

———. 1758. *The natural history of Iceland.* London: A. Linde, D. Wilson, and T. Durham, G. Keith, P. Davey, and B. Law, T. Field, C. Henderson, and J. Staples.

———. 1766. *Description historique, civile et politique de l'Islande, avec des observations critiques sur l'histoire naturelle de cette isle.* 2 vols. Paris: Charpentier.

Hort, A. 1916. *Theophrastus: Enquiry into plants and minor works on odours and weather signs.* 2 vols. New York: G. P. Putnam's Sons.

Houttuyn, M. 1761–85. *Natuurlyke historie, of uitvoerige beschryving der dieren, planten en mineraalen, volgens het samenstel van den Heer Linnaeus.* 3 pts. in 37 vols. Amsterdam: F. Houttuyn, Pt. 1, Dieren: vol. 1, 1761; vol. 2, 1761; vol. 3, 1762; vol. 4, 1762; vol. 5, 1763; vol. 6, 1764; vol. 7, 1764; vol. 8, 1765; vol. 9, 1766; vol. 10, 1766; vol. 11, 1767; vol. 12, 1768; vol. 13, 1769; vol. 14, 1770; vol. 15, 1771; vol. 16, 1771; vol. 17, 1772; vol. 18, 1773. Pt. 2, Planten: vol. 1, 1773; vol. 2, 1774; vol. 3, 1774; vol. 4, 1775; vol. 5, 1775; vol. 6, 1776; vol. 7, 1777; vol. 8, 1777; vol. 9, 1778; vol. 10, 1779; vol. 11, 1779; vol. 12, 1780; vol. 13, 1782; vol. 14, 1783. Pt. 3, Mineraalen: vol. 1, 1780; vol. 2, 1781; vol. 3, 1782; vol. 4, 1784; vol. 5, 1785.

———. 1764. Aanmerkingen over de voortteeling der haaijen en de haaijsen-tasjes, zogenaamd. *Uitgez. Verhandl. Soc. Wetensch. Europa* (Amsterdam) 9:480–87, pl. 62.

———. 1765. Aanmerkingen omtrent eenige vreemde visschen. *Uitgez. Verhandl. Soc. Wetensch. Europa* (Amsterdam) 10:506–16, pl. 67.

———. 1782. Beschryving van eenige Japanse visschen, en andere zee-schepzelen. *Verh. Holl. Maatsch. Wetensch.* (Haarlem) 20 (2): 311–50.

Hughes, G. 1750. *The natural history of Barbados.* London: For the author.

Hulth, J. M. 1907. *Bibliographia Linnaeana: Matériaux pour servir à une bibliographie Linnéenne.* Uppsala: Almqvist och Wiksells.

Humboldt, F. H. A. von. 1811a. Mémoire sur l'Eremophilis et l'Astroblepus, deux nouveaux genres de l'ordre des apodes. In *Voyage aux régions équinoxiales du nouveau continent, fait en 1799–1804,* by F. H. A. von Humboldt and A. J. G. Bonpland, vol. 1, pt. 2, pp. 17–20, pls. 5–6. Recueil d'observations de zoologie et d'anatomie comparée. Paris: F. Schoell et G. Dufour.

———. 1811b. Mémoire sur une nouvelle espèce de Pimelode, jetée par les volcans du royaume de Quito. In *Voyage aux régions équinoxiales du nouveau continent, fait en 1799–1804,* by F. H. A. von Humboldt and A. J. G. Bonpland, vol. 1, pt. 2, pp. 21–25. Recueil d'observations de zoologie et d'anatomie comparée. Paris: F. Schoell et G. Dufour.

———. 1811c. Observations sur l'anguille électrique *(Gymnotus electricus,* Lin.) du nouveau continent. In *Voyage aux régions équinoxiales du nouveau continent, fait en 1799–1804,* by F. H. A. von Humboldt and A. J. G. Bonpland, vol. 1, pt. 2, pp. 49–92, pl. 10. Recueil d'observations de zoologie et d'anatomie comparée. Paris: F. Schoell et G. Dufour.

———. 1826. *Bericht über die naturhistorischen Reisen der Herren Ehrenberg und Hemprich durch Ägypten, Dongola, Syrien, Arabien und den östlichen Abfall des Habessinischen Hochlandes, in den Jahren 1820–1825.* Berlin: Druckerei der Königlichen Akademie der Wissenschaften.

———. 1833. Recherches sur les poissons fluviatiles de l'Amérique équinoxiale. In *Voyage aux régions équinoxiales du nouveau continent, fait en 1799–1804,* by F. H. A. von Humboldt and A. J. G. Bonpland, vol. 2, pt. 2, pp. 145–216. Recueil d'observations de zoologie et d'anatomie comparée. Paris: J. Smith et Gide.

Hunter, J. 1774. Anatomical observations on the torpedo. *Phil. Trans. Roy. Soc. Lond.* 63:481–89, pl. 20.

———. 1775. An account of the *Gymnotus electricus. Phil. Trans. Roy. Soc. Lond.* 65:395–407, pls. 1–4.

———. 1783. Account of the organ of hearing in fish. *Phil. Trans. Roy. Soc. Lond.* 72:379–83.

———. 1786. Account of the organ of hearing in fishes. In his *Observations on certain parts of the animal oeconomy,* 69–75. London.

———. 1792. An account of the organ of hearing in fishes. In his *Observations on certain parts of the animal oeconomy,* 2d ed., 81–87. London: Nicol and Johnson.

Illiger, J. C. W. 1811. *Prodromus systematis mammalium et avium.* Berlin: C. Salfeld.

Imperato, F. 1599. *Dell'historia naturale de Ferrante Imperato Napolitano.* Naples: Costantino Vitale.

———. 1672. *Historia naturale de Ferrante Imperato Napolitano. Nella quale ordinatamente si tratta . . . dedicata all'altezza ser. di Giovan Federico.* Venice: Combi e La Noù.

International Commission on Zoological Nomenclature. 1954. Opinion 212. Designation of the dates to be accepted as the dates of publication of the several volumes of Pallas (P. S.), "Zoographia Rosso-Asiatica." *Opinions and Declarations Rendered by the International Commission on Zoological Nomenclature* 4 (2): 15–24.

Isidore. 1601. *Sancti Isidori Hispalensis episcol, opera omnia que extant.* Paris: Michael Sonnius.

Jacobaeus, O. 1680a. Anatome piscis centrines Italis pesce porco. In *Acta medica et philosophica Hafniensia,* ed. T. Bartholin, 5:251–53. Copenhagen: Petrus Hauboldus.

———. 1680b. Anatome piscis torpedinis motusq; tremuli examen. In *Acta medica et philosophica Hafniensia,* ed. T. Bartholin, 5:253–59. Copenhagen: Petrus Hauboldus.

———. 1680c. De lampetra ejusq; pulmonibus, et anguilla. In *Acta medica et philosophica Hafniensia,* ed. T. Bartholin, 5:259–62. Copenhagen: Petrus Hauboldus.

Jansen, H. J. 1803. *Oeuvres de Pierre Camper, qui ont pour objet l'histoire naturelle, la physiologie et l'anatomie comparée.* 3 vols. Paris: H. J. Jansen.

Jardine, W. 1834. Memoir of Cuvier. In *The natural history of the Felinae, illustrated by thirty-eight plates, coloured, and numerous wood-cuts; with memoir of Cuvier,* 17–58. Edinburgh: W. H. Lizars and Stirling and Kenny.

Jones, H. L. 1917. *The geography of Strabo, with an English translation.* 8 vols. New York: G. P. Putnam's Sons.

Jones, J. 1722. *Oppian's Halieuticks of the nature of fishes and fishing of the ancients in V. books. Translated from the Greek, with an account of Oppian's life and writings, and a catalogue of his fishes.* Oxford: Printed at the Theater.

Jones, W. H. S. 1918. *Pausanias, Description of Greece.* Vol. 1. New York: G. P. Putnam's Sons.

Jones, W. H. S., and H. A. Ormerod. 1926. *Pausanias, Description of Greece.* Vol. 2. New York: G. P. Putnam's Sons.

Jonsell, B. 1984. Daniel Solander—the perfect Linnaean; his years in Sweden and relations with Linnaeus. *Arch. Nat. Hist.* 11 (3): 443–50.

Jonstonus, J. 1649–62. *Historiae naturalis.* 7 pts. Frankfurt am Main: Meriani. Pt. 1, *De piscibus et cetis libri V,* 1649; pt. 2, *De exanguibus aquaticis libri IV,* 1650; pt. 3, *De avibus libri VI,* 1651; pt. 4, *De quadrupedibus libri IV,* 1652; pt. 5, *De insectis libri III,* 1653; pt. 6, *De serpentibus et draconibus libri II,* 1653; pt. 7, *Dendrographias, sive De arboribus et fructicibus libri X,* 1662.

———. 1657. *Historiae naturalis.* 6 pts. in 2 vols. Amsterdam: Joannem Jacobi fil. Schipper, Vol. 1, pt. 1, *De piscibus et cetis libri V;* pt. 2, *De exanguibus aquaticis libri IV;* pt. 3, *De insectis libri III;* pt. 4, *De serpentibus libri II.* Vol. 2, pt. 5, *De quadrupedibus libri IV;* pt. 6, *De avibus libri VI.*

———. 1718. *Theatrum universale omnium animalium piscium, avium, quadrupedum, exanguium, aquaticorum, insectorum, et angium, CCLX. Tabulis ornatum, ex scriptoribus tam antiquis quam recentioribus, Aristotele, Theophrasto, Dioscoride, Aeliano, Oppiano, Plinio, Gesnero, Aldrovando, Wottonio, Turnero, Mouffeto, Agricola, Boetio, Baccio, Ruveo, Schonfeldio, Freygio, Mathiolo, Tabernomontano, Bauhino, Ximene, Bustamantio, Rondeletio, Bellonio, Caesio, Theveto, Margravio, Pisone, et aliis maxima curâ à J. Jonstonio collectum, ac plus quam trecentis piscibus nuperrime ex Indiis orientalibus allatis, ac nunquam antea his terris visis, locupletatum; cum enumeratione morborum, quibus medicamina ex his animalibus petuntur, ac notitiâ animalium, ex quibus vicissim remedia praestantissima possunt capi; cura Henrici Ruysch M.D. Amstelaed. VI. Partibus, duobus tomis, comprehensum.* Amsterdam: 2 vols. R. en G. Wetstein.

Jordan, D. S. 1902. The history of ichthyology, an address by David Starr Jordan. *Proc. Amer. Assoc. Adv. Sci.* 51:427–56.

———. 1905. The history of ichthyology. In *A guide to the study of fishes,* 1:387–428. New York: Henry Holt.

Jurine, L. 1821. Mémoire sur quelques particularités de l'oeil du thon *(Scomber thynnus,* Lin.) et d'autres poissons. *Mém. Soc. Phys. Hist. Nat. Genève* 1:1–18, pl. 1.

———. 1825. Histoire abrégée des poissons du Lac Léman, extraite des manuscrits de feu M. le professor Jurine, et accompagnée de planches dessinées et gravées sous sa direction. *Mém. Soc. Phys. Hist. Nat. Genève* 3 (1): 133–235, 15 pls. [Plates dated 1826.]

Kádár, Z. 1978. *Survivals of Greek zoological illuminations in Byzantine manuscripts, with 232 halftone and 10 colour plates.* Budapest: Akadémiai Kiadó.

Kaempfer, E. 1712. *Amoenitatum exoticarum politico-physico-medicarum fasciculi V, quibus continentur variae relationes, observationes et descriptiones rerum Persicarum et Ulterioris Asiae.* Lemgo: Henricus Wilhelm Meyer.

———. 1727. *The history of Japan, giving an account of the ancient and present state and government of that empire; of its temples, palaces, castles and other buildings; of its metals, minerals, trees, plants, animals, birds and fishes; of the chronology and succession of the emperors, ecclesiastical and secular; of the original descent, religions, customs, and manufactures of the natives, and of their trade and commerce with the Dutch and Chinese. Together with a description of the kingdom of Siam.* 2 vols. London: Impensis Editoris.

———. 1729. *Histoire naturelle, civile, et ecclésiastique de l'empire du Japon.* Composée en Allemand par Engelbert Kaempfer, docteur en médecine à Lemgow; et traduite en françois sur la version angloise de Jean-Gaspar Scheuchzer. 2 vols. The Hague: P. Grosse en J. Neaulme.

———. 1733. *De beschryving van Japan, behelsende een verhaal van den ouden en tegenwoordigen staat en regeering van dat ryk, van deszelfs tempels, paleysen, kasteelen en andere gebouwen; van deszelfs metalen, mineralen, boomen, planten, dieren, vogelen en vissche.* Amsterdam: Jan Roman de Jonge.

Karrer, C. 1978. Marcus Elieser Bloch (1723–1799), sein Leben und die Geschichte seiner Fischsammlung. *Sitzungsber. Ges. Naturf. Freunde Berl.*, n.s., 18:129–49.

———. 1980. Marcus Elieser Bloch und seine "Allgemeine Naturgeschichte der Fische." In *Naturgeschichte der Fische I, Fische Deutschlands, eine Auswahl, mit einem Nachwort von Christine Karrer,* by M. E. Bloch, 1:171–201. Dortmund: Harenberg Kommunikation.

King, J. 1784. *A voyage to the Pacific Ocean, undertaken by the command of His Majesty, for making discoveries in the Northern Hemisphere, to determine the position and extent of the west side of North America, its distance from Asia, and the practicability of a northern passage to Europe.* 3 vols. London: W. and A. Strahan for G. Nicol and T. Cadell.

Klausewitz, W., and J. G. Nielsen. 1965. On Forsskål's collection of fishes in the Zoological Museum of Copenhagen. *Spolia Zool. Mus. Hauniensis* 22:1–29, pls. 1–38.

Klein, J. T. 1740–49. *Historiae piscium naturalis promovendae.* 5 pts. Danzig: Schreiberianis. Pt. 1, De lapillis eorumque numero in craniis piscium, cum praefatione: De piscium auditu, 1740; pt. 2, De piscibus per pulmones spirantibus ad justum numerum et ordinem redigendis, 1741; pt. 3, De piscibus per branchias occultas spirantibus ad justum numerum et ordinem redigendis, 1742; pt. 4, De piscibus per branchias apertas spirantibus ad justum numerum et ordinem redigendis, 1744; pt. 5, De piscibus per branchias apertas spirantibus. Horum series secunda cum additionibus ad missus II. III. IV. et epistola: De cornu piscis carinae navis impacto, 1749.

———. 1743. *Summa dubiorum circa classes quadrupedum et amphibiorum in celebris domini Caroli Linnaei Systemate naturae: sive Naturalis quadrupedum historiae promovendae prodromus cum praeludio de crustatis.* Danzig: Schreiberianis.

Klein, M. 1974. Oken (or Okenfuss), Lorenz. In *Dictionary of Scientific Biography,* 10:194–96. New York: Scribner's.

Knorr, G. W. 1766–67. *Deliciae naturae selectae, oder Auserlesenes Naturalien-Cabinet welches aus den drey Reichen der Natur zeiget, was von curiösen Liebhabern aufbehalten und gesammlet zu werden verdienet; ehemahls herausgegeben von Georg Wolfgang Knorr berühmten Kupferstecher in Nürnberg, fortgesetzet von dessen Erben.* Beschrieben von Philipp Ludwig Statius Müller . . . und in das Französische übersezet von Matthäus Verdier de la Blaquiere. Nuremberg: Matthäus Verdier de la Blaquiere. Vol. 1, 1766; vol. 2, 1767. [Colored plates in pt. 1, with an engraved title page bearing the date 1754.]

Koelreuter, J. G. 1763. Piscium rariorum e museo Petropolitano exceptorum descriptiones. *Novi Comment. Acad. Sci. Impér. Petro.* 8:404–30, pl. 14.

———. 1764. Descriptionis piscium rariorum e museo Petropolitano exceptorum continuatio. *Novi Comment. Acad. Sci. Impér. Petro.* 9:420–70, pls. 9–10.

———. 1770. Descriptio piscis, e gadorum genere, russis nawaga dicti, historico-anatomica. *Novi Comment. Acad. Sci. Impér. Petro.* 14:484–97, pl. 12.

———. 1771. Descriptio piscis, e coregonorum genere, russice sig vocati, historico-anatomica. *Novi Comment. Acad. Sci. Impér. Petro.* 15:504–16.

———. 1772. Observationes splanchnologicae, ad acipenseris rutheni Linn. anatomen spectantes. *Novi Comment. Acad. Sci. Impér. Petro.* 16:511–24.

———. 1773. Observationum splanchnologicarum, ad acipenseris russici et Husonis anatomen, speciatim vero ad ipsorum auditus organum, spectantium, continuatio. *Novi Comment. Acad. Sci. Impér. Petro.* 17:521–39, pls. 10–12.

———. 1774. Descriptio piscis, e coregonorum genere, russice riapucha dicti, historico-anatomica. *Novi Comment. Acad. Sci. Impér. Petro.* 18:503–11.

———. 1775. Observationes in gado lota institutae. *Novi Comment. Acad. Sci. Impér. Petro.* 19:424–34.

———. 1795. Descriptio pleuronect flesi et passeris Linn. historico-anatomica. *Nova Acta Acad. Sci. Impér. Petro.* 9:327–49.

Kohn, A. J. 1992. *A chronological taxonomy of Conus, 1758–1840.* Washington, D.C.: Smithsonian Institution Press.

Kolb, K. 1992. Guillaume Rondelet et les "Libri de piscibus." *Mém. Maît. Hist. Art* (Université de Paris IV—Sorbonne). Vol. 1, text; vol. 2, figures.

Kolb, P. 1719. *Reise an das Capo du Bonne Espérance, oder Das africanische Korgebürge der Guten Hofnung: Nebst einer ausführlichen Beschreibung desselben.* 3 pts. in 1. Nuremberg: Peter Conrad Monath.

———. 1727. *Naaukeurige en uitvoerige beschryving van de Kaap de Goede Hoop.* 2 vols. in 1. Amsterdam: Balthazar Lakeman.

———. 1741. *Description du Cap de Bonne-Esperance; où l'on trouve tout ce qui concerne l'histoire naturelle du pays; la religion, les moeurs et les usages des Hottentots; et l'establissement des Hollandois.* 3 vols. Amsterdam: Jean Catuffe.

Kölpin, A. B. 1770. Anmärkningar vid svård-fiskens, xiphae, anatomie och naturalhistorie. *Kongl. Vetensk. Acad. Handl.* 31:5–16, pl. 2.

———. 1771. Ytterligane anmärkningar, vid svårds-fiskens natural-historia. *Kongl. Vetensk. Acad. Handl.* 32:115–19, pl. 4.

König, E. 1687. Lupi piscis et mugilis ventriculi conformatio. *Misc. Curio. Medico-Physica,* decur. 2, 5 (100): 208–9.

———. 1695. De rana piscatricis anatome. *Misc. Curio. Medico-Physica,* decur. 3, 2 (139): 204–7.

Kramer, G. H. 1756. *Elenchus vegetabilium et animalium per Austriam inferiorem observatorum. Sistens ea in classes et ordines genera et species redacta.* Vienna: Joannis Thomas Trattner.

Kronick, D. A. 1976. *A history of scientific and technical periodicals: The origins and development of the scientific and technical press, 1665–1790.* Metuchen, N.J.: Scarecrow Press.

Kuhl, H. 1820. *Beiträge zur Zoologie und vergleichenden Anatomie.* 2 pts. in 1. Frankfurt am Main: Hermannschen Buchhandlung. [Pt. 2 is by J. C. van Hasselt and H. Kuhl.]

Labat, J. B. 1722. *Nouveau Voyage aux isles de l'Amérique, contenant l'histoire naturelle de ces pays, l'origine, les moeurs, la religion et le gouvernement des habitans anciens et modernes.* 6 vols. Paris: Guillaume Cavelier.

———. 1724. *Nouveau Voyage aux isles de l'Amérique, contenant l'histoire naturelle de ces pays, l'origine, les moeurs, la religion et le gouvernement des habitans anciens et modernes.* 2 vols. The Hague: P. Husson, T. Johnson, P. Gosse, J. van Duren, R. Alberts, en C. Le Vier.

———. 1728. *Nouvelle Relation de l'Afrique Occidentale: contenant une description exacte du Sénégal et des païs situés entre le Cap-Blanc et la riviere de Serrelionne, jusqu'à plus de 300. lieuës en avant dans les Terres. L'Histoire naturelle de ces païs, les differentes nations qui y sont répanduës, leurs religions et leurs moeurs. Avec l'état ancien et present du compagnies qui y font le commerce.* 5 vols. Paris: Theodore Le Gras.

———. 1730. *Voyage du chevalier des Marchais en Guinée, isles voisines, et à Cayenne, fait en 1725, 1726 et 1727.* 4 vols. in 2. Paris: Saugrain.

———. 1731. *Voyage du chevalier des Marchais en Guinée, isles voisines, et à Cayenne, fait en 1725, 1726 et 1727.* 4 vols. Amsterdam: Aux dépens de la Compagnie.

———. 1732. *Relation historique de l'Ethiopie Occidentale: Contenant description des royaumes de Congo, Angole, et Matamba.* Trans. from the Italian by P. Cavazzi. 5 vols. Paris: Charles-Jean-Baptiste Delespine.

———. 1742. *Nouveau Voyage aux isles de l'Amérique, contenant l'histoire naturelle de ces pays, l'origine, les moeurs, la religion et le gouvernement des habitans anciens et modernes.* New ed., enl. 8 vols. Paris: Charles-Jean-Baptiste Delespine.

La Bruyère, J. de. 1699. *The characters, or The manners of the age, with The characters of Theophrastus, translated from the Greek and a prefatory discourse to them, by Monsieur de La Bruyère, to which is added a key to his characters.* London: John Bullord. [Followed by J. de La Bruyère, 1698, *The moral characters of Theophrastus, made English from the Greek, with a prefatory discourse concerning Theophrastus, from the French of Monsieur de La Bruyère,* London: John Bullord.]

Lacepède, B. G. E. 1788–89. *Histoire naturelle des quadrupèdes ovipares et des serpens.* 2 vols. Paris: Hôtel de Thou.

———. 1798–1803. *Histoire naturelle des poissons.* Paris: Plassan. Vol. 1, 1798; vol. 2, 1800; vol. 3, 1801; vol. 4, 1802; vol. 5, 1803.

———. 1799–1804a. *Histoire naturelle des poissons.* Paris: P. Didot. Vol. 1, 1799; vol. 2, 1799; vol. 3, 1799; vol. 4, 1799; vol. 5, 1799; vol. 6, 1799; vol. 7, 1799; vol. 8, 1799; vol. 9, 1799; vol. 10, 1799; vol. 11, 1804; vol. 12, 1804; vol. 13, 1804; vol. 14, 1804.

———. 1799–1804b. *Naturgeschichte der Fische, als eine Fortsetzung von Buffons Naturgeschichte.* Trans. from the French with notes by Ph. Loos. 4 pts. in 2 vols. Berlin: Pauli. Vol. 1, pt. 1, 1799; pt. 2, 1799. Vol. 2, pt. 1, 1803; pt. 2, 1804.

———. 1804a. Mémoire sur plusieurs animaux de la Nouvelle-Hollande dont la description n'a pas encore été publiée. *Ann. Mus. Hist. Nat.* (Paris) 4:184–211, pls. 55–58.

———. 1804b. *Histoire naturelle des cétacés.* Paris: Plassan.

Laet, J. de. 1633. *Novus orbis, seu Descriptionis Indiae Occidentalis, libri XVIII. Authore Joanne de Laet Antwerp. Novis tabulis geographicis et variis animantium, plantarum, fructuumque iconibus illustrati.* Leiden: Elsevier.

Laissus, Y. 1967. *Les Vélins du Muséum.* Conference at the Palais de la Découverte, 7 January 1967. Paris: Université de Paris, Palais de la Découverte.

———. 1973. Note sur les voyages de Jean-Jacques Dussumier (1792–1883). *Ann. Soc. Sci. Nat. Char.-Marit., La Rochelle* 5 (5–9): 387–406.

———. 1978. Catalogue des manuscrits de Philibert Commerson (1727–1773) conservés à la Bibliothèque Centrale du Muséum National d'Histoire Naturelle. *Rev. Hist. Sci.* 31 (2): 131–62.

Lamarck, J. B. P. A. 1815–22. *Histoire naturelle des animaux sans vertèbres, présentant les caractères généraux et particuliers de ces animaux, leur distribution, leurs classes, leurs familles, leurs genres, et la citation des principales espèces qui s'y rapportent; précédée d'une introduction offrant la détermination des caractères essentiels de l'animal, sa distinction du végétal et des autres corps naturels, enfin, l'exposition des principes fondamentaux de la zoologie.* Paris: Verdière. Vol. 1, 1815; vol. 2, 1816; vol. 3, 1816. Deterville et Verdière, vol. 4, 1817; vol. 5, 1818. Published by the author, Vol. 6, 1819; vol. 7, 1822.

Lamorier, L. 1745. Sur un organe particulier du chien de mer. *Hist. Mém. Acad. Roy. Sci.* (Paris) 1742:32–33.

Laundon, J. R. 1981. The date of birth of Sir John Hill. *Arch. Nat. Hist.* 10 (1): 65–66.

Lefebvre de Villebrune, J. B. 1789–91. *Banquet des savans, par Athénée, traduit, tant sur les textes imprimés, que sur plusieurs manuscrits.* Paris: Lamy. Vol. 1, 1789; vol. 2, 1789; vol. 3, 1789; vol. 4, 1789; vol. 5, 1791.

Leguat, F. 1708. *A new voyage to the East-Indies by Francis Leguat and his companions, containing their adventures in two desart islands, and an account of the most remarkable things in Maurice Island, Batavia, at the Cape of Good Hope, and island of St. Helena, and other places in their way to and from the desart isles.* London: R. Bonwicke, W. Freeman, Tim. Goodwin, F. Walthoe, M. Wotton, S. Manship, F. Nicholson, B. Tooke, R. Parker, and R. Smith.

Leigh, C. 1700. *The natural history of Lancashire, Cheshire, and the Peak, in Derbyshire: With an account of the British, Phoenician, Armenian, Gr. and Rom. antiquities in those parts.* Oxford: Guil. Painter.

Lejeune, P., J. M. Boveroux, and J. Voss. 1980. Observation du comportement reproducteur de *Serranus scriba* Linné (Pisces, Serranidae), poisson hermaphrodite synchrone. *Cybium*, ser. 3, 10:73–80.

Lemaire, N. E. 1827–32. *Caii Plinii Secundi Historiae naturalis libri XXXVII.* Paris: Nicolaus Eligius Lemaire. Vol. 1, 1827; vol. 2, 1828; vol. 3, 1827; vol. 4, 1828; vol. 5, 1829; vol. 6, 1829; vol. 7, 1830; vol. 8, 1829; vol. 9, 1831; vol. 10, pt. 1, 1831; vol. 10, pt. 2, 1832.

Lepechin, I. I. 1771–1805. *[Journal of travels through the different provinces of the Russian empire in the years 1768, 1769, 1770, and 1771.]* 4 vols. St. Petersburg: Academie der Wissenschaften. Vol. 1, 1771; vol. 2, 1772; vol. 3, 1780; vol. 4, 1805. [In Russian.]

———. 1774–83. *Tagebuch der Reise durch verschiedene Provinzen des russischen Reiches in den Jahren 1768, 1769, 1770 und 1771.* Trans. from the Russian by Christian Heinrich Hase. Altenburg: Richterischen Buchhandling. Vol. 1, 1774; vol. 2, 1775; vol. 3, 1783.

———. 1795. Varietas acipenseris stellati oppido rara descripta. *Nova Acta Acad. Sci. Impér. Petro.* 9:35–38, pl. A.

Léry, J. de. 1578. *Histoire d'un voyage fait en la terre du Brésil, autrement dite America.* La Rochelle: Antoine Chuppin.

Leske, N. G. 1774. *Ichthyologiae Lipsiensis specimen.* Leipzig: Siegfried Lebrecht Crusius.

Lesson, R. P. 1826–31. Poissons. In *Voyage autour du monde, exécuté par ordre du roi, sur la corvette de Sa Majesté, la Coquille, pendant les années 1822, 1823, 1824 et 1825,* by L. I. Duperrey, vol. 1, pt. 1, chap. 10, pp. 66–238, pls. 1–38. Paris: Arthus Bertrand, Zoologie. [Text dating from 1831; atlas, 1826.]

Lesson, R. P., and P. Garnot. 1826–31. Zoologie. In *Voyage autour du monde, exécuté par ordre du roi, sur la corvette de Sa Majesté, la Coquille, pendant les années 1822, 1823, 1824 et 1825,* by L. I. Duperrey. Paris: Arthus Bertrand. Text: Vol. 1, 1826. Vol. 2, pt. 1, 1831; pt. 2 [assisted by F. E. Guérin-Méneville], 1831. Atlas, 1826.

Lesueur, C. A. 1817a. Description of three new species of the genus *Raia. J. Acad. Nat. Sci. Philad.* 1 (1): 41–45, pls. 1–3.

———. 1817b. A short description of five (supposed) new species of the genus *Muraena,* discovered by Mr. Le Sueur, in the year 1816. *J. Acad. Nat. Sci. Philad.* 1 (1): 81–83, pls. 1–3.

———. 1817c. Description of two new species of the genus *Gadus*. *J. Acad. Nat. Sci. Philad.* 1 (1): 83–85.
———. 1817d. Description of a new species of the genus *Cyprinus*. *J. Acad. Nat. Sci. Philad.* 1 (1): 85–86.
———. 1817e. A new genus of fishes, of the order Abdominales, proposed, under the name of *Catostomus;* and the characters of this genus, with those of its species, indicated. *J. Acad. Nat. Sci. Philad.* 1 (1): 88–96, 102–11, pls. 1–9.
———. 1817f. Descriptions of four new species, and two varieties, of the genus *Hydrargira*. *J. Acad. Nat. Sci. Philad.* 1 (1): 126–34.
———. 1818a. Description of several new species of North American fishes. *J. Acad. Nat. Sci. Philad.* 1 (2): 222–35, pls. 9–10.
———. 1818b. Descriptions of several new species of North America fishes. *J. Acad. Nat. Sci. Philad.* 1 (2): 359–68, pl. 14.
———. 1818c. Description of several new species of the genus *Esox*, of North America. *J. Acad. Nat. Sci. Philad.* 1 (2); 413–22, pl. 17.
———. 1819. Notice de quelques poissons découverts dans les lacs du Haut-Canada, durant l'été de 1816. *Mém. Mus. Nat. Hist.* (Paris) 5:148–61, pls. 16–17.
———. 1821a. Description of a new genus, and of several new species of fresh water fish, indigenous to the United States. *J. Acad. Nat. Sci. Philad.* 2 (1): 2–8, pls. 1–3.
———. 1821b. Description of two new species of *Exocetus*. *J. Acad. Nat. Sci. Philad.* 2 (1): 8–11, pl. 4.
———. 1821c. Observations on several genera and species of fish, belonging to the natural family of the esoces. *J. Acad. Nat. Sci. Philad.* 2 (1): 124–38, pls. 1–2.
———. 1822a. Descriptions of the five new species of the genus *Cichla* of Cuvier. *J. Acad. Nat. Sci. Philad.* 2 (2): 214–21, pl. 1.
———. 1822b. Description of three new species of the genus *Sciaena*. *J. Acad. Nat. Sci. Philad.* 2 (2): 251–56, pl. 1.
———. 1822c. Description of a *Squalus*, of a very large size, which was taken on the coast of New-Jersey. *J. Acad. Nat. Sci. Philad.* 2 (2): 343–52, pl. 1.
———. 1824. Description of several species of the Linnaean genus *Raia*, of North America. *J. Acad. Nat. Sci. Philad.* 4 (1): 100–121, pls. 4–6.
———. 1825a. Description of two new species of the Linnaean genus *Blennius*. *J. Acad. Nat. Sci. Philad.* 4 (2): 361–64.
———. 1825b. Description of a new fish of the genus *Salmo*. *J. Acad. Nat. Sci. Philad.* 5 (1): 48–51, pl. 3.
———. 1825c. Descriptions of four new species of *Muraenophis*. *J. Acad. Nat. Sci. Philad.* 5 (1): 107–9, pl. 4.
———. 1825d. Description of a new species of the genus *Saurus* (Cuvier). *J. Acad. Nat. Sci. Philad.* 5 (1): 118–19, pl. 5.
Levy, H. L. 1976. *Lucian: Seventy dialogues.* Introduction and commentary by Harry L. Levy. Norman: University of Oklahoma Press.
Lichtenstein, M. H. C. 1818. Die Werke von Marcgrave und Piso über die Naturgeschichte brasiliens, erläutert aus den wieder aufgefundenen Originalzeichnungen. I. Säugethiere. *Abhandl. Physik. Kl. Königlich-Preuss. Akad. Wissenschaft.* (Berlin) 1814–15: 201–22.
———. 1819. Die Werke von Marcgrave und Piso über die Naturgeschichte brasiliens, erläutert aus den wieder aufgefundenen Original-Abbildungen. II. Vögel. *Abhandl. Physik. Kl. Königlich-Preuss. Akad. Wissenschaft.* (Berlin) 1816–17: 155–78.
———. 1822a. Die Werke von Marcgrave und Piso über die Naturgeschichte

brasiliens, erläutert aus den wieder aufgefundenen Original-Abbildungen (fortsetzung). III. Amphibien. *Abhandl. Physik. Kl. Königlich-Preuss. Akad. Wissenschaft.* (Berlin) 1820–21: 237–54.

———. 1822b. Die Werke von Marcgrave und Piso über die Naturgeschichte brasiliens, erläutert aus den wieder aufgefundenen Original-Abbildungen (fortsetzung). IV. Fische. *Abhandl. Physik. Kl. Königlich-Preuss. Akad. Wissenschaft.* (Berlin) 1820–21: 267–88.

———. 1829. Die Werke von Marcgrave und Piso über die Naturgeschichte brasiliens, erläutert aus den wieder aufgefundenen Original-Abbildungen. IV. Fische [continued]. *Abhandl. Physik. Kl. Königlich-Preuss. Akad. Wissenschaft.* (Berlin) 1826: 49–65.

Limes, J. M. 1817. *Les halieutiques, traduits du Grec du poëme d'Oppien, ou Il Traite de la pêche et des moeurs des habitans des eaux.* Paris: Lebégue.

Lindsay, A. D. 1954. *Plato's Republic.* New York: E. P. Dutton.

Linnaeus, C. 1735. *Systema naturae, sive Regna tria naturae systematice proposita per classes, ordines, genera, et species.* Leiden: Theodor Haak.

———. 1736a. *Fundamenta botanica, quae majorum operum prodromi instar theoriam scientiae botanices per breves aphorismos tradunt.* Amsterdam: Salomon Schouten.

———. 1736b. *Bibliotheca botanica recensens libros plus mille de plantis huc usque editos, secundum systema auctorum naturale in classes, ordines, genera et species dispositos, additis editionis loco, tempore, forma, lingua. Cum explicatione.* Amsterdam: Salomon Schouten.

———. 1736c. *Musa Cliffortiana florens Hartecampi 1736 prope Harlemum.* Leiden.

———. 1737a. *Critica botanica in quo nomina plantarum generica, specifica, et variantia examini subjiciuntur, selectiora confirmantur, indigna rejiciuntur; simulque doctrina circa denominationem plantarum traditur.* Leiden: Conrad Wishoff.

———. 1737b. *Genera plantarum eorumque characteres naturales secundum numerum, figuram, situm, et proportionem omniun fructificationis partium.* Leiden: Conrad Wishoff.

———. 1737c. *Methodus sexualis sistens genera plantarum secundum mares et feminas in classes et ordines redacta.* Leiden.

———. 1737d. *Flora Lapponica exhibens plantas per Lapponiam crescentes, secundum systema sexuale collectas in itinere impensis Soc. Reg. Litter. et Scient. Sveciae A. 1732 instituto. Additis synonymis, et locis natalibus omnium, descriptionibus et figuris rariorum, viribus medicatis et oeconomicis plurimarum.* Amsterdam: Salomon Schouten.

———. 1737e. *Hortus Cliffortianus, plantas exhibens quas in hortis tam vivis quam siccis Hartecampi in Hollandia coluit vir nobilissimus et generosissimus Georgius Clifford juris utriusque doctor, reductis varietatibus ad species, specibus ad genera, generibus ad classes, adiectis locis plantarum natalibus differentiisque specierum.* Amsterdam.

———. 1738. *Classes plantarum, seu Systemata plantarum omnia a fructificatione desumta, quorum XVI universalia et XIII partialia, compendiose proposita secundum classes, ordines et nomina generica cum clave cujusvis methodi et synonymis genericis.* Leiden: Conrad Wishoff.

———. 1740a. *Systema naturae in quo naturae regna tria secundum classes, ordines, genera, species, systematice proponuntur.* 2d ed., enl. Stockholm: Gottfr. Kiesewetter.

———. 1740b. *Systema naturae, sive Regna tria naturae systematice proposita per classes, ordines, genera et species* [with German subtitle]. Halle: Gedruckt mit Gebauerischen Schriften.

———. 1744. *Systema naturae in quo proponuntur naturae regna tria secundum classes, ordines, genera et species.* 4th ed., rev. and enl. by the author. Paris: Michaelis-Antonii David.

———. 1745. *Öländska och Gothländska resa på rikens högloflige ständers befallning förrättad år 1741. Med anmärckningar uti oeconomien, naturalhistorien, antiquiteter.* Stockholm and Uppsala: Hos. Gottfried Kiesewetter.

———. 1746. *Museum Adolpho-Fridericianum, quod, cum consensu ampliss. Fac. Medicae in Regia Acad. Upsaliensi, sub praesidio viri celeberrimi, D.D. Caroli Linnaei. . . . Speciminis academici loco publico bonorum examini submittit Laurentius Balk Fil.* Stockholm: Laurentius Salvius. [Also published in *Amoenitates academicae,* 1749, 1:277–326.]

———. 1747a. *Systema naturae in quo naturae regna tria, secundum classes, ordines, genera, species, systematice proponuntur. Recusum et societatis, quae impensas contulit, usui accommodatum curante Mich. Gottl. Agnethlero Saxone Transilvano.* Another ed., rev. and enl. Halle: Hala Magdeburgica.

———. 1747b. *Wästgöta-resa, på riksens högloflige ständers befallning förrättad år 1746. Med anmärkningar uti oeconomien, naturkunnogheten, antiquiteter, inwånarnes seder och lefnads-sätt, med tilhörige figurer.* Stockholm: Lars Salvius.

———. 1748a. *Systema naturae, sistens regna tria naturae, in classes et ordines genera et species redacta, tabulisque aeneis illustrata. 6th ed., rev. and enl. Stockholm: Godofr. Kiesewetter.*

———. 1748b. *Systema naturae sistens regna tria naturae in classes et ordines genera et species redacta tabulisque aeneis illustrata.* According to the 6th rev. and enl., Stockholm ed. Leipzig: Godofr. Kiesewetter.

———. 1749. Museum Adolpho-Fridericianum, sub praesidio Dn. D. Caroli Linnaei . . . propositum Laurent. Balk, Fil. In *Amoenitates academicae,* 1:277–327. Stockholm: Laurentius Salvius.

———. 1749–69. *Amoenitates academicae; seu Dissertationes variae physicae, medicae, botanicae, antehac seorsim editae, nunc collectae et auctae, cum tabulis aeneis.* 10 vols. Stockholm: Laurentius Salvius.

———. 1751a. *Philosophia botanica, in qua explicantur fundamenta botanica, cum definitionibus partium, exemplis terminorum, observationibus rariorum, adjectis figuris aeneis.* Stockholm: Godofr. Kiesewetter.

———. 1751b. *Skånska resa, på höga öfwerhetens befallning förrättad år 1749. Med rön och anmärkningar uti oeconomien, naturalier, antiquiteter, seder, lefnadssätt.* Stockholm: Lars Salvius.

———. 1753a. *Museum Tessinianum, opera illustrissimi comitis, Dom. Car. Gust. Tessin, . . . collectum.* Stockholm: Lars Salvius.

———. 1753b. *Species plantarum, exhibentes plantas rite cognitas, ad genera relatas, cum differentiis specificis, nominibus trivialibus, synonymis selectis, locis natalibus, secundum systema sexuale digestas.* 2 vols. Stockholm: Laurentius Salvius.

———. 1753c. *Herr Archiaterns och riddarens D. Caroli Linnaei indelning i Ört-Riket, efter Systema naturae, på Swenska öfwersatt af Johan J. Haartman.* Stockholm: Lars Salvius.

———. 1754a. *Museum S:ae R:ae M:tis Adolphi Friderici . . . in quo animalia rariora imprimis, et exotica: Quadrupedia, aves, amphibia, pisces, insecta, vermes describuntur et determinantur, Latine et Svetice, cum iconibus, jussu Sac. Reg. Maj:tis a Car. Linnaeo, Equ.* Stockholm: Pet. Momma.

———. 1754b. *Specimen academicum, sistens Chinensia Lagerströmiana. Quod, annvente nob:a Facultate Medica. in R. Academia Upsal. praeside viro nobilissimo et experientissimo D:n D. Carolo Linnaeo. . . . Publicae disquisitioni submittit Johannes Laurentius Odhelius, W.G.* Stockholm: Jacob Merckell.

———. 1756. *Systema naturae sistens regna tria naturae in classes et ordines genera et species redacta, tabulisque aeneis illustrata. Accedunt vocabula Gallica.* Ed. much rev. and enl. Leiden: Theodor Haak.

———. 1758–59. *Systema naturae per regna tria naturae, secundum classes, ordines, genera, species, cum characteribus, differentiis, synonymis, locis.* 10th ed. rev. Stockholm: 2 vols. Laurentius Salvius.

———. 1759. Chinensia Lagerströmiana, praeside D.D. Car. Linnaeo, proposita a Johann Laur. Odhelio, W. Gotho. Upsaliae 1754, Decembr. 23. In *Amoenitates Academicae,* 4:230–60. Stockholm: Laurentius Salvius.

———. 1760–70. *Systema naturae per regna tria naturae secundum classes, ordines, genera, species, cum characteribus, differentiis, synonymis, locis.* Praefatus est Joannes Joachimus Langius. Halle: Jo. Jac. Curt. Vol. 1, Regnum animale, 1760; vol. 2, Regnum vegetabile, 1760; vol. 3, Regnum lapideum, 1770.

———. 1762. *Systema naturae per regna tria naturae, secundum classes, ordines, genera, species, cum characteribus, differentiis, synonymis, locis.* 2 vols. Leipzig. [Not seen, publisher unknown; not in BMNH or BL; BMNH Library Cat. (1910, 3:1128) notes that "Linnaeus reckoned this as the 11th edition, and stated that it contains no additions"; cat. of the works of Linnaeus, BMNH, 1933, 11: "this edition may be non-existent."]

———. 1762–63. *Species plantarum, exhibentes plantas rite cognitas, ad genera relatas, cum differentiis specificis, nominibus trivialibus, synonymis selectis, locis natalibus, secundum systema sexuale digestas.* 2d ed. enl. 2 vols. Stockholm: Laurentius Salvius.

———. 1764a. *Museum S:ae R:ae M:tis Ludovicae Ulricae reginae Svecorum, Gothorum, Vandalorumque . . . in quo animalia rariora, exotica, imprimis insecta et conchilia describuntur et determinantur. Prodromi instar editum.* Stockholm: Laurentius Salvius.

———. 1764b. *Museum S:ae R:ae M:tis Adolphi Friderici . . . in quo animalia rariora imprimis, et exotica: Quadrupedia, aves, amphibia, pisces, insecta, vermes describuntur et determinantur, Latine et Svetice, cum iconibus, jussu Sac. Reg. Maj:tis a Car. Linnaeo, Equ. Tomi secundi prodromus.* Stockholm: Laurentius Salvius.

———. 1764c. *Genera plantarum eorumque characteres naturales secundum numerum, figuram, situm, et proportionem omniun fructificationis partium.* 6th ed. rev. and enl. Stockholm: Laurentius Salvius.

———. 1766–68. *Systema naturae per regna tria naturae, secundum classes, ordines, genera, species, cum characteribus, differentiis, synonymis, locis.* 12th ed. rev. Stockholm: Laurentius Salvius. Vol. 1, Regnum animale, 1766; vol. 2, Regnum vegetabile, 1767; vol. 3, Regnum lapideum, 1768.

———. 1767. *Mantissa plantarum Generum editionis VI. (1764) et Specierum editionis II.* Stockholm: Laurentius Salvius.

———. 1767–70. *Systema naturae, per regna tria naturae, secundum classes, ordines, genera, species cum characteribus, differentiis, synonymis, locis.* 13th ed., after the revised 12th, Stockholm ed. Vienna: Joannis Thomae nob. de Trattnern. Vol. 1, Regnum animale, 1767, pt. 1, Mammalia, Aves, Amphibia (Reptilia, Serpentes, and Nantes), and Pisces; pt. 2, Insecta and Vermes. Vol. 2, Regnum vegetabile, 1770; vol. 3, Regnum lapideum, 1770.

———. 1771. *Mantissa plantarum altera Generum editionis VI. et Specierum editionis II. (Mantissa prioris additamenta). Regni animalis appendix. Appendix. Index Mantissae. Index observationum. Index alter. Genera nova, addenda Generibus plantarum editionis sextae (1764).* Stockholm: Laurentius Salvius.

———. 1773–76. *Des Ritters Carl von Linné . . . vollständiges Natursystem nach der zwölften lateinischen Ausgabe und nach Anleitung des holländischen Houttuynischen Werkes, mit einer ausführlichen Erklärung, ausgefertigt von Philipp Ludwig Statius Müller.* 6 vols. and suppl. bound in 8. Nuremberg: Gabriel Nicolaus Raspe. Vol. 1, 1773. Vol. 2, 1773. Vol. 3, 1774. Vol. 4, 1774. Vol. 5, pt. 1, 1774; pt. 2, 1775. Vol. 6, pt. 1, 1775; pt. 2, 1775. Supplement, 1776.

———. 1788–93. *Systema naturae per regna tria naturae, secundum classes, ordines, genera, species, cum characteribus, differentiis, synonymis, locis.* 13th ed., rev. and enl. Cura Jo. Frid. Gmelin. 10 pts. in 3 vols. Leipzig: Georg. Emanuel Beer. Vol. 1, Regnum animale, pt. 1, Mammalia, 1788; pt. 2, Aves, 1789; pt. 3, Amphibia and Pisces, 1789; pt. 4, Insecta, 1790; pt. 5, Insecta, 1790; pt. 6, Vermes, 1791; pt. 7, Index, 1792. Vol. 2, Regnum vegetabile, pt. 1, 1791; pt. 2, 1792. Vol. 3, Regnum lapideum, 1793.

———. 1788–96. *Systema naturae per regna tria naturae, secundum classes, ordines, genera, species; cum characteribus, differentiis, synonymis, locis.* 13th ed., rev. and enl. Cura Jo. Frid. Gmelin. 10 pts. in 3 vols. Lyon: J. B. Delamollière. Vol. 1, Regnum animale, pt. 1, Mammalia, 1788; pt. 2, Aves, 1789; pt. 3, Amphibia and Pisces, 1789; pt. 4, Insecta, 1790; pt. 5, Insecta, 1790; Pt. 6, Vermes, 1791; Pt. 7, Index, 1792. Vol. 2, Regnum vegetabile, in 2 pts., 1792. Vol. 3, Regnum lapideum, 1796.

Löfling, P. 1758. *Iter Hispanicum, eller resa til Spanska Länderna uti Europa och America, förrättad ifrån år 1751 til år 1756, med beskrifningar och rön öfver de märkvärdigaste vaxter, utgifven efter dess frånfälle af Carl Linnaeus.* Stockholm: Lars Salvius.

———. 1776. *Reisebeschreibung nach den spanischen Ländern in Europe und America in den Jahren 1751 bis 1756 nebst Beobachtungen und Anmerkungen über die merkwürdigen Gewächse, herausgegeben Herrn Carl von Linné. . . .* Trans. from the Swedish by Alexander Bernhard Kölpin. Berlin: Gottl. August Lange.

Lonicer, A. 1551. *Naturalis historiae opus novum, in quo tractatur de natura et viribus arborum fruticum, herbarum, animantium'q; terrestrium, volatilium et aquatilium.* 2 vols. in 1. Frankfurt am Main: Christ. Egenolphus.

Lönnberg, E. 1905. *Peter Artedi—a bicentenary memoir written on behalf of the Swedish Royal Academy of Science.* Trans. by W. E. Harlock. Uppsala: Almquist and Wiksells Boktryckeri.

Lorenzini, S. 1678. *Osservazioni intorno alle torpedini fatte da Stefano Lorenzini Fiorentino, e dedicate al serenissimo Ferdinando III, Principe di Toscana.* Florence: Onofri.

———. 1680. Anatomia torpedinis, excerpta ex ejusdem libro inscripto osservationi intorno alle torpedini, Florentiae anno 1678. *Misc. Curio. Medico-Physica,* decur. 1, 9–10 (172): 389–95.

Lott, F. van der. 1762. Kort bericht van den conger-aal, ofte drilvisch. *Verh. Holl. Maatsch. Wetensch. Haarlem* 6 (2): 87–95.

Low, G. 1813. *Fauna Orcadensis, or The natural history of the quadrupeds, birds, reptiles, and fishes, of Orkney and Shetland.* Edinburgh: George Ramsay.

Lucas, P. 1705. *Voyage du Sieur Paul Lucas au Levant, on y trouvera entr'autre une description de la Haute Egypte, suivant le cours du Nil, depuis le Caire jusques aux cataractes, avec une carte exacte de ce fleuve, que personne n'avoit donnée.* 2 vols. The Hague: Guillaume de Voys.

———. 1712. *Voyage du Sieur Paul Lucas, fait par ordre du roy dans la Grèce, l'Asie Mineure, la Macédoine et l'Afrique.* 2 vols. Paris: Nicolas Simart.

———. 1719. *Voyage du Sieur Paul Lucas, fait en M. DCCXIV, etc. par ordre de Louis XIV dans la Turquie, l'Asie, Sourie, Palestine, Haute et Basse Egypte, etc.* 3 vols. Rouen: Robert Machuel.

Ludwig, C. F. 1791–95. *Scriptores neurologici minores selecti, sive Opera minora ad anatomiam physiologiam et pathologiam nervorum spectantia.* Leipzig: Jo. Frid. Iunius, Vol. 1, 1791; vol. 2, 1792; vol. 3, 1793; vol. 4, 1795.

Maar, V. 1910. *Nicolai Stenonis opera philosophia.* 2 vols. Copenhagen: Vilhelm Tryde.

Mabberley, D. J. 1985. *Jupiter Botanicus: Robert Brown of the British Museum.* London: British Museum (Natural History).

Mair, A. W. 1928. *Oppian Colluthus Tryphiodorus.* With an English translation by A. W. Mair, D. Litt. New York: G. P. Putnam's Sons.

Malpighi, M. 1687. Marcelli Malpighii exercitatio epistolica de cerebro. In *M. Malpighi, Opera omnia botanico-medico-anatomica,* 2: 113–24. Leiden: Petrus vander Aa.

Manget, J. 1699. *Bibliotheca anatomica, sive Recens in anatomia inventorum thesaurus locupletissimus.* 2 vols. Geneva: Johan. Anthon. Chouët et David Ritter.

Marcgrave, G. 1648. Historiae rerum naturalium Brasiliae, libri octo: Quorum tres priores agunt de plantis, quartus de piscibus, quintus de avibus, sextus de quadrupedibus, et serpentibus, septimus de insectis, octavus de ipsa regione, et illius incolis, cum appendice de tapuyis, et chilensibus, Joannes de Laet, Antwerpianus, in ordinem digessit et annotationes addidit multas, et varia ab auctore omissa supplevit et illustravit. In *Historia naturalis Brasiliae auspicio et beneficio illustriss. I. Mauritii Com. Nassau illius provinciae et maris summi praefecti adornata. In qua non tantum plantae et animalia, sed et indigenarum morbi, ingenia et mores describuntur et iconibus supra quingentas illustratur,* ed. G. Piso and G. Marcgrave, pt. 2. Leiden and Amsterdam: Haack en Elsevier. [Reprint 1942, São Paulo; Portuguese translation.]

Marolles, M. de. 1680. *Les Quinze Livres des Deipnosophistes d'Athénée, de la ville de Naucrate d'Egypte, écrivain d'une érudition consommée, et presque le plus sçavant des Grecs.* 2 vols. Paris: Jacques Langlois.

Marsden, W. 1783. *The history of Sumatra, containing an account of the government, laws, customs, and manners of the native inhabitants, with a description of the natural productions, and a relation of the ancient political state of that island.* London: For the author.

———. 1788. *Histoire de Sumatra, dans laquelle on traite du gouvernement, du commerce, des arts, des loix, des coutumes et des moeurs des habitans; des productions naturelles, et de l'ancien état politique de cette isle.* 2 vols. Paris: Buisson.

———. 1811. *The history of Sumatra, containing an account of the government, laws, customs, and manners of the native inhabitants, with a description of the natural productions, and a relation of the ancient political state of that island.* London: For the author by J. M'Creery.

Marshall, J. B. 1984. Daniel Carl Solander, friend, librarian and assistant to Sir Joseph Banks. *Arch. Nat. Hist.* 11 (3): 451–56.

Marsigli, L. F. 1718. Letterea del Sig. Conte Luigi Ferdinando Marsilli, scritta al Sig. Antonio Vallisnieri, intorno all'origine dell'anguille. *Gior. Letterati d'Italia* (Venice) 29 (8): 206–17, pl. 3.

———. 1726. *Danubius Pannonico-Mysicus, observationibus geographicis, astronomicis, hydrographicis, historicis, physicis perlustratus et in sex tomos digestus.* 6 vols. The

Hague: P. Gosse, R. Chr. Alberts, P. de Hondt; Amsterdam: Herm. Uytwerf en Franç. Changuion.
Martin, A. R. 1760. Anmärkningar öfver den så kallade spitelska fisk och boskap i Norrige. *Kongl. Svenska Vetensk. Acad. Handling.* (Stockholm) 21:306–11.
———. 1771. Gordier, knut eller tråd-maskar, fundne hos fiskar och människor, med försökte medel at dem fördrifva. *Kongl. Svenska Vetensk. Acad. Handling.* (Stockholm) 31:261–311.
Martini, F. H. W. 1775. Bemerkung von der Schädlichkeit einiger Fische. *Berl. Samml.* 7 (2): 189–92.
———. 1776. Die gefiederte Goldmacher und vom Alter der Fische. *Berl. Samml.* 8 (4): 348–50.
Marwitz, [—] von. 1779. Schreiben des Herrn von Marwitz über die Versetzung einiger Fischarten. *Beschäft. Berl. Ges. Naturf. Fr.* 4:91–94.
Massalien, F. C. 1815. *Dissertatio inauguralis sistens descriptionem oculorum scombri, thynni et sepiae, quam gratiosi medicorum ordinis auctoritate et consensu praeside Carolo Asmund Rudolphi.* Berlin: Joannis Fridericus Starckius.
Massaria, F. 1537. *Francisci Massarii Veneti in nonum Plinii De naturali historia librum castigationes et annotationes.* Basle: Hieronymus Frobenius et Nicolaus Episcopius.
———. 1542. *In nonum Plinii De naturali historia librum castigationes et annotationes.* Paris: Michaëlis Vascosani.
Mattioli, P. A. 1548. *Il Dioscoride dell'eccellente Dottor Medico M.P. Matthioli Siena.* Venice: Vincenzo Valgrisi. [The "Privilegio" is dated 1544.]
———. 1554. *Medici senensis commentarii, in libros sex Pedacii Dioscoridis Anazarbei, de medica materia.* Venice: Vincentius Valgrisius.
———. 1565. *Commentarii in sex libros Pedacii Dioscoridis Anazarbei de medica materia.* Venice: Valgrisi.
Mearns, B., and R. Mearns. 1988. *Biographies for birdwatchers, the lives of those commemorated in western Palearctic bird names.* London: Academic Press.
Meckel, J. F. 1821–33. *System der vergleichenden Anatomie.* Halle: Rengerschen Buchhandlung. Vol. 1, 1821. Vol. 2, pt. 1, 1824; pt. 2, 1825. Vol. 3, 1828. Vol. 4, 1829. Vol. 5, 1831. Vol. 6, 1833.
———. 1828–38. *Traité général d'anatomie comparée.* Traduit de l'Allemand et augmenté de notes par MM. Riester, et Alph. Sanson. Paris: Villeret, vol. 1, 1828; vol. 2, 1828. Paris: Rouen, vol. 3, 1829; vol. 4, 1829; vol. 5, 1829; vol. 6, 1829–30. Paris: Charles Hingray, vol. 7, 1836; vol. 8 (trans. Alph. Sanson and Th. Schuster), 1838; vol. 9 (trans. Th. Schuster), 1837; vol. 10 (trans. Th. Schuster), 1838.
Meidinger, C. von. 1785–94. *Icones piscium Austriae indigenorum quos collegit vivisque coloribus expressos.* Vienna: Wapplerus Bibliopola. Pt. 1, 1785, pls. 1–10; pt. 2, 1786, pls. 11–20; pt. 3, 1788, pls. 21–30; pt. 4, 1790, pls. 31–40; pt. 5, 1794, pls. 41–50.
Merriman, D. 1938. Peter Artedi—systematist and ichthyologist. *Copeia* 1938 (1): 33–39.
———. 1941. A rare manuscript adding to our knowledge of the work of Peter Artedi. *Copeia* 1941 (2): 64–69.
Merry, W. W. 1895. *Homer, Odyssey, books I-XII, with introduction, notes, etc.* 2 pts. in 1. Oxford: Clarendon Press. Pt. 1, introduction and text; pt. 2, notes [and index].
Meyer, J. D. 1748–56. *Angenehmer und nützlicher Zeit-Vertreib mit Betrachtung curioser Vorstellungen allerhand kriechender, fliegender und schwimmender, auf dem Land und im Wasser sich befindender und nährender Thiere, sowohl nach ihrer Gestalt und äus-*

serlichen Beschaffenheit als auch nach der accuratest davon verfertigten Structur ihrer Scelete oder Bein-Körper. Nuremberg: Johann Joseph Fleischmann. Vol. 1, 1748; vol. 2, 1752; vol. 3, 1756.

Mitchill, S. L. 1814. *Report, in part, of Samuel L. Mitchill, M.D., professor of natural history, etc. on the fishes of New-York.* New York: D. Carlisle.

———. 1815. The fishes of New-York, described and arranged. *Trans. Lit. Phil. Soc. N.Y.* 1:355–492, pls. 1–6.

———. 1818. Description of three species of fish. *J. Acad. Nat. Sci. Philad.* 1 (2): 407–12.

———. 1824. Description of an extraordinary fish, resembling the *Stylephorus* of Shaw. *Ann. Lyceum Nat. Hist. N.Y.* 1:82–86.

Molina, G. I. 1782. *Saggio sulla storia naturale del Chili.* Bologna: Stamperia di S. Tommaso d'Aquino.

———. 1789. *Essai sur l'histoire naturelle du Chili.* Traduit de l'italien, et enrichi de notes, par M. Gruvel, D.M. Paris: Née de la Rochelle.

———. 1810. *Saggio sulla storia naturale del Chili di Gio. Ignazio Molina.* 2d ed. Bologna: Fratelli Masi.

Mondini, C. 1783. De anguillae ovariis. *Comment. Bonon. Sci. Inst. Acad.* (Bologna) 6:406–19, pl. 18.

Monod, T. 1963. Achille Valenciennes et *l'Histoire naturelle des poissons. Mém. Inst. Franç. Afr. Noire* 68:9–45.

Monod, T., and J. C. Hureau. 1978. Essai de bibliographie de Risso. *Ann. Mus. Hist. Nat. Nice* 5:159–63.

Monro, A. 1744. *An essay on comparative anatomy.* London: John Nourse.

Monro, A., II. 1783a. *A treatise on comparative anatomy, by Alexander Monro . . . published by his son, Alexander Monro, M.D.* A new edition: with considerable improvements and additions, by other hands. Edinburgh: C. Elliot and G. Robinson.

———. 1783b. *Observations on the structure and functions of the nervous system.* Edinburgh: William Creech and Joseph Johnson.

———. 1785. *The structure and physiology of fishes explained, and compared with those of man and other animals.* Edinburgh: Charles Elliot.

———. 1787. *Vergleichung des Baues und der Physiologie der Fische mit dem Bau des Menschen und der übrigen Thiere durch Kupfer erläutert von Alexander Monro.* Aus dem Englischen übersezt und mit eignen Zusätzen und Anmerkungen von P. Campern vermehrt durch Johann Gottlob Schneider. Leipzig: Weidmanns Erben und Reich.

———. 1797. Observations on the organ of hearing in man and other animals. In his *Three treatises. On the brain, the eye, and the ear. Illustrated by tables,* 177–263. Edinburgh: Bell and Bradfute.

Montagu, G. 1811. An account of five rare species of British fishes. *Mem. Wernerian Nat. Hist. Soc.* 1 (4): 79–101, pls. 2–5.

Monti, C. 1783. De anguillarum ortu et propagatione. *Comment. Bonon. Sci. Inst. Acad.* (Bologna) 6:392–405.

Morel, F. 1591. *Marcelli Sidetae Medici, De remediis ex piscibus. Fragmentum poematis de re medica, è Biblioth. Reg. Medicaea erutum.* Interprete Fed. Morello Par. Prof. Reg. Paris: Lutetaie, Apud Fed. Morellum Typographum Regium, via Iac. ad Insigne Fõtis.

Mortimer, C. 1752. The description of a fish, *Opah guiniensium,* shewed to the Royal

Society by Mr. Ralph Bigland, on March 22, 1749–50. *Phil. Trans. Roy. Soc. Lond.* 46 (495): 518–20, pl. 4.

Morton, J. 1712. *The natural history of Northampton-shire; with some account of the antiquities, to which is annex'd a transcript of Doomsday-book, so far as it relates to that county.* London: R. Knaplock and R. Wilkin.

Müller, O. F. 1776. *Zoologiae Danicae prodromus, seu Animalium Daniae et Norvegiae indigenarum characteres, nomina, et synonyma imprimis popularium.* Copenhagen: Hallageri.

———. 1777–89. *Zoologiae Danicae, seu Animalium Daniae et Norvegiae rariorum ac minus notorum icones.* Copenhagen: Hallageri. Pt. 1, 1777; pt. 2, 1780; pt. 3, 1789.

Muralt, J. von. 1683a. Examen anatomicum mustelae fluviatilis. *Misc. Curio. Medico-Physica,* decur. 2, 1 (46): 124–28.

———. 1683b. Examen anatomicum truttae magnae. *Misc. Curio. Medico-Physica,* decur. 2, 1 (47): 128–29.

Naccari, F. L. 1822. Ittiologia Adriatica, ossia catalogo de' pesci del golfo e lagune di Venezia. *Gior. Fisic. Chim. Stor. Nat.* (Pavia) decad. 2, 5:327–40, 409–18.

———. 1825. Aggiunta all'ittiologia Adriatica. *Gior. Ital. Lett., Soc. Lett. Ital.* (Padua), May–June 1825, 188–92.

Nardo, G. D. 1824. Osservazioni ed aggiunte all'Adriatica ittiologia pubblicata dal sig. Cav. Fortunato Luigi Naccari presentate dal sig. Domenico Nardo al sig. Giuseppe Cernazai di Udine. *Gior. Fisic. Chim. Stor. Nat.* (Pavia), decad. 2, 7:222–34, 249–63, pl. 1.

———. 1827. Prodromus observationum et disquisitionum ichthyologiae Adriaticae. *Isis* 20 (6): 473–88.

Nau, B. S. 1791. Bemerkungen zu des Herrn Prof. Sanders Beyträgen zur Naturgeschichte der Fische im Rhein. *Naturforscher* (Halle) 25:24–34.

Needham, W. 1668. *Disquisitio anatomica de formato foetu.* Amsterdam: Petrus Le Grand.

Nelson, G. 1978. From Candolle to Croizat: Comments on the history of biogeography. *J. Hist. Biol.* 11 (2): 269–305.

Neucrantz, P. 1654. *De harengo, exercitatio medica, in quâ principis piscium exquisitissima bonitas summaque gloria asserta et vindicata.* Lubeck: Gothofredi Jegeri.

Niebuhr, C. 1772. *Beschreibung von Arabien aus eigenen beobachtungen und im Lande selbst gesammleten Nachrichten abgefasset.* Copenhagen: Nicolaus Möller.

———. 1774–78. *Reisebeschreibung nach Arabien und andern umliegenden Ländern.* Copenhagen: Nicolaus Möller. Vol. 1, 1774; vol. 2, 1778. [For vol. 3, see Niebuhr 1837.]

———. 1837. *Reisen durch Syrien und Palästina, nach Cypern, und durch Kleinastien und die Türkey nach Deutschland und Dännemark. Reisebeschreibung nach Arabien und andern umliegenden Ländern.* Vol. 3. Hamburg: Friedrich Perthes. [For vols. 1–2, see Niebuhr 1774–78.]

Nieremberg, J. E. 1635. *Historia naturae, maxime peregrinae, Libri XVI. Distincta.* Anvers: Balthasaris Moreti.

Nieuhof, J. 1682. *Gedenkwaerdige zee- en lant-reize door de voornaemste landschappen van West en Oostindien. Zee en lant-reize door verscheide gewesten van Oostindien, behelzende veele zeltzaame en wonderlijke voorvallen en geschiedenissen.* 2 pts. in 1. Amsterdam: Jacob van Meurs.

Nissen, C. 1951. *Schöne Fischbuch: Kurze Geschichte der ichthyologischen Illustration.* Stuttgart: Lothar Hempe.

———. 1972. Geschichte der zoologischen Buchillustration. In his *Die zoologisches Buchillustration: Ihre Bibliographie und Geschichte*, vol. 2, pt. 2, pp. 73–144. Stuttgart: Anton Hiersemann.

Nollet, J. A. 1746. Mémoire sur l'ouie des poissons, et sur la transmission des sons dans l'eau. *Hist. Mém. Acad. Roy. Sci.* (Paris) 1743:199–224.

O'Brian, P. 1993. *Joseph Banks: A life.* Boston: David R. Godine.

Oken, L. 1807. *Über die Bedeutung der Schädelknochen: Ein Program beim Antritt der Professur an der Gesammt-Universität zu Jena.* Jena: Johann Christian Gottfried Göpferdt.

———. 1809–11. *Lehrbuch der Naturphilosophie.* Jena: Friedrich Frommann., Pt. 1, 1809; pt. 2, 1810; pt. 3, 1811.

———. 1815–16. *Lehrbuch der Naturgeschichte.* Pt. 3. Zoologie. Jena: August Schmid. Text: vol. 1, 1815; vol. 2, 1816. Atlas, 1816.

———. 1821a. *Naturgeschichte für Schulen.* Leipzig: Brockhaus.

———. 1821b. *Esquisse du système d'anatomie, de physiologie et d'histoire naturelle.* Paris: Béchet Jeune.

Olafsen, E., and B. Povelsen. 1772. *Reise igiennem Island, foranstaltet af Videnskabernes Saelskab i Kiøbenhavn, og beskreven af forbemeldte Eggert Olafsen.* 2 vols. Sorøe: Jonas Lindgrens Enke.

———. 1774–75. *Reise durch Island, veranstaltet von der Königlichen Societät der Wissenschaften in Kopenhagen und beschrieben von bemeldtem Eggert Olafsen.* 2 vols. Copenhagen and Leipzig: Heinecke und Faber.

———. 1802. *Voyage en Island, fait par ordre de S. M. Danoise, contenant des observations sur les moeurs et les usages des habitans; une description des lacs, rivières, glaciers, sources chaudes et voleans; des diverses espèces de terres, pierres, fossiles et pétrifications; des animaux, poissons et insects, etc., etc.; avec un atlas.* Trans. from the Danish by Gauthier de la Peyronie, translator of the voyages of Pallas. 5 vols. Paris and Strasbourg: Levrault.

Olearius, A. 1666. *Gottorffische Kunst-kammer, worinnen allerhand ungemeine sachen, so theils die Nature, theils künstliche hände hervor gebracht und bereitet.* Schlesswig: Johan Holwein.

Oliver, S. P. 1909. *The life of Philibert Commerson, D.M., naturaliste du roi: An old-world story of French travel and science in the days of Linnaeus.* Ed. G. F. S. Elliot. London: Murray.

Oppenheimer, J. M. 1936. Guillaume Rondelet. *Bull. Hist. Med.* 4:817–34.

Osbeck, P. 1757. *Dagbok öfver en Ostindisk resa, åren 1750, 1751, 1752, med anmärkningar uti naturkunnigheten främmande. Folkslags språk, seder, hushållning. m. m. Jåmte 12 tabeller och afledne skepps- predikanten Toréns bref.* Stockholm: Lor. Ludv. Grefing.

———. 1765. *Herrn Peter Osbeck . . . Reise nach Ostindien und China. Nebst O. Toreens reise nach Suratte und C. G. Ekebergs Nachricht von der Landwirthschaft der Chineser.* Trans from the Swedish by J. G. Georgi. Rostock: Johann Christian Koppe.

———. 1770. Fragmenta ichthyologiae Hispanicae, collecta a Dn. Petro Osbeck. *Nova Acta Phys.-Med. Nat. Curio.* 4:99–104.

———. 1771. *A voyage to China and the East Indies, by Peter Osbeck, rector of Hasloef and Woxtorp, member of the Academy of Stockholm, and of the Society of Upsal, together with a voyage to Suratte, by Olof Toreen, chaplain of the Gothic Lion East Indiaman, and an account of the Chinese husbandry, by Captain Charles Gustavus Eckeberg.* Translated from the German by John Reinhold Forster, F.A.S., to which are added a faunal and flora Sinensis. 2 vols. London: Benjamin White.

Otto, A. W. 1821. *Conspectus animalium quorundam maritimorum nondum editorum.* Breslau: Typis Universitatis.

———. 1826. Über die Gehörorgane des *Lepidoloprus trachyrhynchus* und *Caelorrhynchus. Zeitschr. Physiol.* 2 (1): 86–96, pl. 6.

Ozeretskovsky, N. Y. 1801. Observata de salmone salare oceani septentrionalis. *Nova Acta Acad. Sci. Impér. Petro.* 12:337–43.

Paget, S. 1897. *John Hunter, man of science and surgeon (1728–1793).* With introduction by Sir James Paget. London: T. Fischer Unwin.

Pallas, P. S. 1766a. *Elenchus zoophytorum sistens generum adumbrationes generaliores et specierum cognitarum succinctas descriptiones cum selectis auctorum synonymis.* The Hague [Hagae-Comitum]: Petrus van Cleef.

———. 1766b. *Miscellanea zoologica quibus novae imprimis atque obscurae animalium species descributur et observationibus iconibusque illustrantur.* The Hague: Petrus van Cleef.

———. 1767. Some further intelligence relating to the jaculator fish, mentioned in the *Philosophical Transactions* for 1764, art. XIV, from Mr. Hommel, at Batavia, together with the description of another species, by Dr. Pallas, F.R.S. in a letter to Mr. Peter Collinson, F.R.S. from John Albert Schlosser, M.D., F.R.S. *Phil. Trans. Roy. Soc. Lond.*, 56:186–88, pl. 8, fig. 6.

———. 1769. *Spicilegia zoologica, quibus novae imprimis et obscurae animalium species iconibus, descriptionibus atque commentariis illustrantur cura P. S. Pallas.* Pt. 7. Berlin: Lange.

———. 1770. *Spicilegia zoologica, quibus novae imprimis et obscurae animalium species iconibus, descriptionibus atque commentariis illustrantur cura P. S. Pallas.* Pt. 8. Berlin: Lange.

———. 1771–76. *Reise durch verschiedene Provinzen des russischen Reichs.* St. Petersburg: Academie der Wissenschaften. Vol. 1, 1771; vol. 2, 1773; vol. 3, 1776.

———. 1784. *Samuel Gottlieb Gmelin's Reise durch Russland zur Untersuchung der drey Natur-Reiche.* Vol. 4. St. Petersburg: Academie der Wissenschaften. [For vols. 1–3, see S. G. Gmelin 1770–74.]

———. 1787–91. *D. Johann Anton Güldenstädt . . . Reisen durch Russland und im caucasischen gebürge.* 2 vols. St. Petersburg: Academie der Wissenschaften.

———. 1810. Labraces, novum genus piscium, oceani orientalis. *Mém. Acad. Impér. Sci. St. Pétersb.* 2:382–98, 22–23 pls.

———. 1811–14. *Zoographia Rosso-Asiatica, sistens omnium animalium in extenso imperio Rossico et adjacentibus maribus observatorum recensionem, domicilia, mores et descriptiones, anatomen atque icones plurimorum.* St. Petersburg: Officina Caes. Academie Scientiarum Impress. Vol. 1, 1811; vol. 2, 1811; vol. 3, 1814. [Dating of volumes after Sherborn 1934.]

Parkinson, S. 1773. *A journal of a voyage to the South Seas, in His Majesty's ship, the Endeavour.* London: Stanfield Parkinson.

Parra, A. 1787. *Descripcion de diferentes piezas de historia natural las mas del ramo maritimo, representadas en setenta y cinco laminas.* Havana: Imprenta de la Capitania General. [Reprint 1989, Havana: Editorial Academia.]

Parsons, J. 1752. Some account of the *Rana piscatrix. Phil. Trans. Roy. Soc. Lond.* 46 (492): 126–31, pl. 3.

Passalacqua, J. 1826. *Catalogue raisonné et historique des antiquités découvertes en Egypte, par M. Joseph Passalacqua, de Trieste.* Paris: Galerie d'Antiquités Egyptiennes.

Paterson, W. 1786. An account of a new electrical fish. In a letter from Lieutenant William Paterson to Sir Joseph Banks, Bart. P.R.S. *Phil. Trans. Roy. Soc. Lond.* 76:382–83, pl. 13.

———. 1787. Traduction de la lettre du Lieutenant William Paterson, à Sir J. M. Joseph Bancks, président de la Société Royale, contenant la description d'un nouveau poisson électrique. *J. Physique* (Paris) 30:196–97.

Paula Schrank, F. von. 1780. Auszug eines Briefes des Herrn Franz von Paula Schrank an die Gesellschaft. *Schr. Berl. Ges. Naturf. Fr.* 1:379–82.

———. 1781. Beytrag zur Naturgeschichte des *Salmo alpinus* Lin. *Schr. Berl. Ges. Naturf. Fr.* 2:297–306.

———. 1783. An den Herrn D. Bloch von Herrn Professor von Paula Schrank zu Burghausen vom 30. November 1782 *[Salmo hucho]. Schr. Berl. Ges. Naturf. Fr.* 4:427–31.

———. 1786. *Baiersche Reise.* Munich: Johann Baptist Stobl.

Pennant, T. 1768–70. *British zoology.* London and Chester: Benjamin White. Vol. 1, Quadrupeds, birds, 1768; vol. 2, Birds, with an essay on birds of passage, 1768; vol. 3, Reptiles, fishes, 1769; vol. 4, Plates and brief explanations, 1770.

———. 1769. *Indian zoology.* London.

———. 1776–77. *British zoology.* London and Warrington: William Eyres for Benjamin White, Vol. 1, Quadrupeds, 1776; vol. 2, Birds, 1776; vol. 3, Reptiles, fishes, 1776; vol. 4, Crustacea, Mollusca, Testacea, 1777.

———. 1784–87. *Arctic zoology.* London: Henry Hughs. Vol. 1, 1784; vol. 2, 1785; supplement, 1787.

———. 1790. *Indian zoology.* 2d ed. London: Printed by Henry Hughs for Robert Faulder.

Pernety, A. J. 1769. *Journal historique d'un voyage fait aux îles Malouïnes en 1763 et 1764, pour les reconnoître, et y former un establissement; et de deux voyages au Détroit de Magellan, avec une relation sur les Patagons.* 2 vols. Berlin: Etienne de Bourdeaux.

Péron, F. 1807–16. *Voyage de découverte aux terres australes, exécuté par ordre de Sa Majesté l'empereur et roi, sur les corvettes le Géographe, le Naturaliste, et la goélette le Casuarina, pendant les années 1800, 1801, 1802, 1803 et 1804; sous le commandement du Capitaine de Vaisseau N. Baudin.* Paris: Imprimerie Impériale. Text: vol. 1, 1807; vol. 2, 1816. Atlas: vol. 1, by C. A. Lesueur and F. P. du Petit, 1807; vol. 2, by L. Freycinet, 1811.

Perrault, C. 1733. Description anatomique d'un renard marin. In C. Perrault 1733–1734, vol. 3, pt. 1, pp. 117–24, pls. 15–16.

———. 1733–34. Mémoires pour servir à l'histoire naturelle des animaux. *Hist. Mém. Acad. Roy. Sci.* (Paris) 1699, vol. 3: pt. 1, 1733; pt. 2, 1733; pt. 3, 1734.

Petit, F. P. du. 1728. Mémoire sur plusieurs découvertes faites dans les yeux de l'homme, des animaux à quatre pieds, des oiseaux et des poissons. *Hist. Mém. Acad. Roy. Sci.* (Paris) 1726:69–83.

———. 1732. Mémoire sur le cristallin de l'oeil de l'homme, des animaux à quatre pieds, des oiseaux et des poissons. *Hist. Mém. Acad. Roy. Sci.* (Paris) 1730:4–26.

Petiver, J. 1764. *Jacobi Petiveri opera, historiam naturalem spectantia, or Gazophylacium, containing several thousand figures of birds, beasts, reptiles, insects, fish, beetles, moths, flies, shells, corals, fossils, minerals, stones, funguses, mosses, herbs, plants, etc. from all nations, on 156 copper-plates, with Latin and English names.* 3 vols. London: John Millan. [Reprints of works of Petiver bound together and issued under a common title page; paging highly variable.]

Petrie, W. M. F. 1892. *Medum.* London: David Nutt.

Peyer, J. C. 1683. Lepusculis, Salmonum extis, intestino coeco, ventriculo anserino, et renibus anserinis. *Misc. Curio. Medico-Physica,* decur. 2, 1 (85): 199–205.

Philes, M. 1730. *De animalium proprietate, ex prima editione Arsenii et libro Oxoniensis restitutus a Joanne Cornelio de Pauw.* Trajecti ad Rhenum: Guilielmus Stouw.

Pies, E. 1981. *Willem Piso (1611–1678). Begründer der kolonialen Medizin und Leibarzt des Grafen Johann Moritz von Nassau-Siegen in Brasilien. Eine Biographie.* Düsseldorf: Interma-Orb.

Pieters, F. F. J. M. 1980. Notes on the menagerie and zoological cabinet of Stadholder William V of Holland, directed by Aernout Vosmaer. *J. Soc. Bibliogr. Nat. Hist.* 9 (4): 539–63.

Pietsch, T. W. 1984. Louis Renard's fanciful fishes. *Nat. Hist.* (New York) 93 (1): 58–67.

———. 1985. The manuscript materials for the *Histoire naturelle des poissons,* 1828–1849: Sources for understanding the fishes described by Cuvier and Valenciennes. *Arch. Nat. Hist.* 12 (1): 59–106.

———. 1986. Fallours, S. Fishes of the Indo-West Pacific: A collection of hand-coloured drawings. *Natural history and travel,* Catalogue 241, Antiquariaat Junk, Amsterdam, item 74, pp. 36–39.

———. 1991. Samuel Fallours and his "sirenne" from the province of Ambon. *Arch. Nat. Hist.* 18 (1): 1–25.

———. 1993. On the three editions of Louis Renard's *Poissons, écrevisses et crabes: Histoire naturelle des plus rares curiositez de la mer des Indes. Arch. Nat. Hist.* 20 (1): 49–68.

———. 1995. *Fishes, crayfishes, and crabs: Louis Renard's natural history of the rarest curiosities of the seas of the Indies.* 2 vols. Baltimore: Johns Hopkins University Press.

Pietsch, T. W., and L. B. Holthuis. 1992. Lamotius and his marine marvels. *Nat. Hist.* (New York) 101 (10): 34–39.

Pietsch, T. W., and D. M. Rubiano. 1988. On the date of publication of the first edition of Louis Renard's *Poissons, écrevisses et crabes: Histoire naturelle des plus rares Curiositez de la mer des Indes. Arch. Nat. Hist.* 15 (1): 63–71.

Piso, G. 1648. De medicina Brasiliensi libri quatuor: I. De aëre, aquis, et locis, II. De morbis endemiis, III. De venenatis et antidotis, IV. De facultatibus simplicium. In *Historia naturalis Brasiliae auspicio et beneficio illustriss. I. Mauritii Com. Nassau illius provinciae et maris summi praefecti adornata. In qua non tantum plantae et animalia, sed et indigenarum morbi, ingenia et mores describuntur et iconibus supra quingentas illustratur,* by G. Piso and G. Marcgrave, pt. 1. Leiden and Amsterdam: Haack en Elsevier.

———. 1658. *De Indiae utriusque re naturali et medica libri quatuordecim, quorum contenta pagina sequens exhibet.* Amsterdam: Ludovic en Daniel Elsevier. [Followed by G. Marcgrave's *Tractatus topographicus,* J. Bontius's *Historiae naturalis et medicae Indiae orientalis,* and G. Piso's *Mantissima aromatica.*]

Plot, R. 1677. *The natural history of Oxford-shire, being an essay toward the natural history of England.* Oxford: At the Theater.

Plumier, C. 1693. *Description des plantes de l'Amérique.* Paris: Imprimerie Royale par les soins de Jean Anisson.

———. 1703. *Nova plantarum Americanarum genera, authore P. Carolo Plumier ordinis Minimorum in provincia Franciae, et apud insulas Americanas botanico regio.* Paris: Joannes Boudot.

———. 1705. *Traité des fougères de l'Amérique. Par le R. P. Charles Plumier, Minime de la province de France, et botaniste du roy dans les isles de l'Amérique.* Paris: Imprimerie Royale, curante Joanne Anisson.

———. 1755–60. *Plantarum Americanarum fasciculus primus, continens plantas, quas olim Carolus Plumierius, botanicorum princeps detexit, eruitque, atque in insulis Antillis ipse depinxit.* Has primum in lucem edidit, concinnis descriptionibus, et observationibus, aeneisque tabulis illustravit Johannes Burmannus, M.D. Amsterdam: Gerard Potvliet en Theodor Haak. Pt. 1, 1755; pt. 2, 1756; pt. 3, 1756; pt. 4, 1756; pt. 5, 1757; pt. 6, 1757; pt. 7, 1758; pt. 8, 1758; pt. 9, 1759; pt. 10, 1760.

Poey y Aloy, F. 1863. Enumeration of the fish described and figured by Parra, scientifically named by Felipe Poey. *Proc. Acad. Nat. Sci. Philad.* ser. 2, 15:174–80.

———. 1865–68. *Repertorio fisico-natural de la isla de Cuba.* Havana: Impr. del Gobierno y Capitania General. Vol. 1, 1865–66; vol. 2, 1866–68.

Pohl, C. E. 1818. *Expositio generalis anatomica organi auditus per classes animalium, accedunt quinque tabulae lithographicae.* Vienna: C. Schaumburg.

Poinsinet de Sivry, L. 1771–82. *Histoire naturelle de Pliny traduite en françois, avec le texte latin rétabli d'après les meilleures leçons manuscrites.* Paris: Veuve Desaint. Vol. 1, 1771; vol. 2, 1771; vol. 3, 1771; vol. 4, 1772; vol. 5, 1772; vol. 6, 1773; vol. 7, 1774; vol. 8, 1776; vol. 9, 1777; vol. 10, 1778; vol. 11, 1778; vol. 12, 1782.

Pontoppidan, E. L. 1752–53. *Det forste forsog paa Norges naturlige historie.* 2 vols. Copenhagen: Rudolph Henrich Rissie.

———. 1753–54. *Versuch einer natürlichen histoire von Norwegen.* 2 vols. Copenhagen: Franz Christian Mumme.

———. 1755. *The natural history of Norway.* 2 pts. in 1. London: A. Linde.

———. 1769. *Versuch einer natürlichen historie Norwegen.* 2 pts. in 1. Flensburg and Leipzig: Johann Christoph Korte.

———. 1977. *Norges naturlige historie.* Facsimile edition. 2 vols. Copenhagen: Rosenkilde and Bagger.

Prévost, A. F. 1768. Voyage au Kamtschatka par la Sibérie. Journal de M. Gmelin, traduit de l'Allemand. In his *Histoire générale des voyages, ou Collection nouvelle de toutes les relations de voyages par mer et par terre qui ont été publiées jusqu'à présent dans les différentes langues de toutes les nations connues,* 18:71–483. Paris: Panckoucke.

———. 1770. Des poissons. In his *Histoire générale des voyages, ou Collection nouvelle de toutes les relations de voyages par mer et par terre qui ont été publiées jusqu'à présent dans les différentes langues de toutes les nations connues,* 19:50–64. Paris: Panckoucke.

Provençal, [J.M.?], and F. H. A. von Humboldt. 1809. Recherches sur la respiration des poissons. *J. Physique* (Paris) 69:261–86.

Quoy, J. R. C., and J. P. Gaimard. 1824. Description des poissons. In *Voyage autour du monde, entrepris par ordre du roi . . . exécuté sur les corvettes de S. M. l'Uranie et la Physicienne, pendant les années 1817, 1818, 1819 et 1820,* by L. de Freycinet, vol. 3, chap. 9, pp. 192–401, pls. 43–65. Paris: Pillet Aîné.

Rackham, H. 1938. *Pliny, Natural history, with an English translation in ten volumes.* Vol. 1. Cambridge: Harvard University Press.

———. 1940. *Pliny, Natural history, with an English translation in ten volumes.* Vol. 3. Cambridge: Harvard University Press.

Rafinesque-Schmaltz, C. S. 1810a. *Caratteri di alcuni nuovi generi e nuove specie di animali e piante della Sicilia, con varie osservazioni sopra i medesimi.* Palermo: Stampe di Sanfilippo.

———. 1810b. *Indice d'ittiologia Siciliana, ossia catalogo metodico dei nomi Latini, Italiani, e*

Siciliani dei pesci, che si rivengono in Sicilia, disposti secondo un metodo naturale, eseguito da un appendice che contiene la descrizione di alcuni nuovi pesci Siciliani, illustrato da due piance. Messina: Giovanni del Nobolo.

———. 1815. *Analyse de la nature, ou Tableau de l'univers et des corps organisés.* Palermo: Aux dépens de l'auteur.

———. 1818. Description of three new genera of fluviatile fish, *Pomoxis, Sarchirus* and *Exoglossum. J. Acad. Nat. Sci. Philad.* 1 (1): 417–22, pl. 17.

———. 1819. Prodrome de 70 nouveaux genres d'animaux découverts dan l'intérieur des Etats-Unis d'Amérique, durant l'année 1818. *J. Physique* (Paris) 88:417–29.

———. 1820a. *Annals of nature, or Annual synopsis of new genera and species of animals, plants, etc., discovered in North America.* By C. S. Rafinesque, professor of botany and natural history in Transylvania University, at Lexington in Kentucky, and member of several learned societies in the United States and in Europe, &c. First annual number, for 1820. Dedicated to Dr. W. E. Leach, of the British Museum, London. Lexington, Ky.: Thomas Smith.

———. 1820b. *Ichthyologia Ohiensis, or Natural history of the fishes inhabiting the river Ohio and its tributary streams, preceded by a physical description of the Ohio and its branches.* Lexington, Ky.: For the author by W. G. Hunt.

Ranzani, C. 1818. Descrizione di un pesce il quale appartiene ad un nuovo genere della famiglia dei tenioidi del Signor G. Cuvier. *Opusc. Sci. Bologna* 2:133–37, pl. 6.

Rathke, M. H. 1822. Bemerkungen über den Bau des *Cyclopterus lumpus* (Lumpfisches, Seehafen). *Deut. Arch. Physiol.* (Meckel) 7 (4): 498–524.

———. 1824. Über den Darmkanal und die Zeugungsorgane der Fische. *Neue. Schr. Naturf. Gesell. Danzig* (Halle) 1 (3): i-vi, 1–211, pls. 1–5.

———. 1825a. Beiträge zur Geschichte der Thierwelt. Pt. 1. Über die Entwickelung der Geschlechtstheile bei den Fische. *Neue. Schr. Naturf. Gesell. Danzig* (Halle) 1 (4): 1–18, pl. 1, figs. 1–7.

———. 1825b. *Bemerkungen über den innern Bau der Pricke oder des Petromyzon fluviatilis des Linnaeus.* Danzig: Wilhelm Theodor Lohde.

———. 1826. Ueber die Herzkammer der Fische. *Arch. Anat. Physiol.* (Meckel) 9 (1): 152–57.

Raven, C. E. 1986. *John Ray: Naturalist.* Cambridge: Cambridge University Press. [Reprint of 2d ed. (1950), with an introduction by S. M. Walters.]

Ray, J. 1686. Praefatio ad lectorem. [Unpaged preface.] In *De historia piscium libri quatuor, jussu et sumptibus Societatis Regiae Londinensis editi. . . . Totum opus recognovit, coaptavit, supplevit, librum etiam primum et secundum integros adjecit Johannes Raius e Societate Regia,* by F. Willughby. Oxford: Theatro Sheldoniano.

———. 1713. *Synopsis methodica piscium.* London: W. Innys.

———. 1978. *Synopsis methodica avium et piscium.* Facsimile edition, edited by W. Derham. New York: Arno Press.

Réaumur, R. A. F. de. 1717. Des effets que produit le poisson appellé en François torpille, ou tremble, sur ceux qui le touchent; et de la cause dont ils dépendent. *Hist. Mém. Acad. Roy. Sci.* (Paris) 1714:344–60, pls. 12–13.

———. 1718. Observations sur la matière qui colore les perles fausses, et sur quelques autres matières animales d'une semblable couleur; à l'occasion de quoi on essaye d'expliquer la formation des écailles des poissons. *Hist. Mém. Acad. Roy. Sci.* (Paris) 1716:229–44.

Recchi, N. A. 1651. *Rerum medicarum Novae Hispaniae thesaurus, seu Plantarum, ani-*

malium, mineralium Mexicanorum historia, ex Francisci Hernandez. Rome: Vitalis Mascardi.

Redouté, P. J. 1812. *Description de l'Egypte, ou Recueil des observations et des recherches qui ont été faites en Egypte pendant l'expédition de l'armée française, publié par ordre de Sa Majesté l'empereur Napoléon le Grand. Antiquités, planches.* Vol. 2. Paris: Imprimerie Impériale, pls. 44–92.

Renard, L. [1719]. *Poissons, écrevisses et crabes de diverses couleurs et figures extraordinaires, que l'on trouve autour des Isles Moluques, et sur les côtes des Terres Australes: Peints d'après nature durant la regence de Messieurs Van Oudshoorn, Van Hoorn, Van Ribeek et Van Zwoll, successivement gouverneurs-généraux des Indes Orientales pour la Compagnie de Hollande. Ouvrage, auquel on a employé près de trente ans, et qui contient un très-grand nombre de poissons les plus beaux et les plus rares de la Mer des Indes: Divisé en deux tomes, dont le premier a été copié sur les originaux de Monsr. Baltazar Coyett, ancien gouverneur des isles de la province d'Amboine, présentement directeur desdites isles et président des commissaires à Batavia; le second tome a été formé sur les recueils de Monsr. Adrien van der Stell, gouverneur regent de ladite province d'Amboine, avec une courte description de chaque poisson. Le tout muni de certificats et attestations authentiques. Histoire des plus rares curiositez de la Mer des Indes.* 2 vols. in 1. Amsterdam: Louis Renard.

———. 1754. *Poissons, écrevisses et crabes, de diverses couleurs et figures extraordinaires, que l'on trouve autour des Isles Moluques, et sur les côtes des Terres Australes: peints d'après nature durant la régence de Messieurs Van Oudshoorn, Van Hoorn, Van Ribeek et Van Zwoll, successivement gouverneurs-généraux des Indes Orientales pour la Compagnie de Hollande. Ouvrage, auquel on a employé près de trente ans, et qui contient un très-grand nombre de poissons les plus beaux et les plus rares de la Mer des Indes: Divisé en deux tomes, dont le premier a été copié sur les originaux de Monsr. Baltazar Coyett, ancien gouverneur et directeur des isles de la province d'Amboine, et président des commissaires à Batavia. Le second tome a été formé sur les recueils de Monsr. Adrien van der Stell, gouverneur regent de la dite province d'Amboine, avec une courte description de chaque poisson. Le tout muni de certificats et attestations authentiques. Donné au public par Mr. Louis Renard, agent de S. M. Brit. à Amsterdam, et augmenté d'une préface par Mr. Arnout Vosmaer.* 2 vols. in 1. Amsterdam: Reinier et Josué Ottens.

Rhanaeus, S. J. 1725. Von einigen merckwürdigen Fischen Curland. IV. Die unvermuthete Karpffen. *Samml. Nat. Med.* (Leipzig), February 1725, 177–78.

Rhodius, J. 1661. Mantissa anatomica. In *Historiarum anatomicarum et medicarum rariorum centuriae V. et VI., accessit viri clarissimi Joannis Rhodii mantissa anatomica,* by T. Bartholin. Copenhagen: Petrus Hauboldus.

Richardson, J. 1823. Notices of the fishes. In *Narrative of a journey to the shores of the Polar Sea, in the years 1819, 20, 21, and 22,* by J. Franklin, app. 6, pp. 705–28, pls. 25–26. London: John Murray.

Richer, J. 1679. *Observations astronomiques et physiques faites en l'isle de Caienne.* Paris: Sebastien Mabre-Cramoisy, Imprimerie Royal.

———. 1729. Observations astronomiques et physiques faites en l'isle de Caienne. *Mém. Acad. Roy. Sci.* (Paris) 7 (1): 233–326.

Richmond, J. A. 1962. *The Halieutica, ascribed to Ovid.* London: Athlone Press, University of London.

Risso, A. 1810. *Ichthyologie de Nice, ou Histoire naturelle des poissons du département des Alpes Maritimes.* Paris: F. Schoell.

———. 1820a. Mémoire sur quelques poissons nouveaux observés dans la mer de Nice. *J. Phys. Chimie Hist. Nat.* 91:241–55.

———. 1820b. Mémoire sur deux nouvelles espèces de poissons du genre Scopèles observées dans la mer de Nice. *Mem. R. Accad. Sci. Torino* 25:262–69, pl. 10, figs. 1–3.

———. 1820c. Mémoire sur un nouveau genre de poisson nommé alépocéphale vivant dans les grandes profondeurs de la mer de Nice. *Mem. R. Accad. Sci. Torino* 25:270–72, pl. 10, fig. 4.

———. 1827a. *Histoire naturelle des principales productions de l'Europe méridionale et particulièrement de celles des environs de Nice et des Alpes Maritimes.* 5 vols. Paris: Levrault.

———. 1827b. Histoire naturelle des poissons de la Méditerranée qui fréquentent les côtes des Alpes Maritimes, et qui vivent dans la golfe de Nice. In his *Histoire naturelle des principales productions de l'Europe méridionale et particulièrement de celles des environs de Nice et des Alpes Maritimes,* 3:97–480, 16 pls. Paris: Levrault.

Rivinus, A. Q. 1687. Observatio anatomica circa poros in piscium cute notandos. *Acta Eruditorum* (Leipzig) 1687: 160–62, pl. 3.

Roberts, T. 1993. The freshwater fishes of Java, as observed by Kuhl and van Hasselt in 1820–23. *Zool. Verh.* (Leiden) 285:1–94.

Rochefort, C. de. 1658. *Histoire naturelle et morale des îles Antilles de l'Amérique. Enrichie de plusieurs belles figures des raretez les plus considérables qui y sont décrites. Avec un vocabulaire Caraïbe.* Rotterdam: Arnould Leers.

Rondelet, G. 1554. *Libri de piscibus marinis, in quibus verae piscium effigies expressae sunt.* Lyons: Matthias Bonhomme. [Followed by G. Rondelet, *Universae aquatilium historiae pars altera, cum veris ipsorum imaginibus,* 1555.]

———. 1555. *Universae aquatilium historiae pars altera, cum veris ipsorum imaginibus.* Lyons: Matthias Bonhomme. [Preceded by G. Rondelet, *Libri de piscibus marinis, in quibus verae piscium effigies expressae sunt,* 1554.]

———. 1558. *L'Histoire entière des poissons, composée premièrement en latin par maistre Guilaume Rondelet docteur regent en medecine en l'Université de Mompelier.* Maintenant traduite en françois sans avoir rien omis estant necessaire à l'intelligence d'icelle. Avec leurs pourtraits au naïf. 2 pts. in 1. Lyons: Matthias Bonhomme.

Rookmaaker, L. C. 1989. *The zoological exploration of southern Africa, 1650–1790.* Rotterdam: A. A. Balkema.

———. 1992. J. N. S. Allamand's additions (1769–1781) to the *Nouvelle edition* of Buffon's *Histoire naturelle* published in Holland. *Bijdr. Dierkunde* 61 (3): 131–62.

Rosenman, H. 1992. *Two voyages to the South Seas: Captain Jules S.-C. Dumont d'Urville.* Translated from the French and retold by Helen Rosenman. Honolulu: University of Hawaii Press.

Rosenthal, F. C. 1811. Ueber das Skelett der Fische. *Arch. Physiol.* (Reil) 10 (2): 340–58.

———. 1812–22. *Ichthyotomical Tafeln.* 4 pts. in 1. Berlin: Realschulbuchhandlung. Pt. 1, 1812; pt. 2, 1816; pt. 3, 1821; pt. 4, 1822.

Rousseau, G. S. 1970. The much-maligned doctor "Sir" John Hill (1707–1775). *J. Amer. Med. Assoc.* 212 (1): 103–8.

———. 1978. John Hill, universal genius manqué: Remarks on his life and times, with a check-list of his works. In *The Renaissance man in the eighteenth century,* by J. A. L. Lemay and G. S. Rousseau, 45–129. Berkeley and Los Angeles: University of California Press.

Roux, C. 1976. On the dating of the first edition of Cuvier's *Règne animal. J. Soc. Bibliogr. Nat. Hist.* 8 (1): 31.

Roux, J. L. F. P. 1825–30. *Ornithologie provençale, ou Description avec figures coloriées de tous les oiseaux qui habitent constamment la Provence, ou qui n'y sont que de passage; suivie d'un abrégé des chasses, de quelques instructions de taxidermie, et d'une table des noms vulgaires.* 2 vols. Paris and Marseille: Levrault, Treuttel et Wurtz, Dufour et d'Occagne, et l'auteur.

Rudolphi, K. A. 1812. *Peter Simon Pallas. Ein biographischer Versuch, vorgelesen in der öffentlichen Sitzung der Königl. Akademie der Wissenschaften den 30ten Januar 1812.* Berlin: Beyträge zur Antropologie und Algemeinen Naturgeschichte.

———. 1826. Ueber den Zitterwels. In his Anatomische Bemerkungen, *Abh. Akad. Wiss.* (Berlin), 1824, pt. 2, pp. 137–44.

Russell, A. 1756. *The natural history of Aleppo, and parts adjacent, containing a description of the city, and the principal natural productions in its neighbourhood; together with an account of the climate, inhabitants, and diseases; particularly of the plague, with the methods used by the Europeans for their preservation.* London: A. Millar.

Russell, P. 1803. *Descriptions and figures of two hundred fishes; collected at Vizagapatam on the coast of Coromandel.* 2 vols. London: W. Bulmer.

Ruysch, H. 1718. Collectio nova piscium Amboinensium partim ibi ad vivum delineatorum partim ex museo Henrici Ruysch M.D. XX. tabulis comprehensa. In *Theatrum universale omnium animalium piscium, avium, quadrupedum, exanguium, aquaticorum, insectorum, et angium, CCLX. Tabulis ornatum, ex scriptoribus tam antiquis quam recentioribus, Aristotele, Theophrasto, Dioscoride, Aeliano, Oppiano, Plinio, Gesnero, Aldrovando, Wottonio, Turnero, Mouffeto, Agricola, Boetio, Baccio, Ruveo, Schonfeldio, Freygio, Mathiolo, Tabernomontano, Bauhino, Ximene, Bustamantio, Rondeletio, Bellonio, Caesio, Theveto, Margravio, Pisone, et aliis maxima curâ à J. Jonstonio collectum, ac plus quam trecentis piscibus nuperrime ex Indiis orientalibus allatis, ac nunquam antea his terris visis, locupletatum; cum enumeratione morborum, quibus medicamina ex his animalibus petuntur, ac notitiâ animalium, ex quibus vicissim remedia praestantissima possunt capi. VI. Partibus, duobus tomis, comprehensum,* by J. Jonstonus, 1:1–40, pls. 1–20. Amsterdam: R. en G. Wetstein.

Rychkov, N. P. 1770–1772. *[Journal of travels through the different provinces of the Russian empire in the years 1769 and 1770.]* 3 pts. in 1. St. Petersburg: Academie der Wissenschaften. Pt. 1, 1770; pt. 2, 1772; pt. 3, 1772. [In Russian.]

———. 1774. *Tagebuch über seine Reise durch verschiedene Provinzen des russischen Reichs in den Jahren 1769, 1770, und 1771.* Trans. from the Russian by Christan Heinrich Hase. Riga: Johann Friedrich Hartknoch.

Salviani, H. 1554–58. *Aquatilium animalium historiae, liber primus, cum eorumdem formis, aere excusis.* Rome: Hippolyto Salviano.

Sander, H. 1779. Von einem merkwürdigen See in der obern Markgrafschaft Baaden. *Beschäft. Berl. Ges. Naturf. Fr.* 4:619–23.

———. 1781. Beyträgen zur Naturgeschichte der Fische im Rhein. *Naturforscher* (Halle) 15:163–83.

Savage, J. J. 1961. *Saint Ambrose, Hexameron, Paradise, and Cain and Abel.* New York: Fathers of the Church.

Scaliger, J. C. 1619. *Aristotelis Historia de animalibus, Julio Caesare Scaligero interprete, cum eiusdem commentariis.* Toulouse: Dominicus et Petrus Bosc.

Scarpa, A. 1789. *Anatomicae disquisitiones de auditu et olfactu.* Ticini: Petrus Galeatius.

Schaeffer, J. C. 1761. *Piscium Bavarico-Ratisbonensium pentas, cum tabulis IV. aeri incisis icones coloribus suis distinctas exhibentibus.* Ratisbon: Impensis Montagii et Typis Weissianis.

Schelhammer, G. C. 1707. *Anatomes xiphiae piscis oceani incolae cultro anatomico, A.MDCCIV. Accedit lumpi et ophidii ejusdem generis examen.* Hamburg: Reumannianis.

Scheltema, P. 1886. *Het leven van Frederik Ruijsch. Academisch proefschrift ter verkrijging van den graad van doctor in de geneeskunde, aan de Rijksuniversiteit te Leiden, op gezag van den rector magnificus Dr. J. P. N. Land, hoogleeraar in de faculteit der letteren en Wijsbegeerte, voor de faculteit te verdedigen op Vrijday 16 April 1886, des Namiddags te 3 Uren, door Pieter Scheltema, geboren te Gouda, arts te Sliedrecht.* Leiden: Sliedrecht, Gebroeders Luijt.

Schierbeek, A. 1967. *Jan Swammerdam (12 February 1637—17 February 1680), his life and works.* Amsterdam: Swets en Zeitlinger.

Schilder, G. G. 1976. *De ontdekkingsreis van Willem Hesselsz. de Vlamingh in de jaren 1696–1697.* The Hague: Nijhoff, Vol. 1, met Inleiding, Journaal en Bijlagen; vol. 2, Bijlagen.

———. 1985. *Voyage to the Great South Land: Willem de Vlamingh, 1696–1697.* Trans. by C. de Heer. Sydney: Royal Australian Historical Society.

Schlosser, J. A. 1765. An account of a fish from Batavia, called jaculator: In a letter to Mr. Peter Collinson, F.R.S. from John Albert Schlosser, M.D., F.R.S. *Phil. Trans. Roy. Soc. Lond.* 54:89–91, pl. 9.

———. 1767. Some further intelligence relating to the jaculator fish, mentioned in the *Philosophical Transactions* for 1764, art. XIV, from Mr. Hommel, at Batavia, together with the description of another species, by Dr. Pallas, F.R.S. in a letter to Mr. Peter Collinson, F.R.S. from John Albert Schlosser, M.D., F.R.S. *Phil. Trans. Roy. Soc. Lond.* 56:186–88, pl. 8, fig. 6.

Schneider, J. G. 1776. *Oppiani poete cilicis, De venatione libri IV et De piscatione libri V, cum paraphrasi Graeca librorum de Aucupio.* Strasbourg: Amandi König.

———. 1784. *Aeliani, De natura animalium libri XVII.* Leipzig: E. B. Schwickerti.

———. 1786. Nachricht von den Originalzeichnungen von Marcgrafs brasilischer Zoologie. *Mag. Naturk. Oecon.* (Leipzig) 3 (2): 270–78.

———. 1811. *Aristotelis De animalibus historiae libri X.* 4 vols. Leipzig: Bibliopolio Hahniano.

———. 1816. *Nicandri Colophonii Theriaca, id est de bestiarum venenis eorumque remediis carmen, cum scholiis Graecis auctioribus eutecnii metaphrasi Graeca editoris Latina. . . .* Leipzig: G. Fleischer.

Schoepff, J. D. 1784a. Der nord-amerikanische Pertsch. *Naturforscher* (Halle) 20:17–25.

———. 1784b. Der gemeine Hecht in Amerika. *Naturforscher* (Halle) 20:26–31.

———. 1788. Beschreibungen einiger nord-amerikanischer Fische, vorzüglich aus den neu-yorkischen Gewässern. *Schr. Berl. Ges. Naturf. Fr.* 8 (15): 138–94.

Scholfield, A. F. 1958. *Aelian on the characters of animals.* With an English translation by A. F. Scholfield. 3 vols. Cambridge: Harvard University Press.

Schonevelde, S. von 1624. *Ichthyologia et nomenclaturae animalium marinorum, fluviatilium, lacustrium, quae in Florentissimis ducatibus Slesvici et Holsatiae et celeberrimo Emporio Hamburgo occurrunt triviales.* Hamburg: Bibliopolio Heringiano.

Schultze, C. A. S. 1818. Ueber die ersten Spuren des Knochensystems und die Entwickelung der Wirbelsäule in den Thieren. *Deut. Arch. Physiol.* (Meckel) 4 (3): 329–402, pl. 4.

Schweighaeuser, J. 1801–7. *Athenaei Naucratitae Deipnosophistarum libri quindecim.* Strasbourg: Societatis Bipontinae. Vol. 1, 1801; vol. 2, 1802; vol. 3, 1803; vol. 4,

1804; vol. 5, 1805. Animadversiones: Vol. 1, 1801; vol. 2, 1802; vol. 3, 1802; vol. 4, 1803; vol. 5, 1804; vol. 6, 1804; vol. 7, 1805; vol. 8, 1805; vol. 9, 1807.

Schwenckfelt, C. 1603. *Theriotropheum Silesiae, in quo animalium, hoc est, quadrupedum, reptilium, avium, piscium, insectorum natura, vis et usus sex libris perstringuntur.* Lignitz: David Albert. [Preceded by C. Schwenckfelt, 1600, *Stirpium et fossilium Silesiae catalogus.* Leipzig: David Albert.]

Scilla, A. 1670. *La vana speculazione disingannata dal senso.* Naples: Andrea Colicchia.

———. 1752. *De corporibus marinus lapidescentibus quae defossa reperiuntur auctore Augustino Scilla, addita dissertatione Fabii Columnae De glossopetris edition altera emendatior.* Rome: Venantius Monaldinus.

Sclatter, W. L. 1947. On the date as from which the names published in Pallas (P. S.) *Zoographia Rosso-Asiatica* are available nomenclaturally. *Bull. Zool. Nomencl.* 1:198–99.

Scopoli, G. A. 1777. *Introductio ad historiam naturalem, sistens genera lapidum, plantarum et animalium hactenus detecta, caracteribus essentialibus donata, in tribus divisa, subinde ad leges naturae.* Prague: W. Gerle.

———. 1786–88. *Deliciae florae et faunae insubricae seu novae, aut minus cognitae species plantarum et animalium quas in insubria Austriaca.* Ticini: Monasterii S. Salvatoris. Pt. 1, 1786; pt. 2, 1786; pt. 3, 1788.

Seba, A. 1734–65. *Locupletissimi rerum naturalium thesauri accurata descriptio, et iconibus artificiosissimis expressio, per universam physices historiam. Opus, cui, in hoc rerum genere, nullum par exstitit. Ex toto terrarum orbe collegit, digessit, descripsit, et depingendum curavit.* Amsterdam: Janssonius van Waesberge, J. Wetstein, en Gul. Smith. Vol. 1, 1734; vol. 2, 1735. Janssonius van Waesberge, vol. 3, 1759. H. C. Arksteum en H. Merkum, en Petrum Schouten, vol. 4, 1765.

Seetzen, U. J. 1794. Verzeichniss der Fische in den Gewässern der Herrschaft Jever. In Uebersicht der neuen zoologischen Entdeckungen, die entweder im Jahr 1793, oder kurz vor dessen Anfange gemacht worden find, ed. F. A. A. Meyer, *Zoolog. Annalen* (Weimar), 1: 173–76, pls. 2–3.

Serres, A. E. R. A. 1824–27. *Anatomie comparée du cerveau, dans les quatres classes des animaux vertébrés, appliquée à la physiologie et à la pathologie du système nerveux.* Paris: Gabon. Text in 2 vols., 1827; atlas, 1824.

Serton, P. 1971. English summary of the introduction. In *François Valentyn, Description of the Cape of Good Hope with the matters concerning it,* ed. P. Serton, R. Raven-Hart, W. J. de Kock, and E. H. Raidt, pt. 1, ser. 2, no. 2., pp. 1–30. Cape Town: Van Riebeeck Society.

Sevastianof, A. 1802. Description de l'*Acarauna longirostris.* Nouveau genre de poisson, appartenant à l'ordre des thoraciques, et qui se trouve dans le musée de notre Académie des Sciences. *Nova Acta Acad. Sci. Impér. Petro.* 13:357–66, pl. 11.

Severinus, M. A. 1645. *Zootomia democritaea: id est, Anatome generalis totius animantium opificii, libris quinque distincta, quorum seriem sequens facies delineabit.* Nuremberg: Volkamer.

———. 1661. *Antiperipatias, hoc est, Adversus Aristoteleos De respiratione piscium diatriba.* Amsterdam: Joannes Janssonius. [Followed by M. A. Severinus, *De piscibus in sicco viventibus commentarius in Theophrasti eresii libellum hujus argumenti,* 1655.]

Shaw, G. 1794. *Zoology of New Holland, by George Shaw, M.D. F.R.S. etc. etc. The figures by James Sowerby, F.L.S.* Vol. 1. London: J. Davis. [Vol. 2 is titled *A specimen of the botany of New Holland by J. E. Smith, 1793–1795,* London: J. Davis; the two volumes were issued separately, and in a single volume with a collective title page: *The zoology and botany of New Holland, the zoology by Shaw, the botany by Smith, 1793–1795.*]

———. 1800–1826. *General zoology, or Systematic natural history by George Shaw, M.D. F.R.S. etc., with plates from the first authorities and most select specimens engraved principally by Mr. Heath.* 14 vols. each in 2 pts. London: George Kearsley. Vol. 1, Mammalia, 1800; vol. 2, Mammalia, 1801; vol. 3, Amphibia, 1802; vol. 4, Pisces, 1803; vol. 5, Pisces, 1804; vol. 6, Insecta, 1806; vol. 7, Aves, 1809; vol. 8, Aves, 1812; vol. 9, Aves, 1815; vol. 10, Aves, 1817; vol. 11, Aves, 1819; vol. 12, Aves, 1824; vol. 13, Aves, 1826; vol. 14, Aves + General Index, 1826.

Shaw, G., and F. P. Nodder. 1789–1813. *The naturalist's miscellany, or Coloured figures of natural objects; drawn and described immediately from nature.* 24 vols. London: Nodder. Vol. 1, 1789–90; vol. 2, 1790–91; vol. 3, 1791–92; vol. 4, 1792–93; vol. 5, 1793–94; vol. 6, 1794–95; vol. 7, 1795–96; vol. 8, 1796–97; vol. 9, 1797–98; vol. 10, 1798–99; vol. 11, 1799–1800; vol. 12, 1800–1801; vol. 13, 1801–2; vol. 14, 1802–3; vol. 15, 1803–4; vol. 16, 1804–5; vol. 17, 1805–6; vol. 18, 1806–7; vol. 19, 1807–8; vol. 20, 1808–9; vol. 21, 1809–10; vol. 22, 1810–11; vol. 23, 1811–12; vol. 24, 1812–13. [Dating of volumes after Sherborn 1895.]

Shaw, T. E. 1932. *The Odyssey of Homer, newly translated into English prose.* New York: Oxford University Press.

Sherborn, C. D. 1895. On the dates of Shaw and Nodder's "Naturalist's miscellany." *Ann. Mag. Nat. Hist.*, ser. 6, 15:375–76.

———. 1934. On the dates of Pallas's "Zoographia Rosso-Asiatica." *Ibis* 4 (1): 164–67.

———. 1947. On the dates of Pallas's "Zoographia Rosso-Asiatica." *Bull. Zool. Nomencl.* 1:199–200.

Sherborn, C. D., and B. B. Woodward. 1893. On the dates of the "Encyclopédie méthodique" (zoology). *Proc. Zool. Soc. Lond.* 1893:582–84.

———. 1899. On the dates of the "Encyclopédie méthodique": Additional note. *Proc. Zool. Soc. Lond.* 1899:595.

———. 1906. On the dates of publication of the natural history portions of the "Encyclopédie méthodique." *Ann. Mag. Nat. Hist.*, ser. 7, 17:577–82.

Sibbald, R. 1684. *Scotia illustrata, sive Prodromus historiae naturalis.* Edinburgh: Jacobi Kniblo, Josuae Solingensis, et Johannis Colmarius, Pt. 1 in 1 vol; pt. 2 in 2 vols: Vol. 1, Historia plantarum in Scotiae; vol. 2, Historia animalium Scotiae.

———. 1773. *Phalainologia nova; sive Observationes de rarioribus quibusdam balaenis in Scotiae littus nuper ejectis: in quibus, nuper conspectae balaenae per genera et species, secundum characteres ab ipsa natura impressos, distribuuntur.* Edinburgh and London: Joannis Redi and Benjamin White.

Silvestre, A. F. de. 1791. Extrait d'un mémoire sur la respiration des poissons, comparée à celle des autres animaux. *Bull. Sci. Soc. Philomat.* (Paris) 1:17–18.

Singer, C. 1959. *A history of biology to about the year 1900: A general introduction to the study of living things.* 3d ed. rev. New York: Abelar-Schuman.

Sloane, H. 1707–25. *A voyage to the islands of Madeira, Barbados, Nièves, St. Christopher's and Jamaica, with the natural history of the herbs and trees, four-footed beasts, fishes, birds, insects, reptiles, etc.* 2 vols. London: Printed by B. M. for the author.

Smith, J. E. 1821. *A selection of the correspondence of Linnaeus, and other naturalists, from the original manuscripts.* 2 vols. London: Longman, Hurst, Rees, Orme, and Brown.

Soemmerring, D. W. 1818. *De oculorum hominis animaliumque sectione horizontali commentatio quam in Georgia Augusta illustris ordinis medici consensu pro obtinendus summis in medicina et chirurgia honoribus.* Göttingen: Vandenhoeck und Ruprecht.

Sonnerat, P. 1776. *Voyage à la Nouvelle Guinée, dans lequel on trouve la description des lieux, des observations physiques et morales, et des détails relatifs à l'histoire naturelle dans le règne animal et le règne végétal.* Paris: Ruault.

———. 1782. *Voyage aux Indes Orientales et à la Chine, fait par ordre du roi, depuis 1774 jusqu'en 1781: dans lequel on traite des moeurs, de la religion, des sciences et des arts des Indiens, des Chinois, des Pégouins et des Madégasses; suivi d'observations sur le Cap de Bonne-Espérance, les isles de France et de Bourbon, les Maldives, Ceylan, Malacca, les Philippines et les Moluques, et de recherches sur l'histoire naturelle de ces pays.* 2 vols. Paris: Auteur, Froulé, Nyon, et Barrois.

Sonnini de Manoncourt, C. N. S. 1803–4. *Histoire naturelle, générale et particulière des poissons; ouvrage faisant suite à l'histoire naturelle, générale et particulière.* Composée par Leclerc de Buffon, et mise dans un nouvel ordre par C. S. Sonnini, avec des notes et des additions. Paris, F. Dufart. Vols. 1–10, 1803; vols. 11–13, 1804.

Spallanzani, L. 1792–97. *Viaggi alle due Sicilie e in alcune parti dell'appennino dell'abbate Lazzaro Spallanzani.* Pavia: Baldassare Comini. Vol. 1, 1792; vol. 2, 1792; vol. 3, 1793; vol. 4, 1793; vol. 5, 1795; vol. 6, 1797.

———. 1798. *Travels in the two Sicilies, and some parts of the Apennines.* 4 vols. London: G. C. and J. Robinson.

———. 1800. *Voyage dans les deux Siciles et dans quelques parties des Apennins . . .* Traduits de l'italien par G. Toscan . . . avec des notes du cit. Faujas-de-St.-Fond. 6 vols. Paris: Maradan.

———. 1803. *Mémoires sur la respiration, par Lazare Spallanzani.* Traduits en français, d'après son manuscrit inédit, par Jean Senebier. Geneva: J. J. Paschoud.

———. 1807. *Rapports de l'air avec les êtres organisés, ou Traité de l'action du poumon et de la peau des animaux sur l'air, comme de celle des plantes sur ce fluide.* 3 vols. Geneva: J. J. Paschoud.

Spinola, M. 1807. Lettre sur quelques poissons peu connus du golfe de Gênes, adressée à M. Faujas-de-Saint-Fond. *Ann. Mus. Hist. Nat.* (Paris) 10:366–80, pl. 10.

Spix, J. B. von. 1815. *Cephalogenesis, sive Capitis ossei structura, formatio et significatio per omnes animalium classes, familias, genera ac aetates digesta, atque tabulis illustrata, legesque simul psychologiae, cranioscopiae ac physiognomiae inde derivatae.* Munich: Franciscus Seraphicus Hübschmannius.

Squire, S. 1744. *Plutarch's treatise of Isis and Osiris, translated into English.* Cambridge: J. Bentham.

Stearn, W. T. 1984. Daniel Carlsson Solander (1733–1782), pioneer Swedish investigator of Pacific natural history. *Arch. Nat. Hist.* 11 (3): 499–503.

Stearns, R. P. 1952. James Petiver, promoter of natural science. *Proc. Amer. Antiqu. Soc.* 62:243–379.

Stejneger, L. 1936. *Georg Wilhelm Steller, the pioneer of Alaskan natural history.* Cambridge: Harvard University Press.

Steller, G. W. 1751. De bestiis marinis. *Novi Comment. Acad. Sci. Impér. Petro.* 2:289–398, pls. 15–16.

———. 1752. Observationes generales universam historiam piscium concernentes. *Novi Comment. Acad. Sci. Impér. Petro.* 3:405–20.

———. 1774. *Beschreibung von dem Lande Kamtschatka dessen einwohnern, deren sitten, nahmen, lebensart und verschiedenen Gewohnheiten herausgegeben.* Frankfurt and Leipzig: Johann Georg Fleischer.

———. 1988. *Journal of a voyage with Bering, 1741–1742.* Trans. by M. A.. Engel and O. W. Frost. Ed. with intro. by O. W. Frost. Stanford: Stanford University Press.

Steno, N. 1664. De anatome rajae epistola. In his *De musculis et glandulis observationum specimen cum epistolis duabus anatomicis,* 48–70. Copenhagen: Matthias Godicchenius.

———. 1667. *Elementorum myologiae specimen, seu Musculi descriptio geometrica, cui accedunt canis carchariae dissectum caput, et dissectus piscis ex canum genere.* Florence: Stella.

———. 1675. Ova viviparorum spectantes observationes factae jussu serenissimi magni ducis hetruriae. In *Acta medica et philosophica Hafniensia,* ed. T. Bartholin, 2:219–31. Copenhagen: Petrus Hauboldus.

Store-Best, L. 1912. *Varro on farming: M. Terenti Varronis Rerum rusticarum libri tres.* Translated, with introduction, commentary, and excursus. London: G. Bell and Sons.

Stoye, J. 1993. *Marsigli's Europe, 1680–1730: The life and times of Luigi Ferdinando Marsigli, soldier and virtuoso.* New Haven: Yale University Press.

Strack, C. F. L. 1816. *Aristoteles Naturgeschichte der Thiere.* Frankfurt am Main: Joh. Christ. Hermannsche Buchhandling.

Stresemann, E. 1951. Date of publication of Pallas' *Zoographia Rosso-Asiatica. Ibis* 93 (2): 217–19.

Strøm, H. 1762–66. *Physist og oeconomisk Beskrivelse over Fogderict Søndmør, beliggende i Bengens Stift, i Norge.* 2 vols. Sorøe: Jonas Lindgren.

Svetovidov, A. N. 1976. On the dates of publication of P. S. Pallas' "Zoographia Rosso-Asiatica." *Zoologicheskii Zhurnal* (Moscow) 55 (4): 596–99.

———. 1981. The Pallas fish collection and the *Zoographia Rosso-Asiatica:* An historical account. *Arch. Nat. Hist.* 10 (1): 45–64.

Swainson, W. 1827. Sketch of the life and character of the late count de Lacépède. *Zool. J.* 3:73–76.

Swammerdam, J. 1737–38. *Biblia naturae, sive Historia insectorum, in classes certas redacta, nec non exemplis, et anatomical variorum animalculorum examine, aeneisque tabulis illustrata.* 2 vols. Leiden: Isaac Severinus, Balduin Vander Aa, en Petrus vander Aa.

———. 1758. *The book of nature, or The history of insects: Reduced to distinct classes, confirmed by particular instances, displayed in the anatomical analysis of many species, and illustrated with copper-plates.* London: C. G. Seyffert.

Sylburg, F. 1587. *Aristotelis De animalium historia libri X.* Frankfurt: Andreae Wecheli Heredes, Claudius Marnius, et Joannes Aubrius.

Taylor, S. 1732. *The history and antiquities of Harwich and Dovercourt, in the county of Essex . . . to which is added a large appendix containing the natural history of the sea-coast and country about Harwich, particularly the cliff, the fossils, plants, threes, birds and fishes, etc. Illustrated with variety of copper plates, by Samuel Dale, author of the Pharmacologia.* London: C. Davis, T. Osborn, and H. Lintot.

Thevet, A. 1558. *Les singularitez de la France Antarctique, autrement nommée Amérique, et de plusieurs terres et isles découvertes de nostre temps.* Anvers: Christophle Plantin.

———. 1584. *Les vrais pourtraits et vies des hommes illustres.* 2 vols. Paris: Veuve J. Kervert et G. Chaudière.

Thompson, D. W. 1947. *A glossary of Greek fishes*. London: Oxford University Press.

Thunberg, C. P. 1789. *Specimen ichthyologicum de muraena et ophichtho, quod venia exp. Fac. Med. Ups. praeside Carol. Pet. Thunberg . . . modeste offert Jonas Nicol Ahl.* Uppsala: Johan. Edman.

———. 1790. Beskrifning på tvånne fiskar ifrån Japan. *Kongl. Svenska Vetensk. Acad. Nya Handling.* (Stockholm) 11:106–10, pl. 3.

———. 1792. Beskrifning på tvånne Japanske fiskar. *Kongl. Svenska Vetensk. Acad. Nya Handling.* (Stockholm) 13:29–32, pl. 1.

———. 1793. Beskrifning på 2:ne nya fiskar af abborrslägtet ifrån Japan. *Kongl. Svenska Vetensk. Acad. Nya Handling.* (Stockholm) 14:55–56, pl. 1.

Tiedemann, F. 1809. *Anatomie des Fischherzens*. Landshut: Joseph Thomann.

Tilesius von Tilenau, W. G. 1809. Description de quelques poissons observée pendant voyage autour du monde. *Mém. Soc. Impér. Sci. Nat.* (Moscow) 2:212–49, pls. 13–17.

———. 1810. Piscium Camtschaticorum descriptiones et icones. *Mém. Acad. Impér. Sci. St. Pétersb.* 2:335–81, pls. 15–21.

———. 1811. Piscium Camtschaticorum descriptiones et icones. *Mém. Acad. Impér. Sci. St. Pétersb.* 3:225–85, pl. 8–13.

———. 1813a. Iconum et descriptionum piscium Camtschaticorum continuatio tertia tentamen monographiae generis Agoni Blochiani sistens. *Mém. Acad. Impér. Sci. St. Pétersb.* 4:406–78, pls. 11–16.

———. 1813b. *Naturhistorische Früchte der ersten kaiserlich-russischen unter dem Kommando des Herrn Krusenstern glücklich vollbrachten Erdumseeglung, gesammelt von Dr. Tilesius, Naturalisten der Expedition.* St. Petersburg.

———. 1820. De piscium Australium novo genere icone illustrato. *Mém. Acad. Impér. Sci. St. Pétersb.* 7:301–10, pl. 9.

Tingbrand, P. 1984. Daniel Solander, Piteå's around-the-world pioneer. *Arch. Nat. Hist.* 11 (3): 489–98.

Treviranus, G. R. 1816–21. *Vermischte Schriften anatomischen und physiologischen Inhalts.* 4 vols. in 2. Göttingen: Johann Friedrich Röwer. Vol. 1, 1816; vol. 2, 1817; vol. 3, 1820; vol. 4, 1821.

Tull, S. 1745. Castration des poissons. *Hist. Mém. Acad. Roy. Sci.* (Paris) 1742:31–32.

———. 1755. An account of Mr. Samuel Tull's method of castrating fish, communicated by W. Watson, F.R.S. *Phil. Trans. Roy. Soc. Lond.* 48:870–74.

Tye, A., and A. Jones. 1993. Birds and birdwatchers in West Africa, 1590–1712. *Arch. Nat. Hist.* 20 (2): 213–27.

Tyson, M. 1772. A letter to the Rev. M. Lort, B.D., F.R.S. containing an account of a singular fish, from the South Seas, by the Rev. Michael Tyson. *Phil. Trans. Roy. Soc. Lond.* 61:247–49, pl. 7, fig. 8.

Urban, I. 1920. *Plumiers Leben und Schriften nebst einem Schlüssel zu seinen Blütenpflanzen.* Repertorium Specierum Novarum Regni Vegetabilis, vol. 5. Berlin: Hrsg. von Friedrich Fedde.

Valenciennes, A. 1822. Sur le sous-genre marteau, *Zygaena*. *Mém. Mus. Hist. Nat.* (Paris) 9:222–28, pls. 11–12.

———. 1824. Description du cernié: *Polyprion cernium*. *Mém. Mus. Hist. Nat.* (Paris) 11:265–69, pl. 17.

———. 1842. Des barbeaux. In *Histoire naturelle des poissons*, by G. Cuvier and A. Valenciennes, 16: 122–205. Paris and Strasbourg: Levrault.

———. 1847. Du genre Ombre (*Umbra*). In *Histoire naturelle des poissons*, by G. Cuvier and A. Valenciennes, 19:538–44, pl. 590. Paris and Strasbourg: Levrault.

———. 1849. Advertissement. In *Histoire naturelle des poissons,* by G. Cuvier and A. Valenciennes, 22: v–vii. Paris and Strasbourg: Levrault.

Valentijn, F. 1724–26. *Oud en nieuw Oost-Indiën, vervattende een naaukeurige en uitvoerige verhandelinge van Nederlands mogentheyd in die gewesten, benevens eene wydluftige beschryvinge der Moluccos, Amboina, Banda, Timor, en Solor, Java, en alle de eylanden onder dezelve landbestieringen behoorende; het Nederlands comptoir op Suratte, en de levens der groote mogols; als ook een keurlyke verhandeling van't wezentlykste, dat men behoort te weten van Choromandel, Pegu, Arracan, Bengale, Mocha, Persien, Malacca, Sumatra, Ceylon, Malabar, Celebes of Macassar, China, Japan, Tayouan of Formosa, Tonkin, Cambodia, Siam, Borneo, Bali, Kaap de Goede Hoop en van Mauritius. Te zamen dus behelzende niet alleen eene zeer nette beschryving van alles, wat Nederlands Oost-Indiën betreft, maar ook't voornaamste dat eenigzins tot eenige andere Europeërs, in die gewesten, betrekking heeft. Met meer dan thien honderd en vyftig prentverbeeldingen verrykt. Alles zeer naaukeurig, in opzigt van de landen, steden, sterkten, zeden der volken, boomen, gewasschen, land- en zeè-dieren, met alle het wereldlyke en kerkelyke, van d'oudste tyden af tot nu toe aldaar voorgevallen, beschreven, en met veele zeer nette daar toe vereyschte kaarten opgeheldert door François Valentyn, onlangs bedienaar des Goddelyken woords in Amboina, Banda, enz. In vyf deelen.* Dordrecht and Amsterdam: Joannes van Braam en Gerard Onder de Linden. Vol. 1, 1724; vol. 2, 1724; vol. 3, 1726, 2 pts. in 1; vol. 4, 1726, 2 pts. in 1; vol. 5, 1726, 2 pts. in 1.

Valentini, M. B. 1720. *Amphitheatrum zootomicum, tabulis aeneis quamplurimis exhibens historiam animalium anatomicam.* 2 pts. in 1. Frankfurt am Main: Haerdes Zunnerianos und Johannem Adamum Jungium.

Vallisneri, A. 1712. Dissertationem de ovario anguillarum. *Misc. Curio. Medico-Physica* centur. 1–2 (appendix): 153–65.

———. 1733. *Opere fisico-mediche stampate e manoscritte del kavalier Antonio Vallisneri raccolte da Antonio suo Figliuolo, corredate d'una prefazione in genere sopra tutte, e d'una in particolare sopra il vocabolario della storia naturale.* 3 vols. Venice: Sebastiano Coleti.

Vayrolatti, F. E., R. G. Gasiglia, J. C. Hureau, and T. Monod. 1978. Biographie. Un pharmacien nicois. Antoine Risso (1777–1845): Sa vie—son oeuvre. *Ann. Mus. Hist. Nat. Nice* 5:7–26.

Vicq-d'Azyr, F. 1776a. Premier mémoire, pour servir à l'histoire anatomique des poissons. *Mém. Math. Phys. Acad. Sci.* (Paris) 7:18–36, pls. 1–2.

———. 1776b. Deuxième mémoire, pour servir à l'histoire anatomique des poissons. *Mém. Math. Phys. Acad. Sci.* (Paris) 7:233–62, pls. 1–5.

———. 1786. Suite des recherches sur la structure du cerveau. Quatrième mémoire. Sur la structure du cerveau des animaux comparé avec celui de l'homme. *Hist. Mém. Acad. Roy. Sci.* (Paris) 1783:468–504, pls. 7–10.

———. 1792. *Encyclopédie méthodique: Système anatomique.* Vol. 2, *Quadrupèdes.* Paris and Liège: Panckoucke et Plomteux.

———. 1805. *Oeuvres de Vicq-d'Azyr, recueillies et publiées avec des notes et un discours sur sa vie et ses ouvrages, par Jacq. L. Moreau (de la Sarthe).* 5 vols. Paris: L. Duprat-Duverger.

Vicq-d'Azyr, F., and H. Cloquet. 1819. *Encyclopédie méthodique: Système anatomique.* Vol. 3. *Mammifères et oiseaux.* Paris: Agasse.

Videler, J. J. 1993. *Fish swimming.* London: Chapman and Hall.

Vincent, W. 1807a. Coast of the Icthyophagi. In his *The commerce and navigation of the ancients in the Indian Ocean*, 1:229–311. London: T. Cadell and W. Davies.

———. 1807b. Agatharchides. In his *The commerce and navigation of the ancients in the Indian Ocean*, 2:25–46. London: T. Cadell and W. Davies.

Vincentius Bellovacensis. 1624. *Bibliotheca mundi Vincentii Burgundi, ex ordine praedicatorum venerabilis Episcopi Bellovacensis, speculum quadruplex, naturale, doctrinale, morale, historiale*. 2 vols. Douai: Baltazaris Belleri.

Viviani, D. 1806. Nomenclature de poissons de la rivière de Gênes et de la Spezzia. In B. Faujas-Saint-Fond, Lettre adressée à M. de Lacepède, sur les poissons du golfe de la Spezzia et de la mer de Gênes. *Ann. Mus. Hist. Nat.* (Paris) 8: 368–70.

Walbaum, J. J. 1783. Naturgeschichte des gelben Kohlmauls. *Schr. Berl. Ges. Naturf. Fr.* 4:147–60.

———. 1784a. Beschreibung der russigen Meerquappe mit einer Bartfaser. *Schr. Berl. Ges. Naturf. Fr.* 5:107–25.

———. 1784b. Beschreibung des breitnasigen Hayes. *Schr. Berl. Ges. Naturf. Fr.* 5:381–93.

———. 1788–93. *Petri Artedi renovati*. 5 pts. in 3 vols. Grypeswaldiae: Ant. Ferdin. Roese. Vol. 1, pt. 1, Bibliotheca ichthyologia, 1788; pt. 2, Philosophia ichthyologia, 1789. Vol. 2, Genera piscium, 1792. Vol. 3, 1793, pt. 1, Synonymia nominum piscium; pt. 2, Descriptiones specierum piscium.

Waldschmidt, W. H. 1700. Lampetrae fluviatilis anatome. *Misc. Curio. Medico-Physica*, decur. 3, 5–6 (231): 545–47.

Wallace, J. 1700. *An account of the Islands of Orkney*. London: Jacob Tonson.

Wallis, J. 1769. *The natural history and antiquities of Northumberland and of so much of the county of Durham as lies between the rivers Tyne and Tweed; commonly called, North Bishoprick*. 2 vols. London: W. and W. Strahan.

Walsh, J. 1774. Of the electric property of the torpedo. In a letter from John Walsh, Esq.; F.R.S. to Benjamin Franklin, Esq.; LL.D., F.R.S. Ac. R. Par. Soc. Ext., etc. *Phil. Trans. Roy. Soc. Lond.* 63:461–80, pl. 19.

Walsh, J. J. 1906–17. *Catholic churchmen in science: Sketches of the lives of Catholic ecclesiastics who were among the great founders in science*. Philadelphia: American Ecclesiastical Review, Dolphin Press. Ser. 1, 1906; ser. 2, 1909; ser. 3, 1917.

Wartmann, B. 1777. Beschreibung und Naturgeschichte des Blaufelchen. *Beschäft. Berl. Ges. Naturf. Fr.* 3:184–213.

———. 1783a. Von dem Rheinanken oder Illanken, *Salmo illanca*. *Schr. Berl. Ges. Naturf. Fr.* 4:55–68.

———. 1783b. Alpforelle aus dem Seealper See, *Salmo alpinus*. *Schr. Berl. Ges. Naturf. Fr.* 4:69–77.

———. 1783c. Vom Herrn D. Wartmann vom 9ten Jan. 1783. *Schr. Berl. Ges. Naturf. Fr.* 4:431–33.

Watson, W. 1779. An account of the blue shark, together with a drawing of the same. *Phil. Trans. Roy. Soc. Lond.* 68:789–90, pl. 12.

Way, A. S. 1934. *Hesiod*. London: Macmillan.

Weber, E. H. 1817. *Anatomia comparata nervi sympathici*. Leipzig: C. H. Reclam.

———. 1820. *De aure et auditu hominis et animalium. Pars I. De aure animalium aquatilium*. Leipzig: Gerhard Fleischer.

Welch, M. A. 1972. Francis Willoughby, F.R.S. (1635–1672). *J. Soc. Bibliogr. Nat. Hist.* 6 (2): 71–85.

Wellisch, H. 1975. Conrad Gessner: A bio-bibliography. *J. Soc. Bibliogr. Nat. Hist.* 7 (2): 151–247.
Wells, E. B. 1981. M. E. Bloch's *Allgemeine Naturgeschichte der Fische:* A study. 7–13. In *History in the service of systematics,* ed. A. C. Wheeler and J. H. Price. Papers from the conference to celebrate the centenary of the British Museum (Natural History) 13–16 April, 1981. Special Publication no. 1. London: Society for the Bibliography of Natural History.
Whalley, P. E. S. 1971. William Arderon, F.R.S., of Norwich, an eighteenth century diarist and letter-writer. *J. Soc. Bibliogr. Nat. Hist.* 6 (1): 30–49.
Wheeler, A. C. 1956. The *Zoophylacium* of Laurens Theodore Gronovius. *J. Soc. Bibliogr. Nat. Hist.* 3 (3): 152–57.
———. 1958. The Gronovius fish collection: A catalogue and historical account. *Bull. Brit. Mus. (Nat. Hist.),* Hist. Ser., 1 (5): 185–249, pls. 26–34.
———. 1961. The life and work of Peter Artedi. In *Petri Artedi Ichthyologia, historiae naturalis classica,* ed. A. C. Wheeler, vii–xxiii. Weinheim: J. Cramer.
———. 1979. The sources of Linnaeus's knowledge of fishes. *Svenska Linnésällsk. Årsskr.* (Uppsala) 1978:156–211.
———. 1984a. Daniel Solander—zoologist. *Svenska Linnésällsk. Årsskr.* (Uppsala) 1982–83:7–30.
———. 1984b. Daniel Solander and the zoology of Cook's voyage. *Arch. Nat. Hist.* 11:505–15.
———. 1985. The Linnaean fish collection in the Linnean Society of London. *Zool. J. Linn. Soc.* 84:1–76.
———. 1986. Catalogue of the natural history drawings commissioned by Joseph Banks on the *Endeavour* voyage 1768–1771 held in the British Museum (Natural History). Part 3: Zoology. *Bull. Brit. Mus. (Nat. Hist.), Hist. Ser.,* 13:1–171.
———. 1987. Peter Artedi, founder of modern ichthyology. *Proc. V Congr. European Ichthyol.* (Stockholm) 1985:3–10.
———. 1988. An appraisal of the zoology of C. S. Rafinesque. *Bull. Zool. Nomencl.* 45 (1): 6–12.
———. 1989. Further notes on the fishes from the collection of Laurens Theodore Gronovius. *Zool. J. Linn. Soc.* 95:205–18.
———. 1991a. The Linnaean fish collection in the Zoological Museum of the University of Uppsala. *Zool. J. Linn. Soc.* 103:145–95.
———. 1991b. *Caroli Linne, Systema naturae editio 12, tomus 1, Regnum animale (1766).* A microfiche reproduction of the author's personal annotated copy from the Linnean Society of London. With an historical introduction by Alwyne Wheeler. London: Natural History Museum in association with the Linnean Society of London.
———. 1993. The zoological manuscripts of Robert Brown. *Arch. Nat. Hist.* 20 (3): 417–24.
White, D. S. 1982. *Patriarch Photios of Constantinople: His life, scholarly contributions, and correspondence together with a translation of fifty-two of his letters.* Brookline, Mass.: Holy Cross Orthodox Press.
White, H. G. E. 1919. *Ausonius.* Vol. 1. New York: G. P. Putnam's Sons.
Whitehead, P. J. P. 1967. The dating of the 1st edition of Cuvier's *Règne animal distribué d'après son organisation. J. Soc. Bibliogr. Nat. Hist.* 4 (6): 300–01.
———. 1969a. Zoological specimens from Captain Cook's voyages. *J. Soc. Bibliogr. Nat. Hist.* 5 (3): 161–201.

———. 1969b. *Forty drawings of fishes made by artists who accompanied Captain James Cook on his three voyages to the Pacific, 1768–71, 1772–75, 1776–78 some being used by authors in the description of new species.* London: Trustees of the British Museum (Natural History).
———. 1976. The original drawings for the *Historia naturalis Brasiliae* of Piso and Marcgrave (1648). *J. Soc. Bibliogr. Nat. Hist.* 7 (4): 409–22.
———. 1978a. The Forster collection of zoological drawings in the British Museum (Natural History). *Bull. Brit. Mus. (Nat. Hist.), Hist. Ser.,* 6 (2): 25–47.
———. 1978b. A guide to the dispersal of zoological materials from Captain Cook's voyages. *Pacific Studies* (Laie, Hawaii) 1978:52–93.
———. 1979a. Georg Markgraf and Brazilian zoology. In *Johan Maurits van Nassau-Siegen, 1604–1679, a humanist prince in Europe and Brazil,* ed. E. van den Boogaart, H. R. Hoetink, and P. J. P. Whitehead, 424–71. The Hague: Johan Maurits van Nassau Stichting.
———. 1979b. The biography of Georg Marcgraf (1610–1643/4) by his brother Christian, translated by James Petiver. *J. Bibliogr. Nat. Hist.* 9 (3): 301–14.
———. 1982. The treasures at Grüssau. *New Scientist* 94 (1302): 226–31.
Whitehead, P. J. P., and M. Boeseman. 1989. *A portrait of Dutch seventeenth century Brazil: Animals, plants and people by the artists of Johan Maurits of Nassau.* New York: North-Holland.
Whitmore, P. J. S. 1967. *The order of Minims in seventeenth-century France.* The Hague: Martinus Nijhoff.
Willughby, F. 1686. *De historia piscium libri quatuor, jussu et sumptibus Societatis Regiae Londinensis editi. . . .* Totum opus recognovit, coaptavit, supplevit, librum etiam primum et secundum integros adjecit Johannes Raius e Societate Regia. Oxford: Theatro Sheldoniano. [Followed by J. Nieuhof, *Appendix ad historium naturalem piscium. Pisces Indiae Orientalis a Joanne Nieuhofs descripti;* followed by *Francisci Willoughby icthyographia ad amplissimum virum d'num Samuelem Pepys, Praesidem Soc:Reg:Londinensis, Concilium, et Socios ejusdem,* a collection of 188 pls. dated 1685.]
———. 1978. *De historia piscium and Ichthyographia.* Facsimile edition. New York: Arno Press.
Wolff, T. 1967. *Danske ekspeditioner på verdenshavene: Danish Expeditions on the seven seas.* Copenhagen: Rhodos International Science and Art.
Wotton, E. 1552. *Edoardi Wottoni Oxoniensis de differentiis animalium libri decem.* Paris: Vascosanus.
Wulff, J. C. 1765. *Ichthyologia, cum amphibiis regni Borussici methodo Linneana disposita.* Regiomonti: Johan. Jacob. Kanter.
Ximenez, F. 1615. *De la naturaleza, y virtudes de las plantas, y animales que estan receuidos en el vso de medicina en la Nueua España, y la methodo, y correccion, y preparacion, que para administrallas se requiere con lo que el doctor Francisco Hernandes escriuio en lengua Latina.* Mexico City: Diego Lopez Daualos.
Yonge, C. D. 1853. *The lives and opinions of eminent philosophers by Diogenes Laërtius.* London: Henry G. Bohn.
———. 1854. *The Deipnosophists, or Banquet of the Learned of Athenaeus.* 2 vols. London: Henry G. Bohn.
Zancaruolo, C. 1560. *Libri di Mons. Paolo Giovio De pesci Romani tradotto in volgare de Carlo Zancaruolo.* Venice: Gualtieri.
Zanobio, B. 1974. Pietro Andrea Mattioli. In *Dictionary of scientific biography,* 9:178–180. New York: Scribner's.

Zorgdrager, C. G. 1720. *Bloeyende opkomst der aloude en hedendaagsche Groenlandsche visschery.* Amsterdam: Joannes Oosterwyk.

———. 1723. *Alte und neue grönlandische Fischerei und Wallfischfang, mit einer kurzen historischen Beschreibung von Grönland, Island, Spizbergen, Nova Zembla, Jan Nayen Island, der Strasse Davis u.a., ausgefertiget durch Abraham Moubach.* Leipzig: Peter Conrad Monath.

———. 1727. *Bloeijende opkomst der aloude en hedendaagsche Groenlandsche visschery.* The Hague: P. van Thol en R. C. Alberts.

———. 1750. *Beschreibung des grönlandischen Wallfischfangs und Fischereii, nebst einer gründlichen Nachricht von dem Bakkeljau- und Stockfischfang bei Terreneuf, und einer kurzen Abhandlung von Grönland, Island, Spizbergen, Nova Zembla, Jan Nayen Island, der Strasse Davis u.a. aus dem Hollandischen übersetzet.* Nuremberg: Georg Peter Monath.

Zuiew, B. 1789. Gymnoti nova species. *Nova Acta Acad. Sci. Impér. Petro.* 5:269–73, pl. 8.

ILLUSTRATION CREDITS

Sources for the illustrations used in this volume are given in the legends for the figures themselves; where more detailed credits are required, they are given below, cited by figure number.

Frontispiece: Bibliothèque Centrale, Muséum National d'Histoire Naturelle, Paris, Inv. no. P.1339. Used with permission.

1. Cuvier and Valenciennes 1828–49, vol. 1; Bibliothèque Centrale, Muséum National d'Histoire Naturelle, Paris. Used with permission.
6. Thévet 1584, 2:612; Bibliothèque Centrale, Muséum National d'Histoire Naturelle, Paris. Used with permission.
9. Salviani 1554–58, frontispiece; Library of Congress, Washington, D.C. Used with permission.
11. Adler 1989, p. 7. Used with permission.
13. Churchill and Churchill 1732, frontispiece; Special Collections Division, University of Washington Libraries, Seattle, UW Neg. no. 15125. Used with permission.
14. Borelli 1743, pl. 14; Health Sciences Library and Information Center, University of Washington, Seattle. Used with permission.
15. Maar 1910, vol. 2, pl. 1; Health Sciences Library and Information Center, University of Washington, Seattle. Used with permission.
18. Willughby 1686, pl. I.5; Library of Congress, Washington, D.C. Used with permission.
20. Catesby 1731–43, vol. 2, pl. 22; Library of Congress, Washington, D.C. Used with permission.
21. Marsigli 1726, vol. 4, pl. 13; Library of Congress, Washington, D.C. Used with permission.
22. Bibliothèque Centrale du Muséum National d'Histoire Naturelle, Paris, MS 339, p. 74. Used with permission.
24. Bibliothèque Centrale, Muséum National d'Histoire Naturelle, Paris, MS 31, fig. 43; Inv. no. 2932. Used with permission.
26. Stedelijk Museum de Lakenhal, Leiden, Inv. no. FU9420049.JvW. Used with permission.
28. Bibliothèque Centrale, Muséum National d'Histoire Naturelle, Paris, Inv. no. P.3467. Used with permission.

29. Bibliothèque Centrale, Muséum National d'Histoire Naturelle, Paris, Inv. no. M.138. Used with permission.
30. Forsskål 1776, pl. 22; Zoological Museum, University of Copenhagen. Used with permission.
31. Zoological Museum, University of Copenhagen. Used with permission.
32. Parra 1787, pl. 6; Allan Hancock Library, University of Southern California, Los Angeles. Used with permission.
33. Bloch 1785–95, vol. 3, title page; Special Collections Department, Smithsonian Institution Libraries, Washington, D.C. Neg. no. 94-4175. Used with permission.
34. Bloch 1785–95, vol. 4, pl. 247; Bibliothèque Centrale, Muséum National d'Histoire Naturelle, Paris. Used with permission.
35. Pallas 1769, vol. 7, pl. 4; Library of Congress, Washington, D.C. Used with permission.
36. Haller 1757, vol. 1, frontispiece; Health Sciences Library and Information Center, University of Washington, Seattle. Used with permission.
37. Bibliothèque Nationale, Paris, Inv. no. 64B.34802. Used with permission.
41. Photographic Services, Smithsonian Institution, Washington, D.C., Neg. no. 34097-N. Used with permission.
46. Hamilton 1822, vol. 2, pl. 14; Library of Congress, Washington, D.C. Used with permission.
48. Oken 1815–16, pt. 3, Zoologie, Atlas, pl. 20; Special Collections Department, Smithsonian Institution Libraries, Washington, D.C. Neg. no. 94-4171. Used with permission.
53. Cuvier and Valenciennes 1828–49, vol. 5, pl. 139; Special Collections Department, Smithsonian Institution Libraries, Washington, D.C. Neg. no. 94-4173. Used with permission.
55. Rijksmuseum van Natuurlijke Histoire, Leiden. Used with permission.
57. Bibliothèque Centrale, Muséum National d'Histoire Naturelle, Paris, Inv. no. P.129. Used with permission.
60. Health Sciences Library and Information Center, University of Washington, Seattle. Used with permission.
63. Cuvier and Valenciennes 1828–49, vol. 3, pl. 65; Special Collections Department, Smithsonian Institution Libraries, Washington, D.C. Neg. no. 94-4176. Used with permission.
64. Cuvier and Valenciennes 1828–49, vol. 8, pl. 214; Special Collections Department, Smithsonian Institution Libraries, Washington, D.C. Neg. no. 94-4178. Used with permission.
65. Cuvier and Valenciennes 1828–49, vol. 2, pl. 9; Special Collections Department, Smithsonian Institution Libraries, Washington, D.C. Neg. no. 94-4174. Used with permission.
66. Cuvier and Valenciennes 1828–49, vol. 7, pl. 173; Special Collections Department, Smithsonian Institution Libraries, Washington, D.C. Neg. no. 94-4177. Used with permission.
67. Cuvier and Valenciennes 1828–49, vol. 15, pl. 429; Houghton Library, Harvard University, Cambridge. Used with permission.

INDEX

Italic figures are references to figures or tables.

Library of Congress Cataloging-in-Publication Data

Cuvier, Georges, baron, 1769–1832.
[Tableau historique des progrès de l'ichtyologie. English]
Historical portrait of the progress of ichthyology : from its origins to our own time / Georges Cuvier ; edited by Theodore W. Pietsch ; translated by Abby J. Simpson.
p. cm.—(Foundations of natural history)
Includes bibliographical references (p.) and index.
ISBN 0-8018-4914-4
1. Ichthyology—History. I. Pietsch, Theodore W. II. Title. III. Series.
QL614.8.C8813 1995
597'.009—dc20 94-38798